Hiroshi Ishii	Tok
Eiji Kamioka	Shib
Le Duy Khanh	Data
Surin Kittitornkun	King ____ ____ ____ute of Technology Ladkrabang, Thailand
Andrea Ko	Corvinus University of Budapest, Hungary
Duc Anh Le	Center for Open Data in the Humanities, Tokyo, Japan
Xia Lin*	Drexel University, USA
Lam Son Le	HCMC University of Technology, Vietnam
Faizal Mahananto	Institut Teknologi Sepuluh Nopember, Indonesia
Clavel Manuel	The Madrid Institute for Advanced Studies in Software Development Technologies, Spain
Nadia Metoui	University of Trento and FBK-Irist, Trento, Italy
Hoang Duc Minh	National Physical Laboratory, UK
Takumi Miyoshi	Shibaura Institute of Technology, Japan
Hironori Nakajo	Tokyo University of Agriculture and Technology, Japan
Nguyen Thai-Nghe	Cantho University, Vietnam
Thanh Binh Nguyen	HCMC University of Technology, Vietnam
Benjamin Nguyen	Institut National des Sciences Appliqués Centre Val de Loire, France
An Khuong Nguyen	HCMC University of Technology, Vietnam
Khai Nguyen	National Institute of Informatics, Japan
Kien Nguyen	National Institute of Information and Communications Technology, Japan
Khoa Nguyen	The Commonwealth Scientific and Industrial Research Organisation, Australia
Le Duy Lai Nguyen	Ho Chi Minh City University of Technology, Vietnam and University of Grenoble Alpes, France
Do Van Nguyen	Institute of Information Technology, MIST, Vietnam
Thien-An Nguyen	University College Dublin, Ireland
Phan Trong Nhan	HCMC University of Technology, Vietnam
Luong The Nhan	University of Pau, France
Alex Norta	Tallinn University of Technology, Estonia
Duu-Sheng Ong	Multimedia University, Malaysia
Eric Pardede	La Trobe University, Australia
Ingrid Pappel	Tallinn University of Technology, Estonia
Huynh Van Quoc Phuong	Johannes Kepler University Linz, Austria
Nguyen Khang Pham	Can Tho University, Vietnam
Phu H. Phung	University of Dayton, USA
Nguyen Ho Man Rang	Ho Chi Minh City University of Technology, Vietnam
Tran Minh Quang	HCMC University of Technology, Vietnam
Akbar Saiful	Institute of Technology Bandung, Indonesia
Tran Le Minh Sang	WorldQuant LLC, USA
Christin Seifert	University of Passau, Germany
Erik Sonnleitner	Johannes Kepler University Linz, Austria

Tran Phuong Thao	KDDI Research, Inc., Japan
Tran Ngoc Thinh	HCMC University of Technology, Vietnam
Quan Thanh Tho	HCMC University of Technology, Vietnam
Michel Toulouse	Vietnamese-German University, Vietnam
Shigenori Tomiyama	Tokai University, Japan
Le Hong Trang	HCMC University of Technology, Vietnam
Tuan Anh Truong	HCMC University of Technology, Vietnam and University of Trento, Italy
Tran Minh Triet	HCMC University of Natural Sciences, Vietnam
Takeshi Tsuchiya	Tokyo University of Science, Japan
Osamu Uchida	Tokai University, Japan
Hoang Tam Vo	IBM Research, Australia
Hoang Huu Viet	Vinh University, Vietnam
Edgar Weippl	SBA Research, Austria
Wolfram Wöß	Johannes Kepler University Linz, Austria
Tetsuyasu Yamada	Tokyo University of Science, Japan
Jeff Yan	Linköping University, Sweden
Szabó Zoltán	Corvinus University of Budapest, Hungary

Additional Reviewers

Pham Quoc Cuong	HCMC University of Technology, Vietnam
Kim Tuyen Le Thi	HCMC University of Technology, Vietnam
Ai Thao Nguyen Thi	Data Security Applied Research Lab, Vietnam
Bao Thu Le Thi	National Institute of Informatics, Japan
Tuan Anh Tran	HCMC University of Technology, Vietnam and Chonnam National University, South Korea
Quang Hai Truong	HCMC University of Technology, Vietnam

Contents

Deep Learning and Applications

Data Analytics and Recommendation Systems

Internet of Things and Applications

Smart City: Data Analytics and Security

Emerging Data Management Systems and Applications

Invited Keynotes

Freely Combining Partial Knowledge
in Multiple Dimensions
(Extended Abstract)

Dirk Draheim$^{(\boxtimes)}$

Large-Scale Systems Group, Tallinn University of Technology,
Akadeemia tee 15a, 12618 Tallinn, Estonia
`dirk.draheim@ttu.ee`

Abstract. F.P. conditionalization (frequentist partial conditionalization) allows for combining partial knowledge in arbitrary many dimensions and without any restrictions on events such as independence or
partitioning. In this talk, we provide a primer to F.P. conditionalization
and its most important results. As an example, we proof that Jeffrey
conditionalization is an instance of F.P. conditionalization for the special case that events form a partition. Also, we discuss the *logics* and the
data science perspective on the matter.

Keywords: F.P. conditionalization · Jeffrey conditionalization
Data science · Statistics · Contingency tables · Reasoning systems
SPSS · SAS · R · Phyton/Anaconda · Cognos · Tableau

1   A Primer on F.P. Conditionalization

In [1] we have introduced F.P.conditionalization (frequentist partial conditionalization), which allows for conditionalization on partially known events. An
F.P. conditionalization $P(A \mid B_1 \equiv b_1, \ldots, B_m \equiv b_m)$ is the probability of an
event A that is conditional on a list of event-probability specifications $B_1 \equiv b_1$
through $B_m \equiv b_m$. A specification pair $B \equiv b^{12}$ stands for the assumption that
the probability of B has somehow changed from a previously given, *a priori*
probability $P(B)$ into a new, *a posteriori* probability b. Consequently, we expect
that $P(B \mid B \equiv b) = b$ as well as $P(A \mid B \equiv P(B)) = P(A)$. Similarly, we expect
that classical conditional probability becomes a special case of F.P. conditional-
ization, i.e., that $P(A|B_1 \cdots B_m)$ equals $P(A \mid B_1 \equiv 100\%, \ldots, B_m \equiv 100\%)$ and,
similarly, $P(A|\overline{B_1} \cdots \overline{B_m})$ equals $P(A \mid B_1 \equiv 0\%, \ldots, B_m \equiv 0\%)$.

But what is the value of $P(A|B_1 \equiv b_1, \ldots, B_m \equiv b_m)$ in general? We have given
a formal, frequentist semantics to it. We think of conditionalization as taking

---

[1] Alternative notations for $B \equiv b$ such as $P(B) \rightsquigarrow b$ or $P(B) := b$ might be considered
more intuitive. We have chosen the concrete notation $B \equiv b$ for the sake of brevity
and readability.

[2] We also use $\mathsf{P}_{B_1 \equiv b_1, \ldots, B_m \equiv b_m}(A)$ as notation for $P(A \mid B_1 \equiv b_1, \ldots, B_m \equiv b_m)$.

© Springer Nature Switzerland AG 2018
T. K. Dang et al. (Eds.): FDSE 2018, LNCS 11251, pp. 3–11, 2018.
https://doi.org/10.1007/978-3-030-03192-3_1

place in chains of repeated experiments, so-called probability testbeds, of sufficient lengths. As a first step, we introduce the notion of F.P. conditionalization bounded by n which is denoted by $\mathsf{P}^n(A \mid B_1 \equiv b_1, \ldots, B_m \equiv b_m)$. We consider repeated experiments of such lengths n, in which statements of the form $B_i \equiv b_i$ make sense frequentistically, i.e., the probability b_i can be interpreted as the frequency of B_i and can potentially be observed. Then we reduce the notion of partial conditionalization to the notion of classical conditional probability, i.e., classical conditional expected value to be more precise. We consider the expected value of the frequency of A, i.e., the average occurrence of A, conditional on the event that the frequencies of events B_i adhere to the new probabilities b_i. Now, we can speak of the b_is as frequencies. Next, we define (general/unbounded) F.P. conditionalization by bounded F.P. conditionalization in the limit.

Definition 1 (Bounded F.P. Conditionalization). Given an i.i.d. sequence (independent and identically distributed sequence) of multivariate characteristic random variables $(\langle A, B_1, \ldots, B_m \rangle_{(j)})_{j \in \mathbb{N}}$, a list of rational numbers $b_1, \ldots, b_m$ and a bound $n \in \mathbb{N}$ such that $0 \leqslant b_i \leqslant 1$ and $nb_i \in \mathbb{N}$ for all b_i in $b_1, \ldots, b_m$. We define the *probability of A conditional on $B_1 \equiv b_1$ through $B_m \equiv b_m$ bounded by n*, which is denoted by $\mathsf{P}^n(A \mid B_1 \equiv b_1, \ldots, B_m \equiv b_m)$, as follows:

$$\mathsf{P}^n(A \mid B_1 \equiv b_1, \ldots, B_m \equiv b_m) = \mathsf{E}(\overline{A^n} \mid \overline{B_1^n} = b_1, \ldots, \overline{B_m^n} = b_m) \qquad (1)$$

Definition 2 (F.P. Conditionalization). Given an i.i.d. sequence of multivariate characteristic random variables $(\langle A, B_1, \ldots, B_m \rangle_{(j)})_{j \in \mathbb{N}}$ and a list of rational numbers $b = b_1, \ldots, b_m$ such that $0 \leqslant b_i \leqslant 1$ for all b_i in b and $lcd(b)$ denotes the smallest $n \in \mathbb{N}$ such that $nb_i \in \mathbb{N}$ for all b_i in $b = b_1, \ldots, b_m$.[3] We define the *probability of A conditional on $B_1 \equiv b_1$ through $B_m \equiv b_m$*, denoted by $\mathsf{P}(A \mid B_1 \equiv b_1, \ldots, B_m \equiv b_m)$, as follows:

$$\mathsf{P}(A \mid B_1 \equiv b_1, \ldots, B_m \equiv b_m) = \lim_{k \to \infty} \mathsf{P}^{k \cdot lcd(b)}(A \mid B_1 \equiv b_1, \ldots, B_m \equiv b_m) \qquad (2)$$

As a first result, we observe that bounded F.P. conditionalization can be expressed more compact, without conditional expectation, merely in terms of conditional probability, i.e., we have that the following holds for any bounded F.P. conditionalization:

$$\mathsf{P}^n(A \mid B_1 \equiv b_1, \ldots, B_m \equiv b_m) = \mathsf{P}(A \mid \overline{B_1^n} = b_1, \ldots, \overline{B_m^n} = b_m) \qquad (3)$$

In most proofs and argumentations we use the more convenient form in Eq. (3) instead of the more intuitive form in Definition 1.

In general, an F.P. conditionalization $\mathsf{P}(A \mid B_1 \equiv b_1, \ldots, B_m \equiv b_m)$ is different from all of its finite approximations of the form $\mathsf{P}^n(A \mid B_1 \equiv b_1, \ldots, B_m \equiv b_m)$. In some interesting special cases, we have that the F.P. conditionalizations are equal to all of their finite approximations; i.e., it is the case if the condition events $B_1 \equiv b_1$ through $B_m \equiv b_m$ are independent or if the condition events form a partition.

---

[3] $lcd(b)$ is the *least common denominator* of $b = b_1, \ldots, b_m$.

The case in which the condition events form a partition is particularly interesting. This is so, because this case makes Jeffrey conditionalization [2–4], valuewise, an instance of F.P. conditionalization as we will discuss further in Sect. 2. In case the conditions events $B_1 \equiv b_1$ through $B_m \equiv b_m$ form a partition, we have that the value of $\mathsf{P}(A \mid B_1 \equiv b_1, \ldots, B_m \equiv b_m)$ is a weighted sum of conditional probabilities $b_i \cdot \mathsf{P}(A|B_i)$, compare with Eq. (5). This is somehow neat and intuitive. Take the simple case of an F.P. conditionalization $\mathsf{P}(A|B \equiv b)$ over a single event B. Such an F.P. conditionalization can be represented differently as an F.P. conditionalization over two partitioning events $B_1 = B$ and $B_2 = \overline{B}$, i.e., $\mathsf{P}(A \mid B \equiv b, \overline{B} \equiv 1 - b)$. Therefore we have that

$$\mathsf{P}(A|B \equiv b) = b \cdot \mathsf{P}(A|B) + (1 - b) \cdot \mathsf{P}(A|\overline{B}) \tag{4}$$

Equation 4 is highly intuitive: it feels natural that the direct conditional probability $P(A|B)$ should be somehow (proportionally) lowered by the new probability b of event B, similarly, we should not forget that the event $\overline{B}$ can also appear, i.e., with probability $1 - b$ and should also influence the final value – symmetrically. So, the b-weighted average of $P(A|B)$ and $P(A|\overline{B})$ as expressed by Eq. (4) seems to be an educated guess. Fortunately, we do not need such an appeal to intuition. In our framework, Eqs. (4) and (5) can be proven correct, as a consequence of probability theory.

Theorem 3 (F.P. Conditionalization over Partitions). *Given an F.P. conditionalization* $\mathsf{P}(A \mid B_1 \equiv b_1, \ldots, B_m \equiv b_m)$ *such that the events* $B_1, \ldots, B_m$ *form a partition, and, furthermore, the frequencies* $b_1, \ldots, b_m$ *sum up to one, we have the following:*

$$\mathsf{P}(A \mid B_1 \equiv b_1, \ldots, B_m \equiv b_m) = \sum_{\substack{1 \leqslant i \leqslant m \\ \mathsf{P}(B_i) \neq 0}} b_i \cdot \mathsf{P}(A \mid B_i) \tag{5}$$

Proof. See [1].

Table 1 summarizes interesting properties of F.P. conditionalization. Proofs of all properties are provided in [1]. Property (a) is a basic fact that we mentioned earlier; i.e., an updated event actually has the probability value that it is updated to. Properties (b) and (c) deal with condition events that form a partition and we have treated them with Theorem 3. Properties (d) and (e) provide programs for probabilities of frequency specifications of the general form $\mathsf{P}(\cap_{i \in I} B_i^n = k_i)$. Having programs for such probabilities is sufficient to compute any F.P. conditionalization. The equation in (d) is called one-step decomposition in [1] and can be read immediately as a recursive programme specification; compare also with the primer on inductive definitions in [5]. Equation (e) provides a combinatorial solution for $\mathsf{P}(\cap_{i \in I} B_i^n = k_i)$. Equation (e) generalizes the known solution for bivariate Bernoulli distributions [6–8] to the general case of multivariate Bernoulli distributions. Property (f) is called conditional segmentation in [1]. Conditional segmentation shows how F.P. conditionalization

Table 1. Properties of F.P. conditionalization. Values of various F.P. conditionalizations $P_{\boldsymbol{B}}(A) = P(A|B_1 \equiv b_1,..., B_m \equiv b_m)$ with frequency specifications of the form $\boldsymbol{B} = B_1 \equiv b_1,..., B_m \equiv b_m$ and condition indices $I = \{1,...,m\}$; probability values (d) and (e) of frequency specifications of the form $P(\bigcap_{i\in I} B_i^n = k_i)$. Proofs of all properties are provided in [1].

	Constraint	F.P. Conditionalization
(a)	b_i belongs to $\boldsymbol{B}$	$P_{\boldsymbol{B}}(B_i) = b_i$
(b)	$m = 1$, $\boldsymbol{B} = (B \equiv b)$	$P_{\boldsymbol{B}}(A) = b \cdot P(A|B) + (1 - b) \cdot P(A|\overline{B})$
(c)	$B_1,..., B_m$ form a partition	$P_{\boldsymbol{B}}(A) = \sum_{i=1}^{m} b_i \cdot P(A \mid B_i)$
(d)	For arbitrary bound n	$P(\underset{i\in I}{\cap} B_i^n = k_i) = \underset{I' \subseteq I}{\sum} P(\underset{i\in I'}{\cap} B_i, \underset{i\notin I'}{\cap} \overline{B_i}) \cdot P(\underset{i\in I'}{\cap} B_i^{n-1} = k_i - 1, \underset{i\notin I'}{\cap} B_i^{n-1} = k_i)$ $\quad \forall i \in I' . k_i \neq 0$ $\quad \forall i \notin I' . k_i \neq n$
(e)	For arbitrary bound n	$P(\underset{i\in I}{\cap} B_i^n = k_i) = \sum \left(\dfrac{n!}{\underset{I' \subseteq I}{\prod} \rho(I')!} \times \underset{I' \subseteq I}{\prod} P(\underset{i\in I'}{\cap} B_i, \underset{i\notin I'}{\cap} \overline{B_i})^{\rho(I')} \right)$ $\quad \rho : \mathbb{P}(I) \to \mathbb{N}_0$ $\quad \forall i \in I . k_i = \sum\{\rho(I') \mid I' \subseteq I \wedge B_i \in I'\}$ $\quad n = \sum\{\rho(I') \mid I' \subseteq I\}$
(f)	–	$P_{\boldsymbol{B}}(A) = \sum P(A|\underset{i\in I}{\cap} \zeta_i) \cdot P(\underset{i\in I}{\cap} \zeta_i | \underset{i\in I}{\cap} B_i \equiv b_i)$ $\quad (\zeta_i \in \{B_i, \overline{B_i}\})_{i\in I}$ $\quad P(\underset{i\in I}{\cap} \zeta_i) \neq 0$
(g)	$B_1,..., B_m$ are independent	$P_{\boldsymbol{B}}(B_1,..., B_k) = b_1 b_2 \cdots b_k$
(h)	$B_1,..., B_m$ are independent	$P_{\boldsymbol{B}}(B_1,..., B_m) = P_{\boldsymbol{B}}(B_1) \cdots P_{\boldsymbol{B}}(B_m)$
(i)	$B_1,..., B_m$ are independent	$P_{\boldsymbol{B}}(A) = \underset{I' \subseteq I}{\sum} \left(P(A|\underset{i\in I'}{\cap} B_i, \underset{i\notin I'}{\cap} \overline{B_i}) \cdot \underset{i\in I'}{\prod} b_i \cdot \underset{i\notin I'}{\prod} (1-b_i) \right)$ $\quad P(\underset{i\in I'}{\cap} B_i, \underset{i\notin I'}{\cap} \overline{B_i}) \neq 0$
(j)	A is independent of $B_1,..., B_m$	$P_{\boldsymbol{B}}(A) = P(A)$
(k)	$B_1 \equiv 100\%,..., B_i \equiv 100\%$ $B_{i+1} \equiv 0\%,..., B_m \equiv 0\%$	$P_{\boldsymbol{B}}(A) = P(A|B_1,..., B_i, \overline{B_{i+1}},..., \overline{B_m})$
(l)	$B_1,..., B_m$ form a partition or $B_1,..., B_m$ are independent $B_1 \equiv P(B_1),..., B_m \equiv P(B_m)$	$P_{\boldsymbol{B}}(A) = P(A)$
(m)	$B_1,..., B_m$ form a partition	$P_{\boldsymbol{B}}(AB_i) = b_i \cdot P(A|B_i)$
(n)	$B_1,..., B_m$ form a partition	$P_{\boldsymbol{B}}(A|B_i) = P(A|B_i)$
(o)	$B_1,..., B_m$ are independent	$P_{\boldsymbol{B}}(A, B_1,..., B_m) = b_1 \cdots b_m \cdot P(A|B_1,..., B_m)$
(p)	–	$P_{\boldsymbol{B}}(A|B_1,..., B_m) = P(A|B_1,..., B_m)$

generalizes Jeffrey conditionalization by dropping the partitioning constraint on events. Conditional segmentation is also often useful as helper Lemma. Properties (g) and (h) are important; they reveal how F.P. conditionalization behaves in case of independent condition events. Property (i) deals with the case that a target event is independent of the condition events. Property (k) has been mentioned earlier; it is about how F.P. conditionalization meets classical conditional probability. Property (l) generalizes the basic fact that $P(A \mid B \equiv P(B)) = P(A)$ to lists of condition events. Properties (m) through (p) all deal with cases, in which condition events also appear, in some way, in the target event. Properties (m) through (p) are highly relevant in the discussion of Jeffrey's probability kinematics and other Bayesian frameworks with possible-world semantics. Actually, property (n) is an F.P. version of what we call Jeffrey's postulate.

Table 2. Properties of F.P. conditional expectations. Values of various F.P. expectations $\mathsf{E}_{\mathsf{P}_B}(\nu \mid A)$, with frequency specifications $\boldsymbol{B} = B_1 \equiv b_1, \ldots, B_m \equiv b_m$ and condition indices $I = \{1, \ldots, m\}$. Proofs of all properties are provided in [1].

	Constraint	F.P. Expectation
(A)	$B_1, \ldots, B_m$ form a partition	$\mathsf{E}_{\mathsf{P}_B}(\nu \mid B_i) = \mathsf{E}(\nu \mid B_i)$
(B)	$m = 1$, $\boldsymbol{B} = (B \equiv b)$	$\mathsf{E}_{\mathsf{P}_B}(\nu \mid A) = \frac{b \cdot \mathsf{P}(A\mid B)\mathsf{E}(\nu\mid AB) + (1-b)\cdot \mathsf{P}(A\mid \overline{B_1})\mathsf{E}(\nu\mid A\overline{B_1})}{b\cdot \mathsf{P}(A\mid B) + (1-b)\cdot \mathsf{P}(A\mid \overline{B})}$
(C)	$B_1, \ldots, B_m$ form a partition	$\mathsf{E}_{\mathsf{P}_B}(\nu \mid A) = \frac{\sum_{i=1}^{m} b_i \cdot \mathsf{P}(A\mid B_i)\cdot \mathsf{E}(\nu \mid AB_i)}{\sum_{i=1}^{m} b_i \cdot \mathsf{P}(A\mid B_i)}$
(M)	$B_1, \ldots, B_m$ form a partition	$\mathsf{E}_{\mathsf{P}_B}(\nu \mid AB_i) = \mathsf{E}(\nu \mid AB_i)$
(N)	$B_1, \ldots, B_m$ form a partition	$\mathsf{E}_{\mathsf{P}_B(\downarrow B_i)}(\nu\mid A) = \mathsf{E}(\nu \mid AB_i)$
(O)	$B_1, \ldots, B_m$ are independent	$\mathsf{E}_{\mathsf{P}_B}(\nu\mid AB_1 \cdots B_m) = \mathsf{E}(\nu \mid AB_1 \cdots B_m)$
(P)	$B_1, \ldots, B_m$ are independent	$\mathsf{E}_{\mathsf{P}_B(\_\mid B_1 \cdots B_m)}(\nu\mid A) = \mathsf{E}(\nu \mid AB_1 \cdots B_m)$

With Table 2 we step from F.P. conditionalization to F.P. conditional expected values, that we also call F.P. conditional expectations or just F.P. expectations for short. Given frequency specifications $\boldsymbol{B} = B_1 \equiv k_1, \ldots, B_m \equiv k_m$, we say that $\mathsf{E}_{\mathsf{P}_B}(\nu \mid A)$ is an F.P. expectation. Here, the event A plays the role of the target event; whereas we consider the random variable ν as rather fixed. This way, each property in Table 1 has a corresponding property in terms of F.P. expectations. Table 2 shows some of them[4]. We do not need an own definition for F.P. expectations. We have that P_B is a probability function, so that the corresponding expected values and conditional expected values[5] are defined and we have that

$$\mathsf{E}_{\mathsf{P}_B}(\nu : \Omega \longrightarrow D \mid A) = \sum_{d \in D} d \cdot \mathsf{P}_B(\nu = d, A) \Big/ \mathsf{P}_B(A) \qquad (6)$$

In Ramsey's subjectivism [9–11] and Jeffrey's logic of decision [4,12] the notion of *desirability* is a crucial concept. Here, the desirability *des A* of an event A is the conditional expected value of an implicitly given utility ν under the condition A, which also explains why F.P. expectations are an important concept.

2 The Logics Perspective

In his *logic of decision* [13], also called *probability kinematics* [13,14], Richard C. Jeffrey establishes Jeffrey conditionalization. Probabilities are interpreted as

---

[4] Rows with same letters in Tables 1 and 2 correspond to each other.

[5] The notation E_P makes explicit that E belongs to the probability space $(\Omega, \Sigma, \mathsf{P})$.

degrees of believe and the semantics of a probability update is explained directly in terms of a *possible world semantics*. Jeffrey denotes *a priori* probability values as $prob(A)$ and *a posteriori* probability values as $PROB(A)$ and maintains the list of updated events $B_1,..., B_m$ in the context of probability statements[6]. It is assumed that in both the worlds, i.e., the *a priori* and the *a posteriori* world, the laws of probability hold. The probability functions $PROB$ and $prob$ are related by a postulate. The postulate deals exclusively with situations, in which the updated events $B_1,..., B_m$ form a partition. Then, it states that conditional probabilities with respect to one of the updated events are preserved, i.e., we can assume that $PROB(A|B_i) = prob(A|B_i)$ holds for all events A and all events B_i from $B_1,..., B_m$ – just as longs as $B_1,..., B_m$ form a partition. Persi Diaconis and Sandy Zabell call this postulate the J-condition [15,16]. Richard Bradley talks about conservative belief changes [17,18]. We call this postulate the probability kinematics postulate, or also just Jeffrey's postulate for short. We say that Jeffrey's postulate is a bridging statement, as it bridges between the *a priori* world and the *a posteriori* world. Next, Jeffrey exploits this postulate to derive Jeffrey conditionalization, also called Jeffrey's rule, compare with Eq. (5). It is crucial to understand, that the F.P. equivalent of Jeffrey's postulate, i.e., $P_B(A|B_i) = P(A|B_i)$[7] does not need to be postulated in the F.P. framework, but is a property that simply holds; i.e., it can be proven from the underlying frequentist semantics.

We have seen that F.P. conditionalization creates a clear link from the Kolmogorov system of probability to one of the important Bayesian frameworks, i.e., Jeffrey's logic of decision. When it comes to Bayesianism, there is no such single, closed apparatus as with frequentism [19–23]. Instead, there is a great variety of important approaches and methodologies, with different flavors in objectives and explications [24–26]. We have de Finetti [27,28] with his Dutch book argument and Ramsey [9,11] with his representation theorem [10]. Think of Jaynes [29], who starts from improving statistical reasoning with his application of maximal entropy [30], and from there transcends into an agent-oriented explanation of probability theory [31]. Also, think of Pearl [32], who eventually transcends probabilistic reasoning by systematically incorporating causality into his considerations [33,34]. Bayesian approaches have in common that they rely, at least in crucial parts, on notions other than frequencies to explain probabilities, among the most typical are degrees of belief, degrees of preference, degrees of plausibility, degrees of validity or degrees of confirmation.

3   The Data Science Perspective

The *data science* perspective is the F.P. perspective *per se*. Current data science has a clear statistical foundation; in practice, we see that data science is

---

[6] Please note, that the notational differences between between Jeffrey conditionalization and F.P. conditionalization are a minor issue and must not be confused with semantical differences – see [1] for a thorough discussion.

[7] With $\boldsymbol{B} = B_1 \equiv PROB(B_1),..., B_m \equiv PROB(B_m)$.

boosted by statistical packages and tools, ranging from SPSS, SAS over R to Phyton/Anaconda. In practice, the more interactive, multivariate data analytics (as represented by business intelligence tools such as Cognos or Tableau) is still equally important in data science initiatives. Again, the findings of F.P. conditionalization are fully in line with the foundations of multivariate data analytics.

An important dual problem to partial conditionalization is about determining the most likely probability distribution with known marginals for a complete set of observations. This problem is treated by Deming and Stephan in [35] and Ireland and Kullback in [36]. Given two partitions of events $B_1,..., B_s$ and $C_1,..., C_t$, numbers of observations n_{ij} for all possible B_iC_j in a sample of size n and marginals $p_{i\star}$ for each B_i in and $p_{\star j}$ for each C_j, it is the intention to find a probability distribution P that adheres to the specified marginals, i.e., such that $\mathsf{P}(B_i) = p_{i\star}$ for all B_i and $\mathsf{P}(C_j) = p_{\star j}$ for all C_j, and furthermore maximizes the probability of the specified joint observation, i.e., that maximizes the following multinomial distribution[8]:

$$\mathfrak{M}_{n,\, \mathsf{P}(B_1C_1),...,\mathsf{P}(B_1C_t),\,...,\, \mathsf{P}(B_sC_1),...,\mathsf{P}(B_sC_t)}(n_{11},..., n_{1t}, \ldots, n_{s1},..., n_{st})$$

Note that the collection of $s \times t$ events B_sB_t form a partition. The observed values n_{ij} are said to be organized in a two-dimensional $s \times t$ *contingency table*. The restriction to two-dimensional contingency tables is without loss of generality, i.e., the results of [35] and [36] can be generalized to multi-dimensional tables. In comparisons with partial conditionalizations, we treat two events B and C as a 2×2 contingency table with partitions $B_1 = B$, $B_2 = \overline{B}$, $C_1 = C$ and $C_2 = \overline{C}$. Now, [35] approaches the optimization by least-square[9] adjustment, i.e., by considering the probability function P that minimizes χ^2, whereas [36] approaches the optimization by considering the probability function P that minimizes the Kullback-Leibler number $I(\mathsf{P}, \mathsf{P}')$[10] with $\mathsf{P}'(B_iC_j) = n_{ij}/n$; compare also with [37,38]. Both [35,39] and [36] use iterative procedures that generates BAN (best approximatively normal) estimators for convergent computations of the considered minima; compare also with [40,41].

4   Conclusion

Statistics is the language of science; however, the semantics of probabilistic reasoning is still a matter of discourse. F.P. conditionalization provides a frequentist semantics for conditionalization on partially known events. It generalizes Jeffrey conditionalization from partitions to arbitrary collections of events. Furthermore, the postulate of Jeffrey's probability kinematics, which is rooted in Ramsey's subjectivism, turns out to be a consequence in our frequentist semantics. F.P. conditionalization is a straightforward, fundamental concept that fits our intuition. Furthermore, it creates a clear link from the Kolmogorov system of probability to one of the important Bayesian frameworks.

---

[8] $\mathfrak{M}_{n,p_1,...,p_m}(k_1, \ldots, k_m) = (n!/(k_1!\cdots k_m!)) \cdot p_1^{k_1} \cdots p_m^{k_m}$.

[9] $\chi^2 = \sum_{i=1}^{s}\sum_{j=1}^{t}(n_{ij} - n \cdot \mathsf{P}(B_iC_j))^2/n_{ij}$.

[10] $I(\mathsf{P}, \mathsf{P}') = \sum_{i=1}^{s}\sum_{j=1}^{t}\mathsf{P}(B_iC_j) \cdot \ln(\mathsf{P}(B_iC_j)/\mathsf{P}'(B_iC_j)))$.

References

1. Draheim, D.: Generalized Jeffrey Conditionalization - A Frequentist Semantics of Partial Conditionalization. Springer, Heidelberg (2017). https://doi.org/10.1007/978-3-319-69868-7. http://fpc.formcharts.org
2. Jeffrey, R.C.: Contributions to the theory of inductive probability. Ph.D. thesis, Princeton University (1957)
3. Jeffrey, R.C.: The Logic of Decision, 1st edn. McGraw-Hill, New York (1965)
4. Jeffrey, R.C.: The Logic of Decision, 2nd edn. University of Chicago Press, Chicago (1983)
5. Draheim, D.: Semantics of the Probabilistic Typed Lambda Calculus - Markov Chain Semantics, Termination Behavior, and Denotational Semantics. Springer, Heidelberg (2017). https://doi.org/10.1007/978-3-642-55198-7
6. Wicksell, S.D.: Some theorems in the theory of probability - with special reference to their importance in the theory of homograde correlations. Svenska Aktuarieforeningens Tidskrift, pp. 165–213 (1916)
7. Aitken, A., Gonin, H.: On fourfold sampling with and without replacement. Proc. R. Soc. Edinburgh 55, 114–125 (1935)
8. Teicher, H.: On the multivariate poisson distribution. Skand. Aktuarietidskr. 37, 1–9 (1954)
9. Ramsey, F.P.: The Foundations of Mathematics and other Logical Essays. Kegan, Paul, Trench, Trubner & Co., Ltd., New York (1931). Ed. by R.B. Braithwaite
10. Ramsey, F.P.: Truth and probability. In: Ramsey, F.P., Braithwaite, R. (eds.) The Foundations of Mathematics and other Logical Essays, pp. 156–198. Kegan, Paul, Trench, Trubner & Co., Ltd., New York (1931)
11. Ramsey, F.P.: Philosophical Papers. Cambridge University Press, Cambridge (1990). Ed. by D.H. Mellor
12. Jeffrey, R.C.: Subjective Probability - the Real Thing. Cambridge University Press, Cambridge (2004)
13. Jeffrey, R.C.: Probable knowledge. In: Lakatos, I. (ed.) The Problem of Inductive Logic, pp. 166–180. North-Holland, Amsterdam, New York, Oxford, Tokio (1968)
14. Levi, I.: Probability kinematics. Br. J. Philos. Sci. 18(3), 197–209 (1967)
15. Diaconis, P., Zabell, S.: Some alternatives to Bayes's rules. Technical report No. 205, Department of Statistics, Stanford University, October 1983
16. Diaconis, P., Zabell, S.: Some alternatives to Bayes's rules. In: Grofman, B., Owen, G. (eds.) Information Pooling and Group Decision Making, pp. 25–38. JAI Press, Stamford (1986)
17. Bradley, R.: Decision Theory with a Human Face. Draft, p. 318, April 2016. http://personal.lse.ac.uk/bradleyr/pdf/DecisionTheorywithaHumanFace(indexed3).pdf (forthcoming)
18. Dietrich, F., List, C., Bradley, R.: Belief revision generalized - a joint characterization of Bayes's and Jeffrey's rules. J. Econ. Theory (forthcoming)
19. Kolmogorov, A.: Grundbegriffe der Wahrscheinlichkeitsrechnung. Springer, Heidelberg (1933). https://doi.org/10.1007/978-3-642-49888-6
20. Kolmogorov, A.: Foundations of the Theory of Probability. Chelsea, New York (1956)
21. Kolmogorov, A.: On logical foundation of probability theory. In: Itô, K., Prokhorov, J.V. (eds.) LNM. Lecture Notes in Mathematics, vol. 1021, pp. 1–5. Springer, Heidelberg (1982). https://doi.org/10.1007/BFb0072897

22. Neyman, J.: Outline of a theory of statistical estimation based on the classical theory of probability. Philos. Trans. R. Soc. Lond. **236**(767), 333–380 (1937)
23. Neyman, J.: Frequentist probability and frequentist statistics. Synthese **36**, 97–131 (1977)
24. Weisberg, J.: Varieties of Bayesianism. In: Gabbay, D., Hartmann, S., Woods, J. (eds.) Handbook of the History of Logic, vol. 10 (2011)
25. Galavotti, M.C.: The modern epistemic interpretations of probability - logicism and subjectivism. In: Gabbay, D., Hartmann, S., Woods, J. (eds.) Handbook of the History of Logic, vol. 10, pp. 153–203. Elsevier, Amsterdam (2011)
26. Weirich, P.: The Bayesian decision-theoretic approach to statistics. In: Bandyopadhyay, P.S., Forster, M.R. (eds.) Philosophy of Statistics. Handbook of Philosophy of Science, vol. 7 (Gabbay, D.M., Thagard, P., Woods, J. general editors). North-Holland, Amsterdam, Boston Heidelberg (2011)
27. de Finetti, B.: Foresight - its logical laws, its subjective sources. In: Kyburg, H.E., Smokler, H.E. (eds.) Studies in Subjective Probability. Wiley, Hoboken (1964)
28. de Finetti, B.: Theory of Probability - A Critical Introductory Treatment. Wiley, Hoboken (2017). First issued in 1975 as a two-volume work
29. Jaynes, E.: Papers on Probability, Statistics and Statistical Physics. Kluwer Academic Publishers, Dodrecht, Boston, London (1989). Ed. by E.D. Rosenkranz
30. Jaynes, E.T.: Prior probabilities. IEEE Trans. Syst. Sci. Cybern. **4**(3), 227–41 (1968)
31. Jaynes, E.T.: Probability Theory. Cambridge University Press, Cambridge (2003)
32. Pearl, J.: Probabilistic Reasoning in Intelligent Systems - Networks of Plausible Inference, 2nd edn. Morgan Kaufmann, San Francisco (1988)
33. Pearl, J.: Causal inference in statistics - an overview. Stat. Surv. **3**, 96–146 (2009)
34. Pearl, J.: Causality - Models, Reasoning, and Inference, 2nd edn. Cambridge University Press, Cambridge (2009)
35. Deming, W.E., Stephan, F.F.: On a least squares adjustment of a sampled frequency table when the expected marginal totals are known. Ann. Math. Stat. **11**(4), 427–444 (1940)
36. Ireland, C.T., Kullback, S.: Contingency tables with given marginals. Biometrika **55**(1), 179–188 (1968)
37. Kullback, S.: Information Theory and Statistics. Wiley, New York (1959)
38. Kullback, S., Khairat, M.: A note on minimum discrimination information. Ann. Math. Stat. **37**, 279–280 (1966)
39. Stephan, F.F.: An iterative method of adjusting sample frequency tables when expected marginal totals are known. Ann. Math. Stat. **13**(2), 166–178 (1942)
40. Neyman, J.: Contribution to the theory of the x^2 test. In: Neyman, J. (ed.) Proceedings of the Berkeley Symposium on Mathematical Statistics and Probability, pp. 239–273. University of California Press, Berkeley, Los Angeles (1946)
41. Taylor, W.F.: Distance functions and regular best asymptotically normal estimates. Ann. Math. Stat. **24**(1), 85–92 (1953)

Risk-based Software Quality and Security Engineering in Data-intensive Environments
(Invited Keynote)

Michael Felderer[1,2]([✉])

[1] University of Innsbruck, Innsbruck, Austria
michael.felderer@uibk.ac.at
[2] Blekinge Institute of Technology, Karlskrona, Sweden

Abstract. The concept of risk as a measure for the potential of gaining or losing something of value has successfully been applied in software quality engineering for years, e.g., for risk-based test case prioritization, and in security engineering, e.g., for security requirements elicitation. In practice, both, in software quality engineering and in security engineering, risks are typically assessed manually, which tends to be subjective, non-deterministic, error-prone and time-consuming. This often leads to the situation that risks are not explicitly assessed at all and further prevents that the high potential of assessed risks to support decisions is exploited. However, in modern data-intensive environments, e.g., open online environments, continuous software development or IoT, the online, system or development environments continuously deliver data, which provides the possibility to now automatically assess and utilize software and security risks. In this paper we first discuss the concept of risk in software quality and security engineering. Then, we provide two current examples from software quality engineering and security engineering, where data-driven risk assessment is a key success factor, i.e., risk-based continuous software quality engineering in continuous software development and risk-based security data extraction and processing in the open online web.

Keywords: Risk assessment · Software quality engineering
Security engineering · Data engineering

1 Introduction

The concept of risk as a measure for the potential of gaining or losing something of value has successfully been applied in software quality and security engineering to support critical decisions.

In software quality engineering, the concept of risk has for instance been applied in risk-based testing, which consider risks of the software product as the guiding factor to steer all phases of a test process, i.e., test planning,

T. K. Dang et al. (Eds.): FDSE 2018, LNCS 11251, pp. 12–17, 2018.
https://doi.org/10.1007/978-3-030-03192-3_2

design, implementation, execution, and evaluation [1–3]. Risk-based testing is a pragmatic, in companies of all sizes widely used approach [4,5] which uses the straightforward idea to focus test activities on those scenarios that trigger the most critical situations of a software system [6]. In general, a risk is an event that may possibly occur and, if it occurs, it has (typically negative) consequences. Therefore, risks are determined by the two factors probability and impact. For testing purposes, the factor probability describes the likelihood that the negative event, e.g., a software failure, occurs and impact characterizes the cost if the failure it occurs in operation. Assessing the risk exposure of a software feature or component requires estimating both factors. Impact can in that context usually be derived from the business value associated to the feature defined in the software requirements specification. Probability is influenced by the implementation characteristics of the feature or component as well as the usage context in which the software system is applied.

In security engineering, the concept of risk in particular and risk management in general receives even more attention than in software quality engineering. Risks are often used as a guiding factor to define security measures throughout the software development lifecycle. For instance, Potter and McGraw [7] consider the process steps creating security misuse cases, listing normative security requirements, performing architectural risk analysis, building risk-based security test plans, wielding static analysis tools, performing security tests, performing penetration testing in the final environment, and cleaning up after security breaches. In security engineering, risk is determined by the probability that a threat will exploit a vulnerability and the impact of the resulting adverse consequence, or loss [8]. A threat is a cyber-based act, occurrence, or event that exploits one or more vulnerabilities and leads to an adverse consequence or loss. A vulnerability is a weakness in an information system, system security procedures, internal controls, or implementation that a threat could exploit to produce an adverse consequence or loss.

The overall risk management comprises the core activities risk identification, risk analysis, risk treatment, and risk monitoring [9]. In the risk identification phase, risk items are identified. In the risk analysis phase, the likelihood and impact of risk items and, hence, the risk exposure is estimated. Based on the risk exposure values, the risk items may be prioritized and assigned to risk levels defining a risk classification. In the risk treatment phase the actions for obtaining a satisfactory situation are determined and implemented. In the risk monitoring phase the risks are tracked over time and their status is reported. In addition, the effect of the implemented actions is determined. The activities risk identification and risk analysis are often collectively referred to as risk assessment while the activities risk treatment and risk monitoring are referred to as risk control.

Several methods to assess software or security risks are available (e.g., RisCal [10] for software risks and the Security Engineering Risk Analysis (SERA) Framework [8] for security risks). In practice, both, in software quality engineering and in security engineering, risks are typically assessed manually, which tends to be subjective, non-deterministic, error-prone and time-consuming.

However, in modern data-intensive environment like open online environments, continuous software development or IoT, the online, system or development environments continuously deliver data, which provides the possibility to automatically assess and utilize software and security risks. In the following two sections, we sketch two examples from software quality engineering and security engineering, where data-driven risk assessment plays a key role, i.e., risk-based continuous software quality engineering and risk-based security data extraction and processing.

2   Risk-Based Continuous Software Quality Engineering

In the data-intensive environment of modern continuous software development based on cloud technologies, system testing and release management merge and have to be performed continuously ranging from automated system testing (for critical system software potentially based on model-based testing), over manual acceptance testing to live online experimentation at runtime. There is unexploited potential to improve system testing, on the one hand by intelligent automation and on the other hand by complementing it with live experimentation. Live experimentation at runtime [11] allows to deploy faster and thus gaining the competitive advantage of giving customers earlier access to new functionality, to reach a larger population than possible with acceptance testing and to check functional as well as non-functional behavior. However, live experimentation can only be implemented for uncritical software components to avoid that critical defects or hazards occur during runtime. Therefore, a suitable software structure and software risk assessment based on automated data analytics (leading to risk analytics) is required to avoid the issue of live experimentation for critical software components prior to sufficient system testing. The three continuous software quality improvement aspects of risk analytics, intelligent test automation and live experimentation are shown together with their characteristics in Fig. 1.

The first aspect is automated software risk analytics. It processes structured, semi-structured and unstructured software product data (e.g., data from source code, test specifications, defects, design models, or requirements specifications), organizational data (e.g., data about the teams developing specific services), process data (e.g., data from the version control system, issue tracking data, or deployment and runtime data), and business data (e.g., data about the business value, market potential or cost of specific software services), which allows to automatically determine probability and impact for risk assessment. The risk information is then applied to perform intelligent test automation to support decisions on what to automate (test-case design, test data generation and test execution of specific components, scenarios or services) and when to automate (in which sequence and iteration) as second aspect. Finally, as a third aspect, if the risk level is moderate, even live experimentation can be performed to test functional and non-functional system properties.

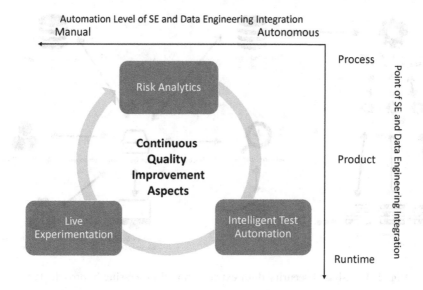

Fig. 1. Risk-based continuous software quality engineering

3   Risk-Based Security Data Extraction and Processing

The proposed approach to risk-based security data extraction and processing in the data-intensive environment of the open online web consists of two major components, i.e., a Security Data Collection and Analysis Component as well as a Security Knowledge Generation Component. The approach was originally presented in [12] and we refer to it here. Figure 2 shows the approach.

The Security Data Collection and Analysis Component is responsible for the data extraction from various data sources, quality assessment of data and data merging in order to provide the data in a processable form. It considers extraction from several online sources including vulnerability knowledge bases like Common Vulnerabilities and Exposures (CVE) [13] or the Malware Information Sharing Platform (MISP) [14], social media like Twitter as well as security forums or websites. Once the data is extracted and available, it must be formatted, the quality assessed and then merged. Because of the type of information being handled and the fact that there are different data fields to deal with, this is a highly complex task. In order to overcome differences, a general format is proposed, which includes information such as name, type, year, target platform, description and reference. It is the basis to automatically assess security risks.

The Security Knowledge Generation Component processes the extracted security information in order to provide it for different roles and various purposes, for instance as knowledge to stakeholders in the agile development process or to generate attack models. For instance, a developer can be provided with a security dashboard showing security risks or concrete guidelines on how code can be secured or security properties can be tested. As for the product owner, they can

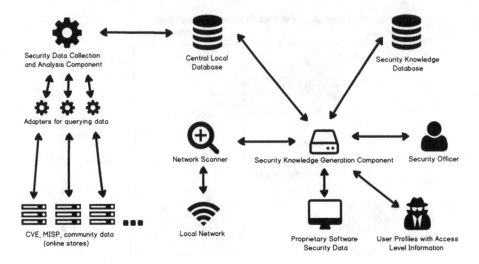

Fig. 2. Risk-based security data extraction and processing approach [12]

receive guidelines on security requirements and risk management. Finally, when developing a safety critical system, a developer who is responsible for the system architecture can be provided with generated attack models annotated with risk information that can be integrated with available system models to perform a combined safety-security analysis [15] or model-based security testing [16].

4   Conclusion

This paper sketched to approaches to automated risk assessment in data-intensive environments, i.e., risk-based continuous software quality engineering in continuous software development as well as risk-based security data extraction and processing in the open online web.

References

1. Gerrard, P., Thompson, N.: Risk-Based E-business Testing. Artech House Publishers, Norwood (2002)
2. Felderer, M., Ramler, R.: Integrating risk-based testing in industrial test processes. Softw. Qual. J. **22**(3), 543–575 (2014)
3. Felderer, M., Schieferdecker, I.: A taxonomy of risk-based testing. Int. J. Softw. Tools Technol. Transf. **16**(5), 559–568 (2014)
4. Felderer, M., Ramler, R.: A multiple case study on risk-based testing in industry. Int. J. Softw. Tools Technol. Transf. **16**(5), 609–625 (2014)
5. Felderer, M., Ramler, R.: Risk orientation in software testing processes of small and medium enterprises: an exploratory and comparative study. Softw. Qual. J. **24**(3), 519–548 (2016)

6. Wendland, M.F., Kranz, M., Schieferdecker, I.: A systematic approach to risk-based testing using risk-annotated requirements models. In: ICSEA 2012, pp. 636–642 (2012)
7. Potter, B., McGraw, G.: Software security testing. IEEE Secur. Priv. **2**(5), 81–85 (2004)
8. Alberts, C., Woody, C., Dorofee, A.: Introduction to the security engineering risk analysis (SERA) framework. Technical report, Carnegie Mellon University Software Engineering Institute, Pittsburgh, Pennsylvania (2014)
9. ISO: ISO 31000 - Risk Management (2018). http://www.iso.org/iso/home/standards/iso31000.htm
10. Haisjackl, C., Felderer, M., Breu, R.: Riscal-a risk estimation tool for software engineering purposes. In: 2013 39th EUROMICRO Conference on Software Engineering and Advanced Applications (SEAA), pp. 292–299. IEEE (2013)
11. Auer, F., Felderer, M.: Current state of research on continuous experimentation: a systematic mapping study. In: EUROMICRO Conference on Software Engineering and Advanced Applications (SEAA 2018). IEEE (2018)
12. Felderer, M., Pekaric, I.: Research challenges in empowering agile teams with security knowledge based on public and private information sources (2017)
13. MITRE: Common vulnerabilities and exposures. https://cve.mitre.org/
14. Andre, D.: Malware information sharing platform. http://www.misp-project.org/
15. Chockalingam, S., Hadžiosmanović, D., Pieters, W., Teixeira, A., van Gelder, P.: Integrated safety and security risk assessment methods: a survey of key characteristics and applications. In: Havarneanu, G., Setola, R., Nassopoulos, H., Wolthusen, S. (eds.) CRITIS 2016. LNCS, vol. 10242, pp. 50–62. Springer, Cham (2017). https://doi.org/10.1007/978-3-319-71368-7_5
16. Felderer, M., Zech, P., Breu, R., Büchler, M., Pretschner, A.: Model-based security testing: a taxonomy and systematic classification. Softw. Test. Verif. Reliab. **26**(2), 119–148 (2016)

Security and Privacy Engineering

A Secure and Efficient kNN Classification Algorithm Using Encrypted Index Search and Yao's Garbled Circuit over Encrypted Databases

Hyeong-Jin Kim, Jae-Hwan Shin, and Jae-Woo Chang[✉]

Department of Computer Engineering, Chonbuk National University, Jeonju, South Korea
{yeon_hui4,djtm99,jwchang}@jbnu.ac.kr

Abstract. Database outsourcing has been popular according to the development of cloud computing. Databases need to be encrypted before being outsourced to the cloud so that they can be protected from adversaries. However, the existing kNN classification scheme over encrypted databases in the cloud suffers from high computation overhead. So we proposed a secure and efficient kNN classification algorithm using encrypted index search and Yao's garbled circuit over encrypted databases. Our algorithm not only preserves data privacy, query privacy, and data access pattern. We show that our algorithm achieves about 17x better performance on classification time than the existing scheme, while preserving high security level.

Keywords: Database outsourcing · Data privacy · Query protection
Hiding data access pattern · kNN classification algorithm · Cloud computing

1 Introduction

Research on preserving data privacy in outsourced databases has been spotlighted with the development of a cloud computing. Since a data owner (DO) outsources his/her databases and allows a cloud to manage them, the DO can reduce the cost of data management by using the cloud's resources. However, because the data are private assets of the DO and may include sensitive information, they should be protected against adversaries including a cloud server. Therefore, the databases should be encrypted before being outsourced to the cloud. A vital challenge in the cloud computing is to protect both data privacy and query privacy. Meanwhile, during query processing, the cloud can derive sensitive information from the actual data items and users by observing data access patterns even if the data and the query are encrypted [1].

Meanwhile, a classification has been widely adopted in various fields such as marketing and scientific applications. Among various classification methods, a kNN classification algorithm is used in various fields because it does not require a time consuming learning process while guaranteeing good performance with moderate k [2]. When a query is given, a kNN classification first retrieves the kNN results for the query. Then, it determines the majority class label (or category) among the labels of kNN results. However, since the intermediate kNN results and the resulting class label are

© Springer Nature Switzerland AG 2018
T. K. Dang et al. (Eds.): FDSE 2018, LNCS 11251, pp. 21–38, 2018.
https://doi.org/10.1007/978-3-030-03192-3_3

closely related to the query, the queries should be more cautiously dealt to preserve the privacy of the users.

However, to the best of our knowledge, a kNN classification scheme proposed by Samanthula [3] is the only work that performs classification over the encrypted data in the cloud. The scheme preserves data privacy, query privacy, and intermediate results throughout the query processing. The scheme also hides data access pattern from the cloud. To achieve this, they adopt SkNN$_m$ [4] scheme among various secure kNN schemes [4–7] when retrieving k relevant records to a query. However, the scheme suffers from high computation overhead because it considers all the encrypted data during the query processing.

To solve the problem, in this paper, we propose a secure and efficient kNN classification algorithm over encrypted databases. Our algorithm can preserve data privacy, query privacy, the resulting class labels, and data access patterns from the cloud. To enhance the performance of our algorithm, we adopt the encrypted index scheme proposed in our previous work [7]. For this, we also propose efficient and secure protocols based on the Yao's garbled circuit [8] and a data packing technique.

The rest of the paper is organized as follows. Section 2 introduces the related work. Section 3 presents our overall system architecture and various secure protocols. Section 4 proposes our kNN classification algorithm over encrypted databases. Section 5 presents the performance analysis. Finally, Sect. 6 concludes this paper with some future research directions.

2   Background and Related Work

2.1   Background

Paillier Crypto System. The Paillier cryptosystem [9] is an additive homomorphic and probabilistic asymmetric encryption scheme for public key cryptography. The public encryption key pk is given by (N, g), where N is a product of two large prime numbers p and q, and g is in $Z_{N^2}^*$. Here, $Z_{N^2}^*$ denotes an integer domain ranging from 0 to N^2. The secret decryption key sk is given by (p, q). Let $E()$ and $D()$ denote the encryption and decryption functions, respectively. The Paillier crypto system provides the following properties. (i) Homomorphic addition: The product of two ciphertexts $E(m_1)$ and $E(m_2)$ results in the encryption of the sum of their plaintexts m_1 and m_2. (ii) Homomorphic multiplication: The bth power of ciphertext $E(m_1)$ results in the encryption of the product of b and m_1. (iii) Semantic security: Encrypting the same plaintexts using the same encryption key does not result in the identical ciphertexts. Therefore, an adversary cannot infer any information about the plaintexts.

Yao's Garbled Circuit. Yao's garbled circuits [8] allows two parties holding inputs x and y, respectively, to evaluate a function $f(x, y)$ without leaking any information about the inputs beyond what is implied by the function output. One party generates an encrypted version of a circuit to compute f. The other party obliviously evaluates the output of the circuit without learning any intermediate values. Therefore, the Yao's garbled circuit provides high security level. Another benefit of using the Yao's garbled circuit is that it can provide high efficiency if a function can be realized with a reasonably small circuit.

Adversarial Models. There are two main types of adversarial models, *semi-honest* and *malicious* [10, 11]. In this paper, we assume that clouds act as insider adversaries with high capability. In the *semi-honest* adversarial model, the clouds honestly follow the protocol specification, but try to use the intermediate data in malicious way to learn forbidden information. In the *malicious* adversarial model, the clouds can arbitrarily deviate from the protocol specification. Protocols against malicious adversaries are too inefficient to be used in practice while protocols under the *semi-honest* adversaries are acceptable in practice. Therefore, by following the work done in [4, 10], we also consider the semi-honest adversarial model in this paper.

2.2 Secure kNN Classification Schemes

To the best of our knowledge, Samanthula proposed a kNN classification scheme (PPkNN) [3], which is the only work that performs classification over the encrypted data. The scheme performs SkNN$_m$ [4] scheme to retrieve k relevant records to a query and determines the class label of the query. The scheme can preserve both data privacy and query privacy while hiding data access pattern. However, the scheme suffers from the high computation overhead because it directly adopts the SkNN$_m$ scheme.

3 System Architecture and Secure Protocols

In this section, we explain our overall system architecture and present generic secure protocols used for our kNN classification algorithm.

3.1 System Architecture

We provide the system architecture of our scheme, which is designed by adopting that of our previous work [7]. Our previous work has a disadvantage that comparison operations cause high overhead by using encrypted binary arrays [7]. To solve this problem, we propose an efficient query processing algorithm that performs comparison operations through yao's garbled circuits [8]. Figure 1 shows the overall system architecture and Table 1 summarizes common notations used in this paper. The system consists of four components: data owner (DO), authorized user (AU), and two clouds (C_A and C_B). The DO stores the original database (T) consisting of n records. A record $t_i (1 \leq i \leq n)$ consists of $(m + 1)$ attributes and $t_{i,j}$ denotes the jth attribute value of t_i. A class label of t_i is stored in $(m + 1)^{th}$ attribute, i.e., $t_{i,m+1}$. We do not consider $(m + 1)^{th}$ attribute when making an index using T. Therefore, the DO indexes on T by using a kd-tree, based on $t_{i,j} (1 \leq i \leq n$ and $1 \leq j \leq m)$. The reason why we utilize a kd-tree (k-dimensional tree) as a space-partitioning data structure is that it not only can evenly partition data into each node, but also is useful for organizing points in a k-dimensional space [14]. When we visit the tree in a hierarchical manner, access patterns can be disclosed. Consequently, we only consider the leaf nodes of the kd-tree and all of the leaf nodes are retrieved once during the query processing step. Let h denote the level of the kd-tree and F be a fan-out which is the maximum number of data to be stored in each node. The total number of leaf nodes is 2^{h-1}. Henceforth, a node refers to a leaf node. The region information of each node is represented as both the lower bound $lb_{z,j}$ and the upper bound $ub_{z,j}(1 \leq z \leq 2^{h-1}, 1 \leq j \leq m)$. Each node stores the identifiers (id) of data located in the node region. Although we consider the kd-tree in this paper, another index structure whose nodes store region information can be applied to our scheme.

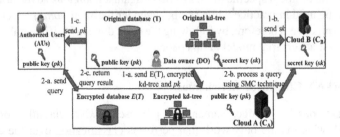

Fig. 1. The overall system architecture

Table 1. Common notations

Notations	Description
$E()$, $D()$	Encryption function and decryption function
t_i, $t_{i,j}$	i^{th} record and j^{th} attribute value of ith record
t'_i	i^{th} extracted record during the index search
q, q_j	a query of a user and j^{th} attribute value of a query q
$node_z$	z^{th} node of the kd-tree
$node_z.t_{s,j}$	j^{th} attribute of sth record stored in z^{th} node of the kd-tree
$lb_{z,j}$, $ub_{z,j}$	j^{th} attribute value of lower/upper bound of zth kd-tree node
r	Random integers

To preserve data privacy, the DO encrypts T attribute-wise by using the public key (pk) of the Paillier cryptosystem [9] before outsourcing the database. Thus, the DO generates $E(t_{i,j})$ for $1 \leq i \leq n$ and $1 \leq j \leq m$. The DO also encrypts the region information of all kd-tree nodes to support efficient query processing. Specifically, $E(lb_{z,j})$ and $E(ub_{z,j})$ are generated with $1 \leq z \leq 2^{h-1}$ and $1 \leq j \leq m$ by encrypting lb and ub of each node attribute-wise. Assuming that C_A and C_B are non-colluding and semi-honest (or honest-but-curious) clouds, they correctly execute the assigned protocols, but an adversary may try to obtain additional information from the intermediate data while executing the assigned protocol. This assumption is not new and has been considered in earlier work [4, 10]. Specifically, because most cloud services are provided by renowned IT companies, collusion between them that would blemish their reputations is improbable [4].

To process kNN classification algorithm over the encrypted database, we utilize a secure multiparty computation (SMC) between C_A and C_B. To do this, the DO outsources both the encrypted database and its encrypted index to a cloud with pk, C_A in this case, but it sends sk to a different cloud, C_B in this case. In addition, the DO outsources the list of encrypted class labels denoted by $E(label_i)$ for $1 \leq i \leq w$ to C_A. The encrypted index includes the region information of each node in cipher-text and the ids of data located in the node in plaintext. The DO also sends pk to AUs to allow them to encrypt a query. At query time, an AU encrypts a query attribute-wise. The encrypted query is denoted by $E(q_j)$ for $1 \leq j \leq m$. C_A processes the query with the help of C_B and sends the query result to the AU.

As an example, assume that an AU has eight data instances as depicted in Fig. 2. Each data t_i is depicted with its class label (e.g., 3 in case of t_6). The data are partitioned into four nodes (e.g., $node_1$– $node_4$) for a kd-tree. The DO encrypts each data instance and the region of each node attribute-wise. For example, t_6 is encrypted as $E(t_6) = \{E(8), E(5), E(3)\}$ because the values of x-axis and y-axis are 8 and 5, respectively, and the class label of t_6 is 3. Meanwhile, the $node_1$ is encrypted as $\{\{E(0), E(0)\}, \{E(5), E(5)\}, \{1, 2\}\}$ because the lb and ub of $node_1$ are $\{0, 0\}$ and $\{5, 5\}$, respectively, and the $node_1$ stores both t_1 and t_2.

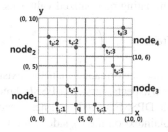

Fig. 2. An example in two-dimensional space

3.2 Secure Protocols

Our kNN classification algorithm is constructed using several secure protocols. In this section, all of the protocols except the SBN are performed with the SMC technique between C_A and C_B. The SBN can be solely executed by C_A. Due to space limitations, we briefly introduce five secure protocols found in the literature [3, 4, 7, 10]. (i) SM (Secure Multiplication) [4] computes the encryption of $a \times b$, i.e., $E(a \times b)$, when two encrypted data $E(a)$ and $E(b)$ are given as inputs. (ii) SBN (Secure Bit-Not) [7] performs a bit-not operation when an encrypted bit $E(a)$ is given as an input. (iii) CMP-S [10] returns 1 if $u < v$, 0 otherwise, when $-r_1$ and $-r_2$ are given from C_A as well as $u + r_1$ and $v + r_2$ are given from C_B. (iv) SMS_n (Secure Minimum Selection) [10] returns the minimum value among the inputs by performing the CMP-S for $n - 1$ times when $E(d_i)$ for $1 \leq i \leq n$ are given as inputs. (v) SF (Secure Frequency) [3] returns $E(f(label_j))$, the number of occurrence of each $E(label_j)$ in $E(c_i)$, when both $E(c_j)$ for $1 \leq i \leq k$ and $E(label_j)$ for $1 \leq j \leq w$ are given as inputs.

Meanwhile, we propose new secure protocols, i.e., ESSED, GSCMP, and GSPE. Contrary to the existing protocols, the proposed protocols do not take the encrypted binary representation of the data, like $E(0)$ or $E(1)$, as inputs. Therefore, our protocols can provide a low computation cost. Next, we propose our new secure protocols.

ESSED Protocol. ESSED (Enhanced Secure Squared Euclidean Distance) computes $E(|X-Y|^2)$ when two encrypted vectors $E(X)$ and $E(Y)$ are given as inputs, where X and Y consist of m attributes. To enhance the efficiency, we pack λ number of σ-bit data instances to generate a packed value. The overall procedure of ESSED is as follows. First, C_A generates random numbers r_j for $1 \leq j \leq m$ and packs them by computing $R = \sum_{j=1}^{m} r_j \times 2^{\sigma(m-j)}$. Then, C_A generates $E(R)$ by encrypting R. Second, C_A calculates $E(x_j-y_j)$ attribute-wise and packs them by computing $E(v) = \prod_{j=1}^{m} E(x_j - y_j)^{2^{\sigma(m-j)}}$. Then, C_A computes $E(v) = E(v) \times E(R)$ and sends $E(v)$ to C_B. Third, assuming that w_j denotes $x_j - y_j + r_j (1 \leq j \leq m)$, C_B acquires $v = [w_1|...|w_m]$ by decrypting $E(v)$. C_B obtains w_j for $1 \leq j \leq m$ by unpacking v through $v \times 2^{-\sigma(m-j)}$. Here, each instance of w_j represents the randomized distance of two input vectors for each attribute. C_B also calculates w_j^2 attribute-wise and stores their sum into d. C_B encrypts d and sends $E(d)$ to C_A. Finally, C_A obtains $E(|X-Y|^2)$ by eliminating randomized values using the following Eq. (1).

$$E(|X-Y|^2) = E(d) \times \prod_{j=1}^{m} \left(E(x_j - y_j)^{-2r_j} \times E(r_j^2)^{-1} \right) \tag{1}$$

Our ESSED achieves better performance than the existing distance computation protocol, DPSSED [10], in two aspects. First, our ESSED requires only one encryption operation on the C_B side while DPSSED needs m times. Second, our ESSED calculates the randomized distance in plaintext on the C_B side while DPSSED computes the sum of the squared Euclidean distances among all attributes over ciphertext on the C_A side. Therefore, the number of computations on encrypted data in our ESSED can be reduced greatly.

GSCMP Protocol. When $E(u)$ and $E(v)$ are given as inputs, GSCMP (Garbled Circuit based Secure Compare) protocol returns 1 if $u \leq v$, 0 otherwise. The main difference between GSCMP and CMP-S is that GSCMP receives encrypted data as inputs while CMP-S receives the randomized plaintext. The overall procedure of the GSCMP is as follows. First, C_A generates two random numbers r_u and r_v, and encrypts them. C_A computes $E(m_1) = E(u)^2 \times E(r_u)$ and $E(m_2) = E(v)^2 \times E(1) \times E(r_v)$. Second, C_A randomly selects one functionality between $F_0 : u > v$ and $F_1 : v > u$. The selected functionality is oblivious to C_B. Then, C_A sends data to C_B, depending on the selected functionality. If $F_0 : u > v$ is chosen, C_A sends $<E(m_2), E(m_1)>$ to C_B. If $F_1 : u < v$ is chosen, C_A sends $<E(m_1), E(m_2)>$ to C_B. Third, C_B obtains $<m_2, m_1>$ by decrypting $<E(m_2), E(m_1)>$ if $F_0 : u > v$ is chosen. If $F_1 : u < v$ is chosen, C_B obtains $<m_1, m_2>$ by decrypting $<E(m_1), E(m_2)>$. Fourth, C_A generates a garbled circuit consisting of two ADD circuits and one CMP circuit. Here, ADD circuit takes two integers u and v as input, and outputs $u + v$ while CMP circuit takes two integers u and v as input, and outputs 1 if $u < v$, 0 otherwise. If $F_0:u > v$ is selected, C_A puts $-r_v$ and $-r_u$ into the 1st and 2nd ADD gates, respectively. If $F_1:u < v$ is selected, C_A puts $-r_u$ and $-r_v$ into the 1st and 2nd ADD gates. Fifth, if $F_0:u > v$ is selected, C_B puts m_2 and m_1 into the 1st and 2nd ADD gates, respectively. If $F_1:u < v$ is selected, C_B puts m_1 and m_2 into the 1st and 2nd ADD gates. Sixth, the 1st ADD gate adds two input values and puts the output $result_1$ into CMP gate. Similarly, the 2nd ADD gate puts the output $result_2$ into CMP gate. Seventh, CMP gate outputs $\alpha = 1$ if $result_1 < result_2$ is true, $\alpha = 0$ otherwise. The output of the CMP is returned to the C_B. Then, C_B encrypts α and sends $E(\alpha)$ to C_A. Finally, only when the selected functionality is $F_0:u > v$, C_A computes $E(\alpha) = SBN(E(\alpha))$ and returns the final $E(\alpha)$. If $E(\alpha)$ is $E(1)$, u is less than v.

GSPE Protocol. GSPE (Garbled circuit based Secure Point Enclosure) protocol returns $E(1)$ when p is inside a *range* or on a boundary of the *range*, $E(0)$ otherwise. GSPE takes an encrypted point $E(p)$ and an encrypted range $E(range)$ as inputs. Here, the *range* consists of the $E(lb_j)$ and the $E(ub_j)$ for $1 \leq j \leq m$. If $E(p_j) \leq E(range.ub_j)$ and $E(p_j) \geq E(range.lb_j)$, the p is inside a *range*. The overall procedure of the GSPE is as follows. First, C_A generates two random numbers ra_j and rb_j for $1 \leq j \leq 2m$. C_A obtains packed values RA and RB by packing ra_j and rb_j, respectively, using the following Eq. (2) for $1 \leq j \leq 2m$.

$$RA = \sum_{j=1}^{2m} ra_j \times 2^{\sigma(2m-j)}, RB = \sum_{j=1}^{2m} rb_j \times 2^{\sigma(2m-j)} \qquad (2)$$

Here, σ means the bit length to represent a data. Then, C_A generates $E(RA)$ and $E(RB)$ by encrypting RA and RB. Second, C_A computes $E(\mu_j) = E(p_j)^2$ and $E(\omega_j) = E(range.lb_j)^2$ for $1 \leq j \leq m$. C_A also computes $E(\delta_j) = E(p_j)^2 \times E(1)$ and $E(\rho_j) = E(range.ub_j)^2 \times E(1)$ for $1 \leq j \leq m$. Third, C_A randomly selects one functionality between $F_0:u > v$ and $F_1:v > u$. Then, C_A performs data packing by using the $E(\mu_j)$ and $E(\rho_j)$, depending on the selected functionality.

- If $F_0: u > v$ is selected, compute

$$E(RA) = E(RA) \times E(\rho_j)^{2^{\sigma(2m-j)}}, E(RB) = E(RB) \times E(\mu_j)^{2^{\sigma(2m-j)}}$$

- If $F_1: v > u$ is selected, compute

$$E(RA) = E(RA) \times E(\mu_j)^{2^{\sigma(2m-j)}}, E(RB) = E(RB) \times E(\rho_j)^{2^{\sigma(2m-j)}}$$

In addition, C_A performs data packing by using the $E(\omega_j)$ and $E(\delta_j)$, depending on the selected functionality. Then, C_A sends packed values $E(RA)$ and $E(RB)$ to C_B.

- If $F_0: u > v$ is selected, compute

$$E(RA) = E(RA) \times E(\delta_j)^{2^{\sigma(2m-j)}}, E(RB) = E(RB) \times E(\omega_j)^{2^{\sigma(2m-j)}}$$

- If $F_1: v > u$ is selected, compute

$$E(RA) = E(RA) \times E(\omega_j)^{2^{\sigma(2m-j)}}, E(RB) = E(RB) \times E(\delta_j)^{2^{\sigma(2m-j)}}$$

Fourth, C_B obtains RA and RB by decrypting $E(RA)$ and $E(RB)$. C_B computes $ra_j + u_j$ $\leftarrow RA \times 2^{-\sigma(2m-j)}$ and $rb_j + v_j \leftarrow RB \times 2^{-\sigma(2m-j)}$ for $1 \leq j \leq 2m$. Here, u_j (or v_j) is one of the $\mu_j, \rho_j, \omega_j,$ and δ_j. Fifth, C_A generates CMP-S circuit and puts $-ra_j$ and $-rb_j$ into CMP-S while C_B puts $ra_j + u_j$ and $rb_j + v_j$ into CMP-S for $1 \leq j \leq 2m$. Once four inputs (i.e., $-ra_j, -rb_j, ra_j + u_j$ and $rb_j + v_j$) are given to CMP-S, the output α'_j is returned to C_B. Then, C_B encrypts α' and sends $E(\alpha')$ to C_A. Sixth, C_A performs $E\left(\alpha'_j\right) = \text{SBN}\left(E\left(\alpha'_j\right)\right)$ for $1 \leq j \leq 2m$ only when the selected functionality is $F_0:u > v$. Then, C_A computes $E(\alpha) = \text{SM}\left(E(\alpha), E\left(\alpha'_j\right)\right)$ where the initial value of $E(\alpha)$ is $E(1)$. Only when all of the $E\left(\alpha'_j\right)$ for $1 \leq j \leq 2m$ are $E(1)$, the value of $E(\alpha)$ remains $E(1)$. Finally, GSPE outputs the final $E(\alpha)$. The p is inside the *range* if the final $E(\alpha)$ is $E(1)$.

SXS$_n$ Protocol. SXS$_n$ (Secure Maximum Selection) returns the maximum value among the inputs when $E(d_i)$ for $1 \leq i \leq n$ are given as inputs. SXS$_n$ can be realized by converting the logic of SMS$_n$ in opposite way. Therefore, we omit the detailed procedure of SXS$_n$ due to the space limitation.

4   KNN Classification Algorithm

In this section, we present our kNN classification algorithm (SkNNC$_G$) which uses the Yao's garbled circuit. Our algorithm consists of four steps; encrypted kd-tree search step, kNN retrieval step, result verification step, and majority class selection step.

4.1   Step 1: Encrypted kd-Tree Search Step

In the encrypted kd-tree search phase, the C_A securely extracts all of the data from a node containing a query point while hiding the data access patterns. To obtain high efficiency, we redesign the index search scheme proposed in our previous work [7]. Specifically, our algorithm does not require operations related to the encrypted binary representation which causes high computation overhead. In addition, we utilize our newly proposed secure protocols based on Yao's garbled circuit.

Algorithm 1: Encrypted kd-tree search

Input : E(q), E(node)
Output : E(cand)   // all the data inside nodes related to a query
C_A : **01:** for $1 \leq z \leq num_{node}$ // $num_{node} = 2^{h-1}$ (h : level of the kd-tree)

   02: $E(\alpha_z) \leftarrow GSPE(E(q), E(node_z))$
   03: $E(\alpha') \leftarrow \pi(E(\alpha))$;   send $E(\alpha')$ to C_B
C_B : **04:** $\alpha' \leftarrow D(E(\alpha'))$; $c \leftarrow$ the number of '1' in α'
   05: create c number of Group
   06: for each NG
   07:   assign a node with $\alpha'=1$ and $num_{node}/c-1$ nodes with $\alpha'=0$
   08: $NG' \leftarrow$ shuffle the *ids* of nodes
   09: send NG' to CA
C_A : **10:** $cnt \leftarrow 0$
   11: $NG^* \leftarrow$ permute node *ids* using π^{-1} for each NG'
   12: for each NG^*
   13:   for $1 \leq s \leq F$
   14:   for $1 \leq i \leq num$ (# nodes in the selected NG^*)
   15: $z = id$ of ith node of NG^*
   16: $E(t'_{i,j}) \leftarrow SM(node_z.t_{s,j}, E(\alpha_z))$ for $1 \leq j \leq m+1$
   17:   for $1 \leq j \leq m+1$
   18: $E(cand_{cnt,j}) \leftarrow \prod_{i=1}^{num} E(t'_{i,j})$
   19: $cnt \leftarrow cnt+1$
   20: return E(*cand*)

The procedure of the encrypted kd-tree search step is shown in Algorithm 1. First, C_A securely finds nodes which include a query by executing $E(\alpha_z) \leftarrow$ $GSPE(E(q), E(node_z))$ for $1 \leq z \leq num_{node}$ where num_{node} means the total number of kd-tree leaf nodes (lines 1–2). Note that the nodes with $E(\alpha_z) = E(1)$ are related to the query, but both C_A and C_B cannot know whether or not the value of each $E(\alpha_z)$ is $E(1)$, because the Paillier encryption provides semantic security. Then, we partially perform the index search algorithm in [7]. Specifically, C_A generates $E(\alpha')$ by permuting $E(\alpha)$ using a random permutation function π and then sends $E(\alpha')$ to C_B (line 3). For example, the output of GSPE is $E(\alpha) = \{E(1), E(0), E(0), E(0)\}$ in Fig. 2 because the q is given inside the $node_1$. Assuming that π permutes data in reverse way, C_A sends the $E(\alpha') = \{E(0), E(0), E(0), E(1)\}$ to C_B.

Third, C_B obtains α' by decrypting $E(\alpha')$ and counts the number of $\alpha' = 1$ and stores it into c. Here, c means the number of nodes that the query is related to (line 4). Fourth, C_B creates c number of node groups. Assuming that NG denotes a node group, C_B assigns to each NG both a node with $\alpha' = 1$ and $num_{node}/c - 1$ nodes with $\alpha' = 0$. Then, C_B obtains NG' by randomly shuffling the *ids* of nodes in each NG and sends NG' to C_A (lines 5–9). For example, C_B can obtain $\alpha' = \{0, 0, 0, 1\}$ which contains one at the fourth position. Because one node group is required, C_B assigns all nodes to one node group and randomly shuffles the *ids* of the nodes, i.e., $NG'_1 = \{2, 1, 3, 4\}$.

Fifth, C_A obtains NG^* by permuting the *ids* of nodes using π^{-1} in each NG' (line 11). Six, C_A gets access to one datum in a node for each NG^* and executes $E\left(t'_{i,j}\right) = \mathrm{SM}\left(E\left(node_z.t_{s,j}\right), E\left(\alpha_z\right)\right)$ for $1 \leq s \leq F$ and $1 \leq j \leq m+1$ where $E\left(\alpha_z\right)$ is the result of GSPE corresponding to $node_z$ (line 12–16). As a result, SM results in $E\left(node_z.t_{s,j}\right)$ only for the data inside the nodes related to the query because their $E\left(\alpha_z\right)$ values are $E(1)$; otherwise SM results in $E(0)$. If a node has the less number of data than F, it performs SM by using $E(max)$, instead of using $E\left(node_z.t_{s,j}\right)$. Here, $E(max)$ is the largest value in the domain. When C_A accesses one datum from every node in a NG^*, C_A performs $E\left(cand_{cnt,j}\right) \leftarrow \prod_{i=1}^{num} E(t'_{i,j})$ where *num* means the total number of nodes in the selected NG^* (line 17–18). As a result, a datum in the nodes related to the query is securely extracted without revealing the data access patterns because the searched nodes are not revealed. By repeating these steps, all of the data in the nodes are safely stored into the $E\left(cand_{cnt,j}\right)$ for $1 \leq i \leq cnt$ and $1 \leq j \leq m+1$ where *cnt* means the total number of data extracted during the index search. For example, C_A obtains $NG^*_1 = \{3, 4, 2, 1\}$ by permuting the $NG'_1 = \{2, 1, 3, 4\}$ using π^{-1}. C_A gains access to $E(t_5)$ in $node_3$, $E(t_7)$ in $node_4$, $E(t_3)$ in $node_2$, and $E(t_1)$ in $node_1$. The results of SM using $E(t_5)$, $E(t_7)$, and $E(t_3)$ are $E(0)$ for all attributes because $E(\alpha_z)$ for the corresponding nodes are $E(0)$. The results are stored into $E\left(t'_1\right), E\left(t'_2\right)$ and $E\left(t'_3\right)$, respectively. However, the results of SM using $E(t_1)$ become $\{E(2), E(1), E(1)\}$ because the values of x-axis and y-axis are 2 and 1, respectively, and the class label of t_1 is 1. The results are stored into $E\left(t'_4\right)$. Thus, the final attribute-wise homomorphic addition of $E(t'_i)$ for $1 \leq i \leq 4$ are $\{E(2), E(1), E(1)\}$. Accordingly, one datum $E(t_1)$ in $node_1$ is securely extracted. By repeating this, the encrypted kd-tree search step can extract all of the data in $node_1$ (e.g., $E(t_1)$ and $E(t_2)$) and finally stores them into $E(cand)$.

4.2   Step 2: kNN Retrieval Step

In the *k*NN retrieval phase, we retrieve the *k* closest data from the query by partially utilizing the S*k*NN$_m$ scheme [4]. However, we only consider $E\left(cand_i\right)$ for $1 \leq i \leq cnt$, which are extracted in the index search phase, whereas the S*k*NN$_m$ considers all the encrypted data. In addition, we utilize our efficient protocols which require relatively low computation costs, instead of using the existing expensive protocols. The procedure of the *k*NN retrieval step is shown in Algorithm 2.

Algorithm 2: kNN retrieval

Input : E(q), E($cand$), k

Output : E(t') // candidate kNN result

C_A : **01:** for $1 \leq i \leq cnt$

    02:   E(d_i) $\leftarrow$ ESSED(E(q), E($cand_i$))

    03: for $1 \leq s \leq k$

    04:   E(d_{min}) $\leftarrow$ SMS$_n$(E(d_1), ..., E(d_{cnt}))

    05:   for $1 \leq i \leq cnt$

    06:   E(τ_i) $\leftarrow$ E(d_{min})$\times$E(d_i)$^{N-1}$;   E(τ'_i) $\leftarrow$ E(τ_i)^{r_i}.

    07:   E(β) $\leftarrow$ π(τ');   send E(β) to C_B

C_B : **08:**   for $1 \leq i \leq cnt$

    09:   if D(β_i)=0 then E(U_i) $\leftarrow$ E(1);   else E(U_i) $\leftarrow$ E(0)

    10:   send E(U) to C_A

C_A : **11:**   E(V) $\leftarrow$ π^{-1}(U)

    12:   for $1 \leq i \leq cnt$

    13:   E($V'_{i,j}$)$\leftarrow$SM(E(V_i),E($cand_{i,j}$)) for $1 \leq j \leq m+1$

    14:   E($t'_{s,j}$) = $\prod_{i=1}^{cnt} E(V'_{i,j})$ for $1 \leq j \leq m+1$

    15:   if $s < k$

    16:   E(d_i)$\leftarrow$SM(E(V_i), E(max))$\times$SM(SBN(E(V_i)), E(d_i))

17: return E(t')

First, using our proposed ESSED, C_A securely calculates the squared Euclidean distances $E(d_i)$ between a query and $E(cand_i)$ for $1 \leq i \leq cnt$ (lines 1–2). Then, instead of using the inefficient SMIN$_n$, C_A performs SMS$_n$ to find the minimum value $E(d_{min})$ among $E(d_i)$ for $1 \leq i \leq cnt$. Second, C_A calculates $E(\tau_i) = E(d_{min}) \times E(d_i)^{N-1}$, i.e., the difference between the $E(d_{min})$ and $E(d_i)$, for $1 \leq i \leq cnt$. Then, C_A computes $E(\tau'_i) = E(\tau_i)^{r_i}$ (lines 3–6). Note that only the $E(\tau'_i)$ corresponding to the $E(d_{min})$ has a value of $E(0)$. C_A obtains $E(\beta)$ by shuffling $E(\tau')$ using a random permutation function π and then sends $E(\beta)$ to the C_B (line 7). For example, because $E(cand) = \{E(t_1), E(t_2)\}$ is given from the index search phase, $E(d_1) = E(4)$ and $E(d_2) = E(5)$. By performing SMS$_n$, $E(d_{min})$ is set as $E(4)$. Then, $E(\tau')$ is computed as $\{E(0), E(-r)\}$. The $E(\tau'_i)$ with $E(0)$ corresponds to the $E(d_{min})$, i.e., $E(t_1)$. Assuming that π permutes data in reverse way, C_A sends the $E(\beta) = \{E(-r), E(0)\}$ to C_B. Third, after decrypting $E(\beta)$, C_B sets $E(U_i) = E(1)$ if $E(\beta_i) = 0$, and sets $E(U_i) = E(0)$ otherwise. After C_B sends $E(U)$ to C_A, C_A obtains $E(V)$ by permuting $E(U)$ using π^{-1} (line 8–11). Then, C_A performs SM protocol by using $E(V_i)$ and $E(cand_{i,j})$ to obtain $E\left(V'_{i,j}\right)$. By computing $E\left(t'_{s,j}\right) = \prod_{i=1}^{cnt} E(V'_{i,j})$ for $1 \leq j \leq m + 1$, C_A can securely extract the datum corresponding to the $E(d_{min})$ (line 12–14). For example, C_B sends $E(U) = \{E(0), E(1)\}$ because the $\beta_2 = 1$. Then, C_A obtains $E(V) = \{E(1), E(0)\}$ by permuting $E(U)$ using π^{-1}. For the x-attribute, C_A performs SM(E($cand_{1,1}$), E(V_1)) = E(2) and SM(E($cand_{2,1}$), E(V_2)) = E(0). By adding the two values, the x-attribute value of E(t_1), i.e., E(2), is securely calculated. Similarly, we can compute E(1), the y-attribute value of $E(t_1)$. Therefore, we can store $E(t_1) = \{E(2), E(1)\}$ into $E(t_1')$ without revealing data access patterns. Finally, to prevent the selected result from being selected in later phase, C_A securely updates the distance of the selected result as $E(max)$ by performing $E(d_i) = \text{SM}(E(V_i), E(max)) \times \text{SM}(\text{SBN}(E(V_i)), E(d_i))$ (line 15-16). This procedure is repeated for k rounds to find the kNN result. For example, in the first round, $E(t_1)$ with distance $E(4)$ is securely selected as the 1NN result and $E(t_2)$ with $E(d_2) = E(5)$ is selected in the second round as the 2NN result.

4.3   Step 3: Result Verification Step

The result of the step 2 may not be accurate because they are retrieved over the partial data being extracted in the step 1. Therefore, the result verification is essential to confirm the correctness of the current query result. Specifically, assuming that $dist_k$ denotes the squared Euclidean distance between the kth closest result, i.e., $E(t_k')$, and the query, the neighboring nodes located within $dist_k$ in the kd-tree need to be searched. For this reason, we use the concept of shortest point (sp) defined in [7]. The sp is a point in a given node whose distance is closest to a given point p as compared with the other points in the node. To find an sp in each node, we use the following properties. (i) If both the lower bound (lb) and the upper bound (ub) of the node are lesser than p, the ub is the sp. (ii) If both the lb and the ub of the region are greater than p, the lb is the sp. (iii) If p is between the lb and the ub of the region, p is the sp. To enhance the efficiency of the result verification algorithm in the previous work [7], we use our newly proposed protocols instead of using the existing expensive protocols.

Algorithm 3: Result verification

Input : E(q), E(node), E(t'), k
Output : E(c)
C_A : **01:** $E(dist_k) = \text{ESSED}(E(q), E(t'_k))$
02: for $1 \leq z \leq num_{node}$
03:   for $1 \leq j \leq m$
04: $E(\psi_1) \leftarrow \text{GSCMP}(E(q_j), E(node_z.lb_j))$
05: $E(\psi_2) \leftarrow \text{GSCMP}(E(q_j), E(node_z.ub_j))$
06: $E(\psi_3) \leftarrow E(\psi_1) \times E(\psi_2) \times \text{SM}(E(\psi_1), E(\psi_2))^{N-2}$ // bit-xor
07: $E(temp) \leftarrow \text{SM}(E(\psi_1), E(node_z.lb_j))$
08: $E(temp) \leftarrow E(temp) \times \text{SM}(\text{SBN}(E(\psi_1)), E(node_z.ub_j))$
09: $E(temp) \leftarrow \text{SM}(E(temp), \text{SBN}(E(\psi_3)))$
10: $E(sp_{z,j}) \leftarrow E(temp) \times \text{SM}(E(\psi_3), E(q_j))$
11: $E(spdist_z) \leftarrow \text{ESSED}(E(q), E(sp_z))$
12: $E(spdist_z) = \text{SM}(E(\alpha_z), E(max)) \times \text{SM}(\text{SBN}(E(\alpha_z)), E(spdist_z))$
13: $E(\alpha_z) \leftarrow \text{GSCMP}(E(spdist_z), E(dist_k))$
14: $E(t'') \leftarrow$ perform 4~20 lines of Algorithm 1
15: $E(t') \leftarrow$ append the $E(t'')$ to $E(t')$
16: $E(result) \leftarrow$ perform Algorithm 2
17: for $1 \leq i \leq k$
18: $E(c_i) \leftarrow E(result_{i,m+1})$
19: return $E(c)$

The procedure of the result verification step is shown in Algorithm 3. First, C_A computes $E(dist_k) = \text{ESSED}(E(q), E(t'_k))$ to calculate the squared Euclidean distance between the query and the kth closest result among E(t'), i.e., the output of the kNN retrieval step (line 1). Second, C_A performs GSCMP by using $E(q_j)$ and $E(node_z.lb_j)$ for $1 \leq z \leq num_{node}$ and $1 \leq j \leq m$ and then stores the result in $E(\psi_1)$. C_A also performs GSCMP by using $E(q_j)$ and $E(node_z.ub_j)$ for $1 \leq z \leq num_{node}$ and $1 \leq j \leq m$ and then stores the result into $E(\psi_2)$. In addition, C_A calculates $E(\psi_3)$ by executing $E(\psi_1) \times E(\psi_2) \times \text{SM}(E(\psi_1), E(\psi_2))^{N-2}$ to obtain the result of bit-xor operation between $E(\psi_1)$ and $E(\psi_2)$ (lines 3–6). Note that "−2" is equivalent to "$N - 2$" under Z_N. Third, C_A securely obtains the shortest point of each node, i.e., $E(sp_{z,j})$, by executing $\text{SM}(E(\psi_3), E(q_j)) \times \text{SM}(\text{SBN}(E(\psi_3)), f(E(lb_{z,j}), E(ub_{z,j})))$ for $1 \leq z \leq num_{node}$ and $1 \leq j \leq m$, where $f(E(lb_j), E(ub_j))$ means $\text{SM}(E(\psi_1), E(lb_{z,j})) \times \text{SM}(\text{SBN}(E(\psi_1)), E(ub_{z,j}))$ (lines 7–10). For example, assuming that the required k is 2, $E(dist_2) = E(5)$ because $E(t_2)$ is the current 2NN. Meanwhile, in Fig. 2, the shortest point of $node_3$ (i.e., sp_3) to the $E(q)$ is computed as follows. Because the x-value of the q is less than the x-values of both lb and ub of $node_3$, the x-value of $E(sp_3)$ is calculated by $E(sp_{3,1}) = E(0) \times E(4) + E(1) \times (E(1) \times E(5) + E(0) \times E(10)) = E(5)$. Similarly, the y-value of $E(sp_3)$ is computed as $E(sp_{3,2}) = E(1)$.

Fourth, C_A calculates $E(spdist_z)$, the squared Euclidean distances between the query and $E(sp_z)$ for $1 \leq z \leq num_{node}$ by using ESSED. In addition, C_A securely updates the

$E(spdist_z)$ of the retrieved nodes into $E(max)$ by computing $E(spdist_z) = \text{SM}(E(\alpha_z), E(max)) \times \text{SM}(\text{SBN}(E(\alpha_z)), E(spdist_z))$ (lines 11–12). Here, $E(\alpha_z)$ is the output of GSPE computed in index search step. Then, C_A performs $E(\alpha_z) = \text{GSCMP}(E(spdist_z), E(dist_k))$ (line 13). The nodes with $E(\alpha_z) = E(1)$ need to be retrieved for query result verification. For example, the initial value of E(spdist) is $(E(0), E(16), E(1), E(26))$ for each node in Fig. 2, and E(spdist) is updated as $(E(max), E(16), E(1), E(26))$. Therefore, the result of GSCMP becomes $E(\alpha) = (E(0), E(0), E(1), E(0))$ because $E(dist_k) = E(5)$. Fifth, C_A securely extracts the data stored in the nodes with $E(\alpha) = E(1)$ by performing the 4–20 lines of the Algorithm 1 and appends them to $E(t')$. Then, C_A executes the kNN retrieval step (Algorithm 2) based on $E(t')$ to obtain the $E(result_i)$ for $1 \le i \le k$ (lines 14–16). Finally, C_A stores $E(result_{i,m+1})$ into $E(c_i)$ for $1 \le i \le k$ to extract the class labels of the kNN results (line 18–19). For example, the final result becomes $E(result) = \{E(t_1), E(t_5)\}$. Because the class labels of both $E(t_1)$ and $E(t_5)$ are 1 in Fig. 2, the final $E(c)$ becomes $(E(1), E(1))$.

4.4   Step 4: Majority Class Selection Step

We securely determine the majority class label among the output of the result verification step, i.e., $E(label)$. The procedure of the result verification step is shown in Algorithm 4. First, C_A performs SF using $E(label_j)$ for $1 \le j \le w$ and $E(c_i)$ for $1 \le i \le k$ to obtain $E(f(label_j))$. Then, C_A finds the maximum value, i.e., $E(f_{max})$, among $E(f(label_j))$ for $1 \le j \le w$ by using SXS_n (line 1–2). Second, C_A securely obtains the class label $E(output)$ corresponding to the $E(f_{max})$ by using the logic similar to 5–10 lines of Algorithm 2. Due to the space limitation, we briefly describe this procedure. C_A calculates $E(\tau_i) = E(f_{max}) \times E(f(label_j))^{N-1}$ for $1 \le i \le w$. Then, C_A computes $E(\tau_i') = E(\tau_i)^{r_i}$ and obtains $E(\beta)$ by shuffling $E(\tau')$ by using π and then sends $E(\beta)$ to the C_B (line 3–5). After decrypting $E(\beta)$, C_B sets $E(U_i) = E(1)$ if $E(\beta_i) = 0$, and sets $E(U_i) = E(0)$ otherwise. After C_B sends $E(U)$ to C_A, C_A obtains $E(V)$ by permuting $E(U)$ using π^{-1} (line 6–9). Then, C_A performs $E(output) = \prod_{j=1}^{w} \text{SM}(E(V_j), E(label_j))$ for $1 \le j \le w$ to obtain the majority class label (line 10–12). For example, $E(output)$ is E(1) because the class label '1' has the maximum occurrence among $E(f(label)) = (E(2), E(0), E(0))$. Third, C_A returns the decrypted result to AU in cooperation with C_B to reduce the computation overhead at the AU side. To do this, C_A computes $E(output) \times E(r)$ by generating a random value r, and then sends the result of $E(output + r)$ to C_B and r to AU (lines 14). C_B decrypts the data sent from C_A and sends the decrypted value (e.g., $output + r$) to AU (lines 15). Finally, AU computes the actual class label by computing $(output + r) - r$ in plaintext (lines 16–17).

Algorithm 4: Majority class selection

Input : E(*label*), E(*c*)

Output : *output*

C_A : **01:** E(*f*(*label*)) ← SF(E(*label*), E(*c*))

 02: E(f_{max}) ← SXS$_n$(E(*f*(*lable$_1$*)), … , E(*f*(*lable$_w$*)))

 03: for $1 \leq j \leq w$

 04:   E(τ_j) ← E(f_{max})×E(*f*(*label*))$^{N-1}$; E(τ'_j) ← E(τ_j)^{r_j}.

 05: E(β) ← $\pi(\tau')$; send E(β) to C_B

C_B : **06:** for $1 \leq j \leq w$

 07:   if D(β_j)=0 then E(U_j) ← E(1);   else E(U_j) ← E(0)

 08: send E(*U*) to C_A

C_A : **09:** E(*V*) ← π^{-1}(*U*)

 10: for $1 \leq j \leq k$

 11:   E(V'_j)←SM(E(V_j), E(*label$_j$*))

 12: E(*output*) = $\prod_{j=1}^{w}$ SM(E(V_j), E(*label$_j$*))

 14: send E(*output*) × E(*r*) to C_B;   send *r* to AU

C_B : **15:** decrypt E(*output+r*); send *output+r* to AU

AU: **16:** *output* = (*output* + *r*) − *r*

 17: return *output*

---

5   Performance Analysis

In this section, we compare our SkNNC$_G$ (secure kNN classification algorithm using the Yao's garbled circuit) with PPkNN [3] that is the only existing work to perform classification over encrypted databases in the cloud. To measure the performance gains of using our newly proposed protocols, we also compare our scheme with SkNNC$_I$ (secure kNN classification algorithm with secure index) that performs classification based on the existing expensive secure protocols, instead of using our newly proposed protocols. Therefore, we can see that the performance gap between SkNNC$_I$ and PPkNN comes from the use of secure index search scheme. We implemented three schemes by using C++ and evaluate their performances in terms of classification time under different parameters settings. The parameters used for our performance analysis are shown in Table 2. We used the Paillier cryptosystem to encrypt a database for all of the schemes. Our experiments were performed on a Linux machine running Ubuntu 14.04.2 with an Intel Xeon E3-1220v3 4-Core 3.10 GHz and 32 GB RAM. We conducted performance analysis by using the real Chess dataset because it is considered as an appropriate dataset for classification [15]. It consists of 28,056 records with six attributes and their class labels.

Table 2. Experimental parameters

Parameters	Values	Default value
Total number of data (n)	4k, 8k, 12k, 16k, 20k, 24k, 28k	28k (28,056)
Level of kd-tree (h)	5, 6, 7, 8, 9	7
Required k (k)	5, 10, 15, 20	10
Encryption key size (K)	512	512

In Fig. 3, we measure the performance of SkNNC$_I$ and our SkNNC$_G$ by varying the level of kd-tree because PPkNN does not use the secure index. The classification times of both schemes are decreased as h changes from 5 to 7 while the classification time increase as h changes from 7 to 9. This is because as h increases, the total number of leaf nodes grows, thus requiring more GSPE and SPE [7] executions for SkNNC$_G$ and SkNNC$_I$, respectively. Whereas, as h increases, the number of data in the node decreases, thus requiring less computation cost for distance calculation. However, our SkNNC$_G$ outperforms SkNNC$_I$ because our scheme uses both efficient secure protocols based on the Yao's garbled circuit and the data packing technique.

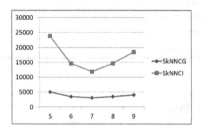

Fig. 3. Performance of varying h

Figure 4(a) shows the performance of three schemes by varying the n. As the n becomes larger, the query processing time of PPkNN linearly increases because it considers all of the data. Although the overall query processing times of SkNNC$_I$ and SkNNC$_G$ are increased as the n increases, they are less affected by n than PPkNN. Overall, our SkNNC$_G$ shows 17.1 and 4.7 times better performance than PPkNN and SkNNC$_I$, respectively. Due to the index-based data filtering, both SkNNC$_G$ and SkNNC$_I$ shows better performance than PPkNN. However, our SkNNC$_G$ outperforms

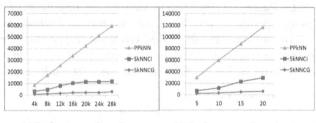

(a) Performance of varying n (b) Performance of varying k

Fig. 4. Classification time for varying n

$SkNNC_I$ because our algorithm can reduce the computation cost by using the Yao's garbled circuit and the data packing technique.

Figure 4(b) shows the performance of three schemes by varying the k. As the k becomes larger, the query processing times of three schemes increase because the larger k requires more executions of expensive protocols, e.g., SMS_n in case of our $SkNNC_G$ and $SMIN_n$ in case of both PPkNN and $SkNNC_I$, to retrieve the more kNN results. Overall, our $SkNN_G$ shows 17.7 and 4.2 times better performance than PPkNN and $SkNNC_I$, respectively, due to the same reasons described for Fig. 3.

6 Conclusion

Databases need to be encrypted before being outsourced to the cloud, due to its privacy issues. However, the existing kNN classification scheme over encrypted databases in the cloud has a problem that it suffers from high computation overhead. Therefore, in this paper we proposed a new secure and efficient kNN classification algorithm over encrypted databases. Our algorithm not only preserves data privacy and query privacy, but also conceals resulting class labels and data access pattern. In addition, our algorithm can support efficient kNN classification by using an encrypted index search scheme, the Yao's garbled circuit and a data packing technique. We showed from our performance analysis that the proposed algorithm showed about 17 times better performance on classification time than the existing PPkNN scheme, while preserving high security level.

As a future work, we plan to expand our algorithm to the distributed cloud computing environment. We also plan to study on data clustering and association rule mining over encrypted database for cloud computing.

Acknowledgment. This work was supported by Basic Science Research Program through the National Research Foundation of Korea (NRF) funded by the Ministry of Education (grant number 2016R1D1A3B03935298). This work was partly supported by Institute for Information & communications Technology Promotion (IITP) grant funded by the Korea government (MSIP) (No. R0113-17-0005, Development of an Unified Data Engineer-ing Technology for Large-scale Transaction Processing and Real-time Complex Analytics).

References

1. Vimercati, S., Foresti, S., Samarati, P.: Managing and accessing data in the cloud: privacy risks and approaches. In: CRiSIS, pp. 1–9 (2012)
2. Riley, J.W., Alfons, C., Fredäng, E., Lind, P.: Nearest Neighbor Classifiers (2009)
3. Samanthula, B., Elmehdwi, Y., Jiang, W.: K-nearest neighbor classification over semantically secure encrypted relational data. TKDE 27(5), 1261–1273 (2015)
4. Elmehdwi, Y., Samanthula, B.K., Jiang, W.: Secure k-nearest neighbor query over encrypted data in outsourced environments. In: ICDE, pp. 664–675 (2014)
5. Hu, H., Xu, J., Ren, C., Choi, B.: Processing private queries over untrusted data cloud through privacy homomorphism. In: ICDE, pp. 601–612 (2011)

6. Zhu, Y., Xu, R., Takagi, T.: Secure k-NN computation on encrypted cloud data without sharing key with query users. In: Security in cloud computing, pp. 55–60 (2013)
7. Kim, H., Kim, H., Chang, J.: A kNN query processing algorithm using a tree index structure on the encrypted database. In: Big Data and Smart Computing (BigComp), pp. 93–100 (2016)
8. Yao, A.C.C.: How to generate and exchange secrets. In: Foundations of Computer Science, pp. 162–167 (1986)
9. Paillier, P.: Public-key cryptosystems based on composite degree residuosity classes. In: EUROCRYPT, pp. 223–238 (1999)
10. Liu, A., Zhengy, K., Liz, L., Liu, G., Zhao, L., Zhou, X.: Efficient secure similarity computation on encrypted trajectory data. In: ICDE, pp. 66–77 (2015)
11. Huang, Y., Evans, D., Katz, J., Malka, L.: Faster secure two-party computation using garbled circuits. In: USENIX Security, vol. 201, no. 1 (2011)
12. Domingo-Ferrer, J.: A provably secure additive and multiplicative privacy homomorphism*. In: Chan, A.H., Gligor, V. (eds.) ISC 2002. LNCS, vol. 2433, pp. 471–483. Springer, Heidelberg (2002). https://doi.org/10.1007/3-540-45811-5_37
13. Samanthula, B.K., Chun, H., Jiang, W.: An efficient and probabilistic secure bit-decomposition. In: ASIACCS, pp. 541–546 (2013)
14. Samet, H.: Foundations of Multidimensional and Metric Data Structures. Morgan Kaufmann (2006)
15. http://archive.ics.uci.edu/ml/

A Security Model for IoT Networks

Alban Gabillon[1](✉) and Emmanuel Bruno[2]

[1] Université de la Polynésie Française, BP 6570, 98702 Punaauia, Faa'a,
French Polynesia
alban.gabillon@upf.pf
[2] Université de Toulon, CNRS, LIS, UMR 7020, 83957 La Garde, France
emmanuel.bruno@univ-tln.fr

Abstract. The MQTT (Message Queuing Telemetry Transport) protocol is
becoming the main protocol for the Internet of Things (IoT). In this paper, we
define a highly expressive ABAC (Attribute-Based Access Control) security
model for the MQTT protocol. Our model allows us to regulate not only publications and subscriptions but also distribution of messages to subscribers. We
can express various types of contextual security rules, (temporal security rules,
content-based security rules, rules based on the frequency of events etc.).

Keywords: Security policy · MQTT · ABAC · IoT · First-order logic

1 Introduction

The MQTT (Message Queuing Telemetry Transport) protocol is becoming the main
protocol behind pub-sub networks for the Internet of Things, that is, in networks
implementing the publication-subscription paradigm. The MQTT protocol is an ISO
standard (ISO/IEC PRF 20922) [2] and the 3.1 version became an OASIS specification
in 2013 [3]. Basically, the MQTT protocol works as follows: publishers post messages
to logical channels called topics; subscribers receive messages published to the topics
to which they subscribed; the MQTT broker routes messages from publishers to
subscribers.

The MQTT protocol supports very few security features. It includes a MQTT client
identification mechanism and supports the basic login/password authentication scheme.
Consequently, there have been several papers aiming at defining security solutions for
the MQTT protocol or more generally for the pub-sub pattern. These papers address
various issues like how to implement a security policy regulating publications and
subscriptions [4–6], how to distribute the evaluation and the enforcement of the security
policy at the edge of the IoT network [7, 8], how to distribute and synchronize the
security policy between different pub-sub architectures [9] or how to protect the confidentiality of the messages from the broker or the pub-sub architecture itself [10, 11].
Although these issues are all very important, we noticed that none of these papers fully
addressed the definition of a security model allowing to express security policies for
regulating IoT messages. Some of the papers [5, 6] mention that they are using the
ABAC (Attribute-Based Access Control) model [12] for expressing the security policy
controlling publications and subscriptions. However, they do not go much into details

© Springer Nature Switzerland AG 2018
T. K. Dang et al. (Eds.): FDSE 2018, LNCS 11251, pp. 39–56, 2018.
https://doi.org/10.1007/978-3-030-03192-3_4

and do not elaborate on the expressive power of the security policy. In this paper, we define a highly expressive ABAC model for regulating IoT messages in a MQTT network. We believe that the definition of such a security model (which does not contradict the solutions proposed by the aforementioned papers) has been missing in the literature related to security solutions for pub-sub architectures. Our model allows us to regulate not only publications or subscriptions but also *distribution* of messages by the broker to subscribers. Our model supports positive and negative authorizations and allows us to express various types of context-based policies, including policies based on the *frequency* of events. This paper is an extension of a short paper we previously published [1]. In [1], we could only draft our model. In this paper we give a complete definition, including the administration model. We also provide an extensive review of the literature on security models for IoT networks.

The remainder of this paper is organized as follows: In Sect. 2, we define our model. In Sect. 3, we present our security administration model. In Sect. 4, we sketch our secure MQTT broker prototype based on our model. In Sect. 5, we review related works before the conclusion in Sect. 6.

2 ABAC Model

Some papers [4, 5] mention that they are using the ABAC (Attribute-Based Access Control) model [11] for expressing the security policy controlling publications and subscriptions in a pub-sub network. However, these papers do not go much into details and do not elaborate on the expressive power of the security policy. Moreover, none of these papers address the security administration issue. Our aim in this paper is to define a security model which can be seen as a *profile* of the ABAC model for pub-sub networks based on MQTT. We first identify some requirements specific to IoT security policies. Then we make some assumptions on the IoT network and on some security aspects that we shall not cover. Finally, we devise our model starting from the requirements we identified.

2.1 Requirements

- Our model should offer the possibility to regulate not only publications and subscriptions but also *distribution* of messages by the broker to subscribers. Controlling distribution of messages is essential to regulate the various flows of messages coming from the broker. Solely controlling subscriptions is too coarse grained to achieve that task.
- Our model should allow for various types of dynamic and contextual authorization rules i.e. authorization rules whose outcome (permit ort deny) depend on some contextual conditions applying to the nodes, the messages (including the content of the messages) or the environment. In particular, *authorization rules based on the frequency of events* should be supported since controlling the rate at which a node may send or receive messages is important in many IoT applications.

2.2   Assumptions

- For the sake of simplicity, we assume a pub-sub architecture with only one MQTT broker. Since we focus on the expressive power of the security policy, we do not investigate issues like distributing and synchronizing the security policy between different bridged brokers or evaluating the security policy at the edge of the network [7–9].
- We assume the broker to be trusted i.e. we do not investigate solutions to protect the confidentiality of the messages from the broker [10, 11].
- We do not investigate authentication techniques. We believe that standard authentication techniques can be used to authenticate both nodes and attributes.
- Finally, we assume TLS/SSL is used at the transport layer between all nodes of the IoT network. Most existing MQTT servers support the use of TLS/SSL.

2.3   Language

We use first-order logic with equality to define our model, i.e. we define a logical language allowing us to represent nodes, attributes, events (like publications, subscriptions, messages distribution) and authorization rules. Note, however, that the reader who is not familiar with logic should be able to understand the main principles of our model since we translate in plain English each logical formula.

Although, we define our own logical language, we wish to make it clear that this paper is *not* about a new logic-based policy language. To specify our model, we could use XACML [13] (but it would be unreadable by a human), or an existing logical language like SecPAL [14]. However, we prefer defining our own language so that we can restrict ourselves to Horn clauses which can easily be read by a human and for which there exists efficient resolution methods.

Constants

Constants of our language are string expressions. They are node identifiers such as *sensor1*, *user1* etc. or the special string *broker* referring to the MQTT broker.

Topics are defined by path expressions (written as strings) such as *temperatures/sensor1*. Several topics can be referenced by using wildcards # and +. For examples, *temperatures/#* addresses any topic having *temperature* as path root and *home/+/temperatures* addresses topics such as *home/room1/temperature*, *home/room2/temperature* etc. See [3] for more details about the use of wildcards in MQTT topics.

Note that, to lighten the notations, we omit the quotation marks for the strings.

Variables

Variables are written in capitalized letters like in Prolog. Our language includes the anonymous variable _ which means *anything*. If variable S contains a string value, then we assume this value can be referred to in a path expression. For example, if S contains the string *sensor1* then *temperatures/S* represents the topic *temperatures/sensor1*.

In this paper, to distinguish variables from constants, we constrain ourselves to consider only constants written as strings of lowercase characters.

Predicates

Authorizations can be derived from a set of facts $\mathcal{F}$ and from a set of logical rules $\mathcal{R}$. Set $\mathcal{F}$ keeps track of registered nodes and events (publications, subscriptions and distributions) whereas set $\mathcal{R}$ records the nodes hierarchy.

Set $\mathcal{F}$ includes instances from the following node predicates:

Registering a node creates an instance of one of these node predicates (Table 1).

Table 1. Node predicates

Predicate	Meaning
$node(N)$	N is an IoT node
$broker(N)$	N is the broker
$sensor(N)$	N is a sensor
$client(N)$	N is a client

Set $\mathcal{R}$ includes the following rules:

$$node(N) \leftarrow broker(N) \tag{1}$$

$$node(N) \leftarrow sensor(N) \tag{2}$$

$$node(N) \leftarrow client(N) \tag{3}$$

These three rules can be used to derivate that the broker or a sensor or a client is also a node. These rules define a roles hierarchy that could be expanded according to the needs of the application.

Set $\mathcal{F}$ also includes instances from the following event predicates:

Publishing a message creates an instance of the *hasPublished/3* predicate. Subscribing to a topic creates an instance of the *hasSubscribed/3* predicate. Delivering a message creates an instance of the *hasDelivered/3* predicate. As we shall see in Sect. 2.3, recording these events allows us in particular to express security rules controlling the frequency of publishing/delivering messages.

As we said previously, topics are path expressions possibly written with wildcards. Therefore, set $\mathcal{F}$ also includes instances from the following topic predicate (Table 3):

Table 2. Event predicates

Predicate	Meaning
$hasPublished(N, T, D)$	At time D, node N has published a message in topic T
$hasSubscribed(N, T, D)$	At time D, node N has subscribed to topic T
$hasDelivered(T, N, D)$	At time D, the broker has delivered a message from topic T to node N

Table 3. Matching predicate

Predicate	Meaning
$addresses(T, T')$	Topic T addresses topic T'

For example, fact $addresses(temperature/*, temperature/sensor1)$ belongs to $\mathcal{J}$. For the sake of simplicity, we do not give the logical rules allowing us to derive instances of the *addresses/2* predicate.

Functions

Functions of our language represent attributes. They are either,

- Functions applying to messages or
- Functions for evaluating temporal conditions or any other contextual conditions.

Lists of functions in Tables 4 and 5 are not exhaustive and can be extended depending on the needs.

Table 4. Message attribute functions

Function	Purpose
$length(M)$	Returns the length of the message M
$retained(M)$	Returns true if the message M is retained, false else
$value(M)$	Returns the content of the message M
$encoding(M)$	Returns the character encoding of the message M
$ciphered(M)$	Returns true if the message M is encrypted[a], false else

[a]Encrypting a message means encrypting the payload of the MQTT packet transporting the message. This should not be confused with encrypting the whole communication between nodes at the transport layer by means of TLS/SSL.

Table 5. Contextual functions

Function	Purpose
$time()$	Returns the current time
$date()$	Returns the current date
$latency()$	Returns the network's latency
$bandwidth()$	Returns the network's bandwidth

2.4   Security Policy

Actions

We define the three *compound terms* to represent the following three actions:
Variables represent action parameters. Note that there is no QoS parameter for the deliver operation. This is because the QoS used by the broker to deliver a message to

Table 6. Actions

Term	Action
$publish(M, T, Q)$	Publishing message M in topic T at QoS Q
$subscribe(T, Q)$	Subscribing to topic T at QoS Q
$deliver(M, T, N)$	Delivering message M from topic T to node N

node N is the QoS chosen by node N when it subscribed to topic T. This means that if, in our security policy, we need to restrict the QoS used by the broker to deliver messages, then it should be done during the subscription step.

Contextual Authorization Rules

We consider positive authorizations and negative authorizations represented by the two following predicates (Table 7):

Table 7. Authorizations

Predicate	Meaning
$allow(N, A)$	Node N is allowed to perform action A
$deny(N, A)$	Node N is denied to perform action A

Variable A contains any of the three compound terms of Table 6. Note that if A is a *deliver* action then we assume that N cannot be different from *broker*.

The security policy $\mathcal{P}$ regulates publish, subscribe and deliver operations. It consists of a set of authorization rules. Any authorization rule is an instance of one of the following rule templates:

$$allow(N, A) \leftarrow conditions \tag{4}$$

$$deny(N, A) \leftarrow conditions \tag{5}$$

Symbol *conditions* stands for a possibly empty conjunction of *contextual conditions* on nodes, topics, QoS, messages and the environment. Here are a few examples of authorization rules:

$$deny(sensor1, publish(\_, alarms/sensor1, \_)) \\ \leftarrow time() > 8 \wedge time() < 20 \tag{6}$$

Rule 6 denies *sensor1* to publish messages (whichever the QoS is), in topic *alarms/sensor1* during day time.

$$allow(N, subscribe(alarms/\#, \_)) \\ \leftarrow guest(N) \tag{7}$$

Rule 7 allows guest nodes to subscribe to the alarms hierarchy of topics. Here we assume *guest/1* is a role predicate expanding the hierarchy defined in Sect. 2.2.

Regarding the delivering operation, we should first note that the normal MQTT behavior is to deliver messages from topic T to the nodes which subscribed to topic T. This can be expressed by the following *default policy rule*:

$$allow(broker, deliver(\_, T, N))$$
$$\leftarrow hasSubscribed(N, T, \_) \tag{8}$$

Rule 8 allows the broker to deliver any messages from topic T to the nodes which subscribed to topic T. However, this default policy can be *overridden* in some specific cases (see Sect. 2.4 for conflicts resolution between rules):

$$deny(broker, deliver(M, alarms/\#, N))$$
$$\leftarrow guest(N) \wedge value(M) =' failure' \tag{9}$$

Rule 9 overrides rule 8 and denies the broker to deliver failure messages from the alarms hierarchy of topics to guest nodes. Rule 9 is an example of a *content-based* authorization rule.

In rules 7 and 9, there is a path expression referring to the set of topics alarms/#. Therefore, we need to include in set $\mathcal{P}$ some rules to derive instances of predicates *allow/2* and *deny/2* addressing any subset of a set of topics expressed by means of wildcards:

$$allow/deny(N, publish(M, T', Q))$$
$$\leftarrow allow/deny(N, publish(M, T, Q)) \wedge addresses(T, T') \tag{10}$$

Rule 10 says that if publication is allowed/denied for a set of topics T then publication is also allowed/denied for each subset T' of T. We could write similar rules for the *subscribe/3* and *deliver/3* predicates.

For example, since *addresses(alarms/#,alarms/sensor1)* is true, then *allow(subscribe(user1,alarms/sensor1,1))* can be derived from *allow(subscribe(user1, alarms/#,1))*.

Controlling the Frequency of Events

Our experience has shown us that in some applications being able to control the frequency of publications, subscriptions and messages distribution is important. Consider for example an online trading broker. An online trading broker is a pub-sub service where clients may send trade orders and receive various tips and hints related to the stock market. Assume that the online broker sells standard accounts and premium accounts. Premium account holders receive more hints and tips per day than standard account holders. Moreover, premium account holders can send more trading orders per day than standard account holders. In such a scenario, we would need to express authorization rules controlling the frequency of publications (e.g. trade orders) and the frequency of messages (e.g. hints and tips) delivered by the broker. Another obvious use of having authorization rules based on the frequency of publications would be to mitigate the effects of compromised sensors involved in DDOS attacks against the pub-sub architecture.

To define authorization rules allowing us to express conditions on the frequency of events, we define the following high-order predicate (Table 8):

Table 8. Frequency predicate

Predicate	Meaning
$freq(E, F, I)$	F is the instant frequency of repeating event E per unit of time I

Frequencies are always evaluated at the time the policy is evaluated. This explains why instances of the *freq/3* predicates represent *instant* frequencies.

Variable E refers to any formula instance of the three event predicates *hasPublished/3*, *hasSubscribed/3* and *hasDelivered/3* defined in Sect. 2.2, with the last variable referring to the timestamp of the event always equal to the anonymous variable _.

Here are two examples of frequencies:

$$freq(hasPublished(sensor1, alarms/sensor1, \_), 5, 24) \tag{11}$$

Formula 11 says that the instant frequency of publications made by *sensor1* in topic *alarms/sensor1* is 5 in the last 24 h.

$$freq(hasPublished(\_, alarms/\#, \_), 152, 24) \tag{12}$$

Formula 12 says that the instant frequency of publications (made by all sensors) in topics hierarchy *alarms/#* is 152 in the last 24 h.

Note that, by defining the high-order predicate *freq/3*, we are no longer in strict first-order logic. However, computing instances of the *freq/3* predicate can easily be done by using some aggregate predicate which would be implemented in many inference engines. For example, the rule below is the SWI Prolog [15] definition of the *freq/3* predicate for the *hasPublished/3* predicate. It uses the Prolog built-in *aggregate_all/3* predicate:

$$freq(hasPublished(N, T, \_), F, I))$$
$$\leftarrow aggregate\_all(count, (hasPublished(N, T, D) \wedge (time() - D) < I)), F) \tag{13}$$

Basically, Prolog rule 13 counts the number of instances of the *hasPublished/3* predicate referring to node N and topic T with a timestamp not older than I hours.

The following rules are examples of authorization rules regulating the frequency of publications and messages distribution:

$$allow(sensor1, publish(\_, alarms/sensor1, \_))$$
$$\leftarrow freq(hasPublished(sensor1, alarms/sensor1, \_), F, 24) \wedge F < 5 \tag{14}$$

Rule 14 allows *sensor1* to publish messages in topic *alarms/sensor1* as long as it does not post more than 5 alert messages per 24 h.

$$
\begin{aligned}
&deny(broker, deliver(\_, alarms/sensor1, N)) \leftarrow guest(N) \\
&\wedge freq(hasDelivered(alarms/sensor1, N, \_), F, 24) \wedge F > 1
\end{aligned}
\tag{15}
$$

Rule 15 denies the broker to deliver to guest nodes more than one alert message per 24 h from topic *alarms/sensor1*.

2.5 Conflict Resolution Policy

Since our authorization model allows for positive and negative authorizations, conflicts between rules may arise. For example, consider the following two rules:

$$
deny(N, subscribe(\_, \_)) \leftarrow sensor(N)
\tag{16}
$$

$$
allow(N, subscribe(N/\#, \_)) \leftarrow sensor(N)
\tag{17}
$$

Rule 16 says that subscriptions are forbidden for sensors while rule 17 says that sensors can subscribe (at any QoS) to topic for which the path root corresponds to their identifier. Clearly these two rules conflict whenever a sensor subscribes to a topic for which the path root corresponds to the sensor identifier.

There are many possible solutions to solve conflicts between authorization rules. The XACML standard [13] enumerates several *combining algorithms* to solve conflicts between rules (deny overrides, permit overrides, first applicable overrides, permit unless deny, deny unless permit etc.). We can use any of these algorithms depending on our needs. Regarding the small example above, the *permit overrides* algorithm would allow a node subscribing to a topic for which the path root corresponds to the node identifier.

3 Security Administration Model

3.1 Principles

Definition of a security model must include the definition of a model for administering the security policy. To introduce our model, let us first consider the scheme depicted in Fig. 1.

Sensors (*S1* and *S2*) sends messages to *Analytics* through topic *A*. *Monitor* sends commands to sensors through topic *B*. Monitor *owns* sensors *S1* and *S2* and created topics *A* and *B*. This scenario suggests us that *Monitor* could be the administrator defining the security policy regulating messages going through channels *A* and *B*. Of course, this is not the only possible scenario. The IoT network could be more centralized; topics *A* and *B* could also be shared by other applications and sensors etc. Nevertheless, decentralizing the security administration should be possible even if the network contains only one broker. Moreover, to give flexibility, delegation of rights should also be supported.

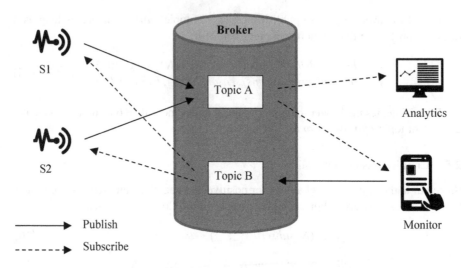

Fig. 1. IoT network

In our model, security administration is *topic-based*. We state that there is *at least* one security administrator for each topic. A security administrator for a given topic *T* is responsible for defining the security policy regulating publications/subscriptions to topic *T* and distribution of messages from topic *T*. There is also one *Root Administrator* (*RA*) who can administrate the security policy for all topics.

More precisely, the *RA* can perform the following tasks:

- Administrate (i.e. define the security policy for) all or some topics.
- Grant to a node the admin privilege on a given topic (with possibly the right to transfer this right).
- Revoke from a user the admin privilege on a topic.

Each administrator node which is admin for a given topic *T* can define the security policy for that topic *T*. If it has also been granted the right to transfer this right, then it may also grant to another node the right to administrate topic *T*. We believe this administration scheme is flexible enough to support various cases of application.

In the following sections we show how we extend our logical language to define our administration policy.

3.2 Constants

We define the following constant to represent the *RA*: *root*

3.3 Function

We define the following function which returns the topic addressed by an authorization rule *R* (instance of either template 4 or template 5) (Table 9).

Table 9. Topic function

Function	Purpose
$topic(R)$	Returns the topic appearing in the action parameter A of rule R

Recall that action A is represented by one of the compound term *publish/3*, *subscribe/2* or *deliver/3* defined in Table 6.

3.4 Predicate

We extend set $\mathcal{F}$ with instances of the following event *rights delegation* predicate (Table 10):

Table 10. Rights delegation predicate

Predicate	Meaning
$hasGranted(N,T,N',G)$	Node N has granted the *admin privilege* on topic T to node N' with or without the *grant option* (depending on the value of the Boolean variable G)

Granting an admin right *creates* a new instance of this predicate. Revoking the right *deletes* the corresponding instance of this predicate. The grant option is similar to the grant option of the SQL grant statement [16].

3.5 Security Administration Policy

Let TA be an administrator node of topic T. Admin TA can add and delete authorization rules addressing topic T in the security policy $\mathcal{P}$. If admin TA holds the grant option on topic T, then it can also grant and revoke the admin rights on topic T to some others nodes.

Actions

We define four compound terms to represent the four following actions (Table 11):

Table 11. Security administration actions

Term	Action
$ruleAdd(R)$	Adding rule R in policy $\mathcal{P}$
$ruleDel(R)$	Deleting rule R from policy $\mathcal{P}$
$grant(T,N,G)$	Granting the admin privilege on topic T to node N with or without (depending on the value of G) the grant option
$revoke(T,N)$	Revoking the admin privilege on topic T from node N

Note that the grant option cannot be granted nor revoked separately. The same principle applies in the SQL delegation scheme.

Security Administration Rules

The security administration policy $\mathcal{A}$ is *mandatory* and consists of the five following administration rules:

$$hasGranted(root, \#, root, true) \tag{18}$$

Rule 18 says that the RA has granted to himself the admin option on the whole topics hierarchy with the admin option.

$$allow(N, ruleAdd(R)) \\ \leftarrow hasGranted(\_, T, N, \_) \land addresses(T, T') \land topic(R) = T' \tag{19}$$

Rule 19 says that if node N was granted the admin privilege on topic T, then it can add authorization rules referring to topic T (or to a subset of topics T if T represents a set of topics).

$$allow(N, ruleDel(R)) \\ \leftarrow hasGranted(\_, T, N, \_) \land addresses(T, T') \land topic(R) = T' \tag{20}$$

Rule 20 says that if node N was granted the admin privilege on topic T, then it can delete authorization rules referring to topic T (or to a subset of topics T if T represents a set of topics).

$$allow(N, grant(T', \_, \_)) \\ \leftarrow hasGranted(\_, T, N, true) \land addresses(T, T') \tag{21}$$

Rule 21 says that if node N was granted the admin privilege on topic T with the grant option then it can grant the admin option on topic T (or on a subset of topics T if T represents a set of topics).

$$allow(N, revoke(T', N')) \\ \leftarrow hasGranted(N, T, N', \_) \tag{22}$$

Rule 22 says that if node N has granted to node N' the admin privilege on topic T then it can revoke this privilege from node N'. In other words, only the node which transferred a privilege can revoke it. Moreover, as we said previously revoking an admin privilege deletes the corresponding instance of the *hasGranted/4* predicate. Since the grantee N' might also have transferred this right to some other nodes, revocation would also delete all the instances of *hasGranted/4* corresponding to the delegation chain originating from N'. This mechanism is usually referred to as *cascade revocation*.

Finally, let us mention the following two points:

- The combining algorithm enforced in a is obviously *deny unless permit* [13], i.e. the default policy is deny. This default policy is overridden by rules 18 to 22.
- If T represents a set of topics, then it is not possible to revoke the admin privilege for a subset of T. One would have to revoke the admin privilege for the set T and then grant again the admin privilege on a subset of T.

4 Prototype

4.1 Architecture

This paper is more about the model than the implementation. Nevertheless, we have implemented a proof-of-concept prototype depicted in Fig. 2. Our prototype is built according to the XACML architecture [13]. We use the EMQ[1] MQTT broker written in Erlang/OTP for which we have developed the MQTTsec plugin acting as a Policy Enforcement Point (PEP). The Policy Information Point (PIP) contains OWL2 ontologies representing nodes, topics and events. The Policy Administration Point (PAP) contains a set of SWRL [17] rules representing the security policy. The MQTTsec manager, written as a Java Web Application, acts as a Policy Decision Point (PDP).

First, the PEP intercepts an event (publication, subscription or distribution of a message). It then submits the event to the PDP. The PDP loads the security policy from the PAP and queries the PIP to retrieve the necessary attribute values. Then it runs an OWL2 [18] inference engine which applies the conflict resolution policy and eventually issues a decision (allow/deny). Whether the request is authorized or not, it is always recorded in the PIP as a new instance of the *event* class of our ontology model. If the request has been authorized then the corresponding instance is tagged as *allowed*, *denied* otherwise. It should be noted that instances of the event predicates defined in Table 2 correspond to the *allowed* events recorded in the PIP. In our prototype, we also keep track of the denied requests for traceability purpose.

4.2 Proof Graphs

We should also mention that the inference engine can show the logical reasoning that led to the decision producing a proof graph of the decision. This feature can be very useful for debugging security policies, auditing, or devising new conflict resolution algorithms. Basically, it works as follows:

- Policy rules and captured events (publication, subscription or message distribution) are represented in the OWL language.
- The inference engine computes a list of possible authorization values and shows the derivation steps for each value.

---

[1] http://emqtt.io/.

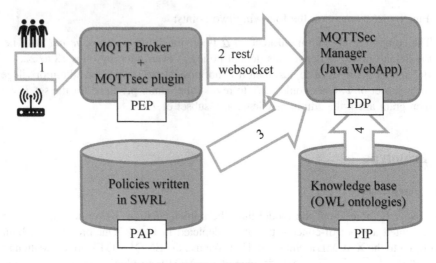

Fig. 2. MQTTsec broker prototype

- If the list is empty, then the default policy is applied
- If there are conflicting values in the list (at least one permit and one deny) then the conflict solver computes the final decision according to the predefined conflict resolution algorithm and shows the derivation steps.
- Whether rejected or accepted, the event is timestamped and added to the PIP.

5 Related Works

As we already said, to our knowledge, there is no paper directly addressing the definition of a security policy model for IoT messages. Nevertheless, in this section we review security works related to pub-sub architectures and MQTT protocol.

In [4], the authors define the *secKit* tool integrated in the IOT network simulator developed as part of the FP7 iCore project [19]. This tool is used to define security policies protecting the data exchanged between the different components (virtual objects, composite virtual objects) which abstract the IoT network. *secKit* is based on a collection of models for modeling objects, data, time, roles, activities, interactions, risk, contexts, trust management and so on. It was implemented as a Mosquitto plugin [20]. Authorizations rules can be positive or negative and include obligations. Authorization rules are Event Condition Action (ECA) rules [21]. This formalism makes it possible to express contextual and dynamic authorizations but requires the implementation of an event manager capable of intercepting all events. Many aspects related to this tool are not clearly defined. Authors claim that the tool supports many features, but they do not elaborate on the features (expressive power, risks, trust management, obligations, conflicts resolution etc.). Therefore, it is very difficult to have a clear view of the model supported by *secKit*. Note also that the tool *secKit* seems to be abandoned. The source code of an alpha version can be downloaded from gitHub (https://github.com/iot-icore/iCore-security-toolkit) but has not been modified for 3 years.

In [5], the authors describe a NOS (NetwOrked Smart object) middleware, located between the objects and the MQTT client. NOS intercepts the messages intended to be published, normalizes them (i.e. extracts metadata) and according to a security policy implementing the ABAC model decides whether to grant the publication of the message. If the message is authorized, it is encrypted by means of a temporary key corresponding to the subject (topic) where the message is to be published. Once encrypted the message is published by the MQTT client. Customers wishing to subscribe to a topic, contact the NOS which according to the security policy will issue them or not the key to decipher the messages of the subject. This approach frees the MQTT broker from the evaluation and enforcement of the authorization policy. It requires, however, to set up a key management mechanism and offers a rather coarse level of granularity since the object of protection is not the message but the subject (thus including all the messages published in the subject). NOS has been implemented using the node.js platform and the objects transmit their messages via the http protocol. Currently NOS is not available for download. In [9], the same authors improve their architecture by proposing a solution to distribute and synchronize the security policies hosted by the NOS of several IoT networks. Their synchronization protocol uses the MQTT protocol.

In [6], the authors propose a solution to implement the ABAC model in a federation of IoT platforms. Their solution decouples the authorization process of the authentication process. An application first connects to an authorization manager to obtain a set of tokens. Each token represents an attribute of the application. Once in possession of its tokens the application that needs to access a resource turns to an authorization manager who will accept or reject access to the requested resource based on the tokens presented. If the application wants to access a resource belonging to a foreign IoT platform, then it must present its tokens to the authentication manager of the foreign platform to obtain foreign tokens. These foreign tokens are computed by means of a tokens conversion function (little information is given on this function in the article). Once in possession of the foreign tokens, the application can then turn to the foreign authorization manager who will accept or reject access. The authors suggest to implement the tokens either in the form of google macaroons [22] or in the form of JWT [23] tokens (Json Web Tokens).

In [24], the authors implement the RBAC model in a pub-sub network. They consider the privilege of logging in, the privileges of adding a topic and deleting it, and the privileges of publishing and subscribing. They define a solution to disseminate security policy [25] in a network of partially trusted brokers so that access control decisions are taken and enforced at the earliest. Their solution is implemented in the Hermès broker [26] using the http protocol.

In [27], the authors propose to adapt the OAuth protocol [28] to the case of an IoT network where resources are discovered and exposed according to the IETF standard [29]. The data (resources) produced by the sensors of the low power IoT network are transmitted to the gateway between the low power network and the IP network. This schema ignores the MQTT protocol, resources being logged and exposed at the gateway level. Within this gateway, an authorization server is installed. According to the OAuth protocol, a third-party application that wants to access a resource exposed by the gateway, (i) requests access delegation from the owner of the resource by asking him to authenticate, (ii) obtains an access token to the resource provided by the authorization server and (iii) presents the access token to the authorization server to access the resource. The authors do not mention any case study that could benefit from such a security architecture.

There is a general trend in the world of IoT which consists, for scaling purposes, of moving the processing and controls at the edge of the IoT network towards the objects themselves [30]. Thus, there are several approaches [7, 8] for moving security-related services and processes to end-of-network gateways or servers. In [7], the authors even use the concept of *sticky policy* [31] which implies that data owners encapsulate the policy protecting the data with the data itself.

There are several articles that deal with security issues in pub-sub networks, without specifically considering the MQTT protocol. In [10], the authors attempt to identify the security problems specific to this type of network. They are particularly interested in protecting the confidentiality of messages and subscriptions in case the pub-sub infrastructure is not trusted. They suggest some lines of research such as the use of numerical calculation on encrypted data [32, 33]. In order to protect the content of the messages and subscription schemes of the brokers, the authors in [11] propose a technique based on the encryption preserving the asymmetric scalar product [34].

In [35], the authors propose a model that implements the reliable distribution service, that is, the messages received by a client do not depend on the connection location, the network latency, or the possible points of network failure. These messages depend only on the customer's subscription filter and her access rights, which are uniformly enforced throughout the pub-sub network.

6 Conclusion

In this paper, we have defined a model to express security policies for a pub-sub architecture consisting of a single MQTT broker. The most important contributions of our paper are the followings:

- Our model allows us to regulate not only publications and subscriptions but also distribution of messages. To our knowledge, this feature has not been addressed in any other paper related to IoT security.
- Our model is an interpretation of the ABAC model for the pub-sub architecture with some unique features like the possibility to control the frequency of events.
- We have developed a prototype based on OWL2 and SWRL showing the feasibility of our approach.

Regarding future works, we are planning to investigate the following issues:

- We will extend our model to the case of a pub-sub architecture consisting of several *bridged* brokers. In such a scenario, we might need to apply the solution presented in [9] to synchronize the security policy at every node of the pub-sub architecture.
- We will also consider an IoT network consisting of a TCP/IP network hosting the pub-sub architecture coupled with a Low Power Wide Area Network (LPWAN) hosting the sensors. In such a scenario, we might also need to implement solutions

proposed by others [7, 8] to move, for scaling purposes, the security controls at the various gateways between the TCP/IP network and the LPWAN network.

- We are also planning to include the possibility to declare obligations in the security policy.
- Finally, we will update and improve our prototype to turn it into a scalable secure broker engine

References

1. Gabillon, A., Bruno, E.: Regulating IoT messages. Presented at the 14th International Conference on Information Security Practice and Experience (ISPEC 2018) - Short Paper, Tokyo (2018)
2. ISO/IEC 20922:2016 - Information Technology – Message Queuing Telemetry Transport (MQTT) v3.1.1. https://www.iso.org/standard/69466.html. Accessed 12 Jan 2018
3. Banks, A., Gupta, R.: MQTT Version 3.1.1. OASIS Stand., vol. 29 (2014)
4. Neisse, R., Steri, G., Fovino, I.N., Baldini, G.: SecKit: a model-based security toolkit for the internet of things. Comput. Secur. **54**, 60–76 (2015)
5. Rizzardi, A., Sicari, S., Miorandi, D., Coen-Porisini, A.: AUPS: an open source AUthenticated publish/subscribe system for the internet of things. Inf. Syst. **62**, 29–41 (2016)
6. Sciancalepore, S., et al.: Attribute-based access control scheme in federated IoT platforms. In: Podnar Žarko, I., Broering, A., Soursos, S., Serrano, M. (eds.) InterOSS-IoT 2016. LNCS, vol. 10218, pp. 123–138. Springer, Cham (2017). https://doi.org/10.1007/978-3-319-56877-5_8
7. Sicari, S., Rizzardi, A., Miorandi, D., Coen-Porisini, A.: Security towards the edge: sticky policy enforcement for networked smart objects. Inf. Syst. **71**, 78–89 (2017)
8. Phung, P.H., Truong, H.-L., Yasoju, D.T.: P4SINC-an execution policy framework for IoT services in the edge. In: 2017 IEEE International Congress on Internet of Things (ICIOT), pp. 137–142 (2017)
9. Sicari, S., Rizzardi, A., Miorandi, D., Coen-Porisini, A.: Dynamic policies in internet of things: enforcement and synchronization. IEEE Internet Things J. **4**(6), 2228–2238 (2017)
10. Wang, C., Carzaniga, A., Evans, D., Wolf, A.L.: Security issues and requirements for internet-scale publish-subscribe systems. In: 2002 Proceedings of the 35th Annual Hawaii International Conference on System Sciences, HICSS, pp. 3940–3947 (2002)
11. Choi, S., Ghinita, G., Bertino, E.: A privacy-enhancing content-based publish/subscribe system using scalar product preserving transformations. In: Bringas, P.G., Hameurlain, A., Quirchmayr, G. (eds.) DEXA 2010. LNCS, vol. 6261, pp. 368–384. Springer, Heidelberg (2010). https://doi.org/10.1007/978-3-642-15364-8_32
12. Yuan, E., Tong, J.: Attributed based access control (ABAC) for web services. In: 2005 Proceedings of IEEE International Conference on Web Services, ICWS 2005 (2005)
13. Moses, T., et al.: Extensible access control markup language (xacml) version 2.0. Oasis Stand., vol. 200502 (2005)
14. Becker, M.Y., Fournet, C., Gordon, A.D.: SecPAL: design and semantics of a decentralized authorization language. J. Comput. Secur. **18**(4), 619–665 (2010)
15. Wielemaker, J., Ss, S., Ii, I.: SWI-Prolog 2.7-Reference Manual (1996)
16. Date, C.J., Darwen, H.: A Guide to the SQL Standard, vol. 3. Addison-Wesley, New York (1987)

17. Horrocks, I., et al.: SWRL: a semantic web rule language combining OWL and RuleML. W3C Memb. Submiss. **21**, 79 (2004)
18. WOW Group, et al.: OWL 2 Web Ontology Language Document Overview (2009)
19. Giaffreda, R.: iCore: a cognitive management framework for the internet of things. In: Galis, A., Gavras, A. (eds.) FIA 2013. LNCS, vol. 7858, pp. 350–352. Springer, Heidelberg (2013). https://doi.org/10.1007/978-3-642-38082-2_31
20. Light, R.: Mosquitto-an open source mqtt v3. 1 broker. URL Httpmosquitto Org (2013)
21. Han, W., Lei, C.: A survey on policy languages in network and security management. Comput. Netw. **56**(1), 477–489 (2012)
22. Birgisson, A., Politz, J.G., Erlingsson, U., Taly, A., Vrable, M., Lentczner, M.: Macaroons: cookies with contextual caveats for decentralized authorization in the cloud. In: NDSS (2014)
23. Jones, M., Bradley, J., Sakimura, N.: JSON web token (JWT) (2015)
24. Belokosztolszki, A., Eyers, D.M., Pietzuch, P.R., Bacon, J., Moody, K.: Role-based access control for publish/subscribe middleware architectures. In: Proceedings of the 2nd international workshop on Distributed event-based systems, pp. 1–8 (2003)
25. Singh, J., Vargas, L., Bacon, J., Moody, K.: Policy-based information sharing in publish/subscribe middleware. In: 2008 IEEE Workshop on Policies for Distributed Systems and Networks, pp. 137–144 (2008)
26. Hermes. http://hermes-pubsub.readthedocs.io/en/latest/. Accessed 05 Nov 2017
27. Sciancalepore, S., Piro, G., Caldarola, D., Boggia, G., Bianchi, G.: OAuth-IoT: an access control framework for the Internet of Things based on open standards. In: 2017 IEEE Symposium on Computers and Communications (ISCC), pp. 676–681 (2017)
28. Hardt, D.: The OAuth 2.0 authorization framework (2012)
29. Shelby, Z.: Constrained RESTful environments (CoRE) link format. Internet Engineering Task Force IETF, vol. RFC6690 (2012)
30. Hu, Y.C., Patel, M., Sabella, D., Sprecher, N., Young, V.: Mobile edge computing—a key technology towards 5G. ETSI White Pap. **11**(11), 1–16 (2015)
31. Pearson, S., Casassa-Mont, M.: Sticky policies: an approach for managing privacy across multiple parties. Computer **44**(9), 60–68 (2011)
32. Abadi, M., Feigenbaum, J., Kilian, J.: On hiding information from an oracle. In: Proceedings of the Nineteenth Annual ACM Symposium on Theory of Computing, pp. 195–203 (1987)
33. Feigenbaum, J.: Encrypting problem instances. In: Williams, H.C. (ed.) CRYPTO 1985. LNCS, vol. 218, pp. 477–488. Springer, Heidelberg (1986). https://doi.org/10.1007/3-540-39799-X_38
34. Wong, W.K., Cheung, D.W., Kao, B., Mamoulis, N.: Secure kNN computation on encrypted databases. In: Proceedings of the 2009 ACM SIGMOD International Conference on Management of Data, New York, NY, USA, pp. 139–152 (2009)
35. Zhao, Y., Sturman, D.C.: Dynamic access control in a content-based publish/subscribe system with delivery guarantees. In: 26th IEEE International Conference on Distributed Computing Systems (ICDCS 2006), p. 60 (2006)

Comprehensive Study in Preventive Measures of Data Breach Using Thumb-Sucking

Keinaz Domingo$^{(\boxtimes)}$, Bryan Cruz, Froilan De Guzman$^{(\boxtimes)}$, Jhinia Cotiangco, and Chistopher Hilario

University of the East, Caloocan, Philippines
keinazd6@gmail.com, bryancruz014@gmail.com,
froilan.deguzman@ue.edu.ph, jhiniaaa04@yahoo.com,
chris.hilario0108@gmail.com

Abstract. This research presents a method of data breach that is known as thumb-sucking. Through the insertion of a flash drive into a computer's USB port, it allows for the extraction of the user's private information. In order to protect against the possibility of data breach in the form of thumb-sucking, the researchers have devised preventive measures in order to protect a user from malicious hackers. These measures include testing on various Windows operating systems that are in use, mainly Windows 7, Windows 8.1, Windows 10 and some Linux platform. The choice of conducting tests on Windows is due to the problem that not all users of Windows are tech savvy enough to avoid their data being accessed without permission.

Keywords: Data breach · Thumb-sucking · Preventive measures

1 Introduction

In an age of information where what one is searching for is just a click away, it's also no wonder that there are people that wish to gain access to an individual's private information through illegal means. The sharing of data has become an essential means of communication. Although data sharing has become a necessity, many are still not aware of the possible dangers that it poses, especially owners of small establishments that may fall prey to this danger. [12] Since innovations in technology are constantly progressing and more data is being displayed in public, this constitutes a new risk for IT departments. [1] Although it is not only data on the internet that needs to be protected, but as well as data that's saved on personal storage devices. Portable devices are convenient for the purpose of business; however, it is also because of their portable nature that they are more likely to be stolen or even misplaced [3].

This act of data being leaked and becoming privy to the eyes of others is known as a data breach. A data breach is said to occur when private data is exposed to outside parties, whether it be unintentional or not. [7] Data breaches and illegal access to private information happens very often, although what makes this alarming is not the amount of common people being affected, but rather how hackers are targeting corporations that

© Springer Nature Switzerland AG 2018
T. K. Dang et al. (Eds.): FDSE 2018, LNCS 11251, pp. 57–65, 2018.
https://doi.org/10.1007/978-3-030-03192-3_5

have a considerable amount of value, not to mention that their methods of attack are improving [4].

Numerous ideas have been formulated in order to deal with the problem of data breaches. These include methods such as imposing restrictions on companies that use wireless networks, contributing to government projects to improve security, and as well as making personal improvements in the security of one's network. [6] Various governments have also taken to devising countermeasures against data breach, and making them known to the public. One method is through the restriction. Creating separate networks that do not have access to each other can help by limiting access from the outside, and firewalls can be implemented in order to further improve security [8].

2   Literature Review

According to study, mobile devices have also become a viable source for hackers to commit their attacks. This is because a large portion of the world is now covered by mobile networks, which only continue to increase in coverage over the years. [5] This means that the amount of personal information that a hacker potentially has access to can only increase as well, putting even more data at risk.

Another study conducted a survey on data breaching. According to their finds, organizations whether large or small are more prone to being targeted by outside parties, the attacks of which were able to compromise a great portion of these organizations. [9] The idea that most organizations suffer from being targets of data breach is definitely food for thought.

A different source explains that the employees of a company could actually be seen as a threat to security, explaining that the mistakes that an employee makes accounts to up to 95% of incidents. [10] This means to say that most data breaches could possibly result from human error, which is not a far off explanation given the fact that not all employees are trained properly. Even a portable device such as a flash drive can be used to commit data theft. A study on data breach says that portable devices such as hard drives, laptops and flash drives being stolen or lost has also been a contributor to the leakage of data. [2] This is because a simple portable device could be used by a hacker to infiltrate and retrieve an organization or individual's information, and very easily at that if they can manage to retrieve its contents.

Data breaches aren't completely restricted to physical storage, because with the advent of cloud computing, it allows for the possibility of storing data within a cloud on the internet. Although the technology is not without any risks, as the cloud can be used as a medium for hackers to commit their crimes. In this way, hackers are able to retrieve personal data from websites that store their information on the cloud, such as common social media sites like Facebook, Youtube, Twitter, and many others. [15] Although there are ways to protect a cloud against data breach, which involves using technology such as a cloud antivirus that scans files such as documents whenever a network cloud receives it [17].

While on the discussion of the internet and the cloud, a paper discussed the benefits of using an ANS (anonymizing network system). The way that an ANS works is by making the user anonymous on the internet, protecting any personal information that could be stolen and used against the user. [11] This could help against data breaches by giving a user the benefit of anonymity. Through this anonymity, a potential hacker will not be able to infer any information from the user.

Another study claims that the traditional methods of defense are no longer enough to protect against targeted attacks. [16] There is some truth to it, because a persistent hacker could gain access to an organization's information with enough time and resources.

Because the protection of data is an important matter, other researchers have devoted their time to making a model that presents an attacker's possible methods of attack. The model which has been named Cyber Kill Chain is described as a model with 7 layers or phases of a cyber-attack. With the information that was presented in the paper, it could be seen as a valuable resource for those who are aiming to create new methods of defense against cyber-attacks, as this lets them gain understanding as to how an attacker plans out their attack [13].

Another paper by other researchers used a system called TrustBox. The system functions by preventing data breaches when a user is connected to a network, but it also provides protection even if a user is not currently connected to any networks. [14] Through the use of that system, sensitive data could not be leaked accidentally, and access was only granted by an authentication scheme. The implementation of a similar system to various organizations could possibly lessen the amount of data breach that happens.

One paper discussed how anti-viruses are vulnerable during updates, because they are either partly or totally deactivated during this time. The reasoning behind this is because a malware installer is able to monitor or even trigger an antivirus into installing an update, which leaves the system open to attacks in this brief window. [18] Through this method, the malware would gain access to the computer. Another paper on anti-viruses discussed a way of using an anti-virus to assist an attack, which is called an anti-virus assisted attack. This method of attack works by utilizing byte patterns in order to slip through an anti-virus undetected. [19] In theory, the anti-virus itself would be a means for an attack to get through.

3   Thumb-Sucking

Thumb-sucking is a method to obtain information from a computer by inserting a thumb drive into a USB port. For the specificity of testing, the researchers used a password recovery tool named WebBrowserPassView that was installed on a flash drive. It retrieves passwords by revealing the cached contents of the user's web browsers. It is normally used for recovering lost passwords, but for the purpose of this study, the researchers have used it as a way to access passwords that have been stored in a web browser cache.

The thumb-sucking procedure is shown in Fig. 1, the process of which the software goes through. The process of the attack is as follows:

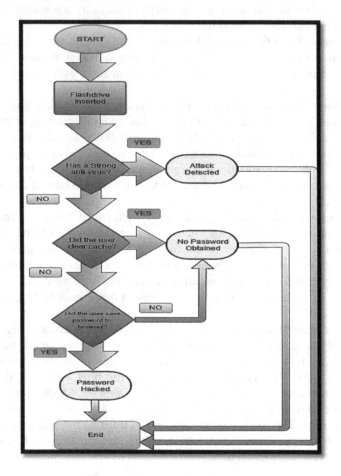

Fig. 1. Diagram of thumb-sucking process

1. Flash drive inserted – The beginning of the process starts with the insertion of the flash drive containing WebBrowserPassView into an open USB port on the target computer. Once it is finished, it moves onto the next step.
2. Has a strong anti-virus? – From here, the results can vary in two ways. If the computer has a strong anti-virus, it can detect the attack, rendering it useless and ending the process. If the anti-virus is weak, the diagram continues.
   Attack detected – If detected by an anti-virus, the attack is halted.
3. Did the user clear cache? – If the software bypasses the anti-virus, it begins to scan the caches of the web browsers. If the cache has been cleared recently, or it contains no available information, then no password is obtained and the process ends. However, if it finds that there is content in the cache, it moves to the next step.

No password obtained – If the cache has been cleared recently, no passwords are obtained by the software.

4. Did the user save password to browser? – If the user did not save any passwords to the browser that was used, then no password is obtained and the process ends. In the event that the software detects any passwords in the browser cache, then the password is obtained and saved by the software.

No password obtained – If no passwords have been saved to the browser, then no information is obtained.

Password hacked – If passwords have been saved to the browser, then the software detects the passwords and retrieves them from the cache.

4 Testing and Discussion

In order to help protect against data breach, the researchers have conducted tests in order to observe the vulnerability of a user's passwords being retrieved by an outside party through thumb-sucking. To begin with, the researchers first tested v1.85 of WebBrowserPassView on a laptop running the latest version of Windows 10. As shown in the figure that follows, the attack was detected by Windows 10 before it could retrieve any passwords from the caches of the browsers, and was subsequently prevented from doing so (Fig. 2).

Fig. 2. The attack is detected by Windows

Next, we tried using v1.86 of WebBrowserPassView. Surprisingly enough, the software was not detected and it was able to obtain the passwords from various browsers, as shown in the figure below (Fig. 3).

Fig. 3. Password obtained

With this information, we can assume that having the latest version of your operating system installed, which in this case, was Windows 10, will decrease the risk of your passwords being hacked by password recovery tools. The reason why is because these updates might be able to fix potential vulnerabilities in earlier versions of the operating system, which hacking tools will be able to take advantage of in order to obtain your private information.

After testing on Windows 10, the researchers conducted another test on Windows 7 using the same tool. Upon opening the application, it was instantly detected by Microsoft Security Essentials. This test was done with the Windows Firewall turned off. After the attack was detected, it was quarantined and removed from the computer by Microsoft Security Essentials. Any subsequent attempts to download the application resulted in it being removed once it was extracted from its .zip file.

The researchers attempted another test with the anti-virus deactivated, and this time, the application opened and managed to obtain passwords from Firefox and Google Chrome's browser caches. Passwords and usernames obtained were blurred in order to protect personal information (Fig. 4).

Fig. 4. Password obtained from Firefox and Google Chrome

Tests were conducted on Windows 8.1, and the application was able to access passwords that were stored on Firefox. WebBrowserPassView also allows for the viewing of detailed information of each individual instance that is retrieved from the browser caches. It is able to show the type of username and password field used for that specific instance (Fig. 5).

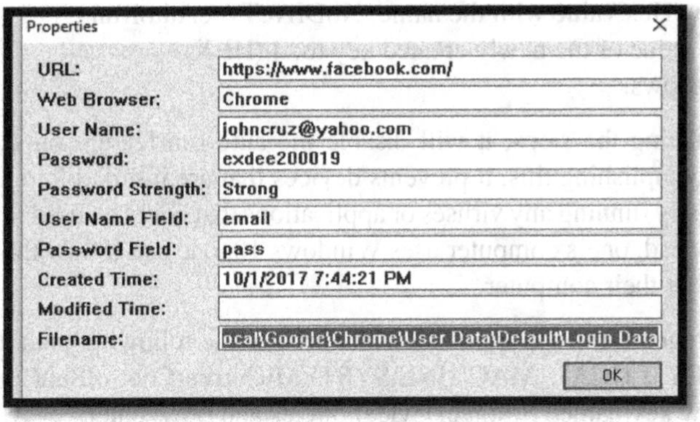

Properties	×
URL:	https://www.facebook.com/
Web Browser:	Chrome
User Name:	johncruz@yahoo.com
Password:	exdee200019
Password Strength:	Strong
User Name Field:	email
Password Field:	pass
Created Time:	10/1/2017 7:44:21 PM
Modified Time:	
Filename:	ocal\Google\Chrome\User Data\Default\Login Data

OK

Fig. 5. Detailed account information

After testing, the researchers were able to compile the information obtained into a table. In some Linux environment such CentOs, Debian, Fedore and Ubuntu, and Kali Linux the thumb-sucking tool is tested (Table 1).

Table 1. Testing results

Operating system	Tool version	Penetrated	Secured
Windows 7	V1.85	No	Yes
Windows 7	V1.86	No	Yes
Windows 8.1	V1.85	No	Yes
Windows 8.1	V1.86	Yes	No
Windows 10	V1.85	No	Yes
Windows 10	V1.86	Yes	No
CentOs	V1.86	No	Yes
Debian	V1.86	No	Yes
Fedora	V1.86	No	Yes
Ubuntu	V1.86	No	Yes
Kali	V1.86	No	Yes

Although this kind of attack is easy to avoid for users who are aware of the possibility of being the target of such an attack, since the application needs to be inserted with a flash drive into a computer's USB port. But if a hacker were to modify the application to run automatically upon insertion of the flash drive, it could endanger a person's private information. To prevent that from happening, one can disable the autorun feature on removal devices on the Windows Registry. To do so, the following steps need to be followed on Windows 7:

1. Search for regedit and run it. Doing so opens the Registry Editor.
2. Go to the key that is shown as follows: HKLM\Software\Microsoft\Windows \CurrentVersion\Policies\Explorer
3. Create a new Hex value with the name: NoDriveTypeAutoRun
4. Change the value of the newly created key to: 4 (HEX)
5. Restart Windows.

After completing the steps, it will disable the auto-run feature on all removable devices. By accomplishing this, it prevents devices that are inserted into the computer from automatically running any viruses or applications that could be used to tamper with your files. If instead, one's computer uses Windows 10, one can follow the steps below in order to protect their computer.

1. Search for regedit, run the application and browse the following path that is shown below: HKEY_LOCAL_MACHINE\SYSTEM\CurrentControlSet\Control
2. Create a new key named StorageDevicePolicies and press enter.
3. Select the newly created key, right-click on the right side and create a new DWORD (32-bit) value.
4. Name the new DWORD WriteProtect and change its value from 0 to 1

Upon completion of the steps, outsiders who connect a USB device to the user's computer or laptop will be denied copy privileges, which will protect the user's information since the outsider cannot obtain any of the user's information.

Ensuring that one constantly clears their cache from time-to-time can also prevent a person's information from being hacked. On Windows, CTRL + SHIFT + DELETE will open the Window. On the prompt that opens, you can select a time range of cached data to be cleared as well as specify what will be deleted. This will remove potential data that can be retrieved by a possible attacker.

If one uses Firefox, they can customize the configuration of the browser to never remember any browsing, download, search or form history, as well as clear history every time Firefox closes. By doing this, it ensures that no data enters the cache, and with the lack of data, it prevents a data breach from happening.

Although if the user instead makes use of Google Chrome, then one can click CTRL + H in order to open the cache. From there, various settings can be chosen in order to decide what data will be cleared or not.

5 Conclusion

The researchers believe that it is possible for a user to avoid a data breach simply through being aware. This awareness will make them more conscious of their actions, since human error is a common reason why information often gets leaked to outside entities. Knowing the proper steps of preventing a data breach can go a long way for a person without any professional training, as they are now aware of the possible consequences that can arise from being careless. Since the thumb-sucking flash drive only uses a simple method of attack, it is quite possible for a user who is aware of the risks to avoid the attack. The attack relies heavily on cached information not being cleared in order to obtain any information. Therefore, if the user constantly clears their cache, then the risk of their information being sucked significantly decreases. The use of a strong anti-virus is also another layer of defense against such attacks, because it allows for the early detection of an attack. If a user wishes to further enhance the protection of their information, they can set their browser cache to not remember any information, thereby limiting the information that a hacker can potentially retrieve from their computer. Using the Registry tool can also aid in protecting one's information by modifying the registries.

Having access to tools that can aid a user in protecting against a data breach is also viable. Software such as anti-viruses can allow for the early detection of possible threats to a system.

References

1. Five Solutions for Improving Your Organization's Data Security. Kanguru, Secure. Anytime. Anywhere (2016)
2. Huq, N.: Follow the data: analyzing breaches by industry. TrendLabs Research Paper (2015)
3. Protecting the confidentiality of personal data, guidance note. Department of Finance (2008)
4. Data Protection and Breach. Online Trust Alliance (2015)
5. Internet Security Threat report. Symantec Corporation (2014)
6. Crews, C.W. Jr., Oberwetter, B.: Preventing identity theft and data security breaches: the problem with regulation. Competitive Enterprise Institute (2006)
7. Cheng, L., Liu, F., Yao, D.: Enterprise data breach: causes, challenges, prevention, and future directions. WIREs Data Min. Knowl. Discov. 7, 1–14 (2017)
8. Basic Cybersecurity Measures. WaterISAC (2015)
9. Vaizey, E.: 2015 Information Security Breaches Survey. HM Government (2015)
10. Investing in Human Firewall. Lawroom (2014)
11. Clark, J, Stavrou, A.: Breaching & Protecting an Anonymizing Network System. In: Annual Symposium on Information Assurance (2011)
12. Frost, S.: Could a data breach cause your business to fail? Faronics
13. Yadav, T., Rao, A.M.: Technical aspects of cyber kill chain. In: Abawajy, J.H., Mukherjea, S., Thampi, S.M., Ruiz-Martínez, A. (eds.) SSCC 2015. CCIS, vol. 536, pp. 438–452. Springer, Cham (2015). https://doi.org/10.1007/978-3-319-22915-7_40
14. Schmidt, M., Fahl, S., Schwarzkopf, R., Freisleben, B.: TrustBox: a security architecture for preventing data breaches
15. Nandhakumar, C., Ranjithprabhu, K., Raja, M.: Counter measures for data breach in cloud computing. Int. J. Res. Comput. Appl. Robot. 2, 124–129 (2014)
16. Balazs, Z., Effitas, M.: Breach detection system testing methodology
17. Chamorro, E., Han, J., Beheshti, M.: The design and implementation of an antivirus software advising system. In: Ninth International Conference on Information Technology (2012)
18. Min, B., Varadharajan, M., Tupakula, U., Hitchens, M.: Antivirus security: naked during updates. Softw. – Pract. Exp. Wiley Online Library (2013)
19. Wressnegger, C., Freeman, K., Yamaguchi, F., Rieck, K.: Automatically inferring malware signatures for anti-virus assisted attacks. In: Asia CCS (2017)

Intrusion Prevention Model for WiFi Networks

Julián Francisco Mojica Sánchez[1], Octavio José Salcedo Parra[1,2(✉)], and Alberto Acosta López[2]

[1] Department of Systems and Industrial Engineering, Faculty of Engineering, Universidad Nacional de Colombia, Bogotá D.C., Colombia
{jfmojicas,ojsalcedop}@unal.edu.co
[2] Faculty of Engineering, Intelligent Internet Research Group, Universidad Distrital "Francisco José de Caldas", Bogotá D.C., Colombia
{osalcedo,aacosta}@udistrital.edu.co

Abstract. WIFI technology has consolidated in the market with a wide range of devices connected at different distances, frequencies and with different characteristics. Despite progress in the creation of new devices and improvements in technology, although security has some protective measures is lagging behind and more vulnerabilities are being discovered every day of the networks with these devices. Therefore, it is essential the research and development of security measures and tools to ensure information security, this document proposes a game theory model that could be the basis of an algorithm for the prevention of intrusions in WIFI networks.

Keywords: Intrusion prevention system · Game theory · Vulnerabilities · WIFI

1 Introduction

WIFI technology has been imposed in recent years over other types of connection in local networks due to its versatility and low cost compared to wired networks, this has allowed the creation of different network configurations from small home and office networks up to business networks, with a considerable amount of devices sharing information at all time. This versatility has been achieved leaving behind security, therefore it is necessary to work on tools and protocols that increase the security of WIFI networks [1].

Currently, the security protocols designed by IEEE for WIFI networks are the 802.11 family, where the most used form of encryption due to its efficiency is WPA2. In spite of the efforts made by IEEE, the advance in new devices has been much greater than the advance in the creation, implementation and updating of security protocols so that each time new attacks are created taking advantage of this gap. Intrusion prevention systems are used in large companies and it is not feasible to implement them in public networks with a large number of connected devices sharing information at all times and in different locations. As for office and home networks, router security mechanisms are obsolete against various types of attacks [2].

© Springer Nature Switzerland AG 2018
T. K. Dang et al. (Eds.): FDSE 2018, LNCS 11251, pp. 66–73, 2018.
https://doi.org/10.1007/978-3-030-03192-3_6

This document proposes a game theory model that could be the basis of an algorithm for the prevention of intrusions in WIFI networks, based on two intrusion detection models.

2 Related Works

In 2013 Manshaei, Zhu, Alpcan, Bacar and Hubaux conduct a review of research on privacy and security in communication and computer networks that have as their focus game theory. They present a review of the different works found in the literature, the way in which IDS are configured, Networked IDS, where in the network operate different IDS independently and the security of each subsystem they protect individually depends on the performance of the other IDS, Collaborative Intrusion Detection System Networks, in this case in the network operate different IDS collaborative way, i.e. share the knowledge of new attacks that detect, but the system may be compromised if the control of an IDS is taken by an attacker and finally the response to intrusions, where they expose a system of response to intrusions based on Stackelberg stochastic game called Response and Recovery Engine (RRE) [3].

In 2016 Sharma, Moon, Moon and Park designed a DFA-AD (distributed framework architecture for the detection of advanced persistent threats), in which one of the 4 traffic classification modules was Dynamic bayesian game model Based, in this case the game model is dynamic since each player selects his behavior depending on the current state of the system and the information he possesses. The attackers identify users and special targets of the system in an exhaustive way, therefore, the attackers have more data about the module than the protectors, which creates a system in incompleteness and asymmetry [4].

In 2016 Wang, Du, Yang, Zhu, Shen and Zhang propose an attack-defense game model to detect malicious nodes in Embedded Sensor Networks (ESNs) using a repeat game approach, define the reward function that attackers and defenders will receive for their actions. To fix detection errors and detection absences they use a game tree model. They demonstrate that the game model does not have a Nash balance of pure strategy but mixed, where the nodes are changing due to the strategies of attackers and defenders so that they are in dynamic balance, in this balance is made use of limited resources and security protection is provided at the same time. Finally they perform simulations of the proposed model from which they conclude that they can reduce energy consumption by 50% compared to the existing All Monitor (AM) model and improve the detection percentage from 10% to 15% compared to the existing Cluster Head (CH) model [5].

3 Game Theory Models

Intrusion prevention can be understood as an attack-defense scenario, in which the network security manager decides whether or not it is necessary to implement the intrusion prevention system, because such operation has a cost that would not be necessary if the network is not being attacked. The game consists of a defender (the one in charge of starting or not the IPS) and an attacker (which seek to enter the network and take advantage of the intrusion), this was taken from the model proposed by Wang in 2016 but limited to a single attacker and defender, as it took multiple attackers and defenders in multiple nodes and time periods [5].

As for the defender, he has two strategies (UD): defend or not defend and in the case of the attacker (UA): attack or not attack. Carrying out these strategies has rewards and costs that will determine the way the two actors act. These costs and rewards are defined below:

- Cost of starting the IPS Cm
- Average loss when the system is attacked Ci
- Cost to attack by the attacker Ca
- Cost of not attacking by the attacker Cw
- Payment to the defender for taking an action strategy defensive Ui
- Payment to the attacker for taking an action strategy offensive Ua

Now it can be understood that the reward of the attackers Pa is equal to the average losses when the system is attacked, that is:

$$P_a = C_i \tag{1}$$

Now it is necessary to define when it is profitable for the attacker to perform the attack:

$$C_w < P_a - C_a \tag{2}$$

The above equation means that the attacker will perform an attack when its reward minus the cost of attacking greater than the cost of not attacking.

On the other hand, the attacker will not make an attack when the cost of starting the IPS is much lower than the loss average when the system is attacked, because in this case surely the defender would have started the IPS, therefore this will be in operation and the attack will be detected and the attacker isolated from the network.

From this it is possible to define the reward matrix as:

$$\begin{bmatrix} P_a - C_a, U_i - C_i & -U_a, U_i - C_m \\ C_w, U_i & C_w, U_i - C_m \end{bmatrix}$$

Where the columns correspond to the strategies of the defender, that is, not defend and defend; and the rows do reference to the attacker's strategies, that is, attack and not attack. It is now necessary to analyze Nash's balance by analyzing each actor's payoffs depending on the strategy taken by the other actor. When the defender does not defend the attacker has two possible strategies, but by Eq. (1) we know that:

$$Cw < Pa - Ca \tag{3}$$

Therefore, the attacker will always choose to attack.

Secondly when the defender defends the same mind the attacker can choose between attacking and not attacking, but how:

$$U_i - C_a < U_i - C_m \tag{4}$$

The defender will always choose to defend.

Second, when the attacker decides not to attack

$$U_i > U_i - C_m \tag{5}$$

Therefore, the defender will always choose not to defend.

From analyzing the previous strategies that the actors will take depending on the behavior of the other it can be said that there is no pure Nash balance, since there is no place in the matrix in which both actors are satisfied with their reward. Because there is no pure Nash balance it is necessary to analyze if the game model is in mixed Nash balance, for this you define the probability that the attacker attacks σ and the probability that the defender defends δ.

The attacker's mixed strategy is

$$U_A = \left(P_a - C_a\right)(1 - \delta)\sigma + \left(-U_a\right)\delta\sigma + C_\omega(1 - \sigma) \tag{6}$$

The defender's mixed strategy is

$$U_I = \left(U_i - C_i\right)(1 - \delta)\sigma + \left(U_i - C_m\right)\delta + U_i(1 - \delta)(1 - \sigma) \tag{7}$$

$$U_I = U_i - C_i\delta\sigma - C_i\sigma - C_m\delta \tag{8}$$

Now using the extreme value method to solve the strategy of the Nash mixed model, Eqs. (6) and (8) are derived with respect to σ and δ respectively.

$$\frac{\partial U_A}{\partial \sigma} = \left(P_a - C_a\right)(1 - \delta) + \left(-U_a\right)\delta - C_\omega = 0 \tag{9}$$

$$\frac{\partial U_I}{\partial \delta} = C_i\sigma - C_m = 0 \tag{10}$$

From Eq. (9) it is possible to find σ

$$\delta = \frac{P_a - C_a - C_\omega}{P_a - C_a + U_a} \tag{11}$$

From Eq. (10) it is possible to find σ

$$\sigma = \frac{C_m}{C_i} \tag{12}$$

To analyze the Nash equilibrium by mixed strategy, it is possible to start assuming that the probability of attacking *sigma* is high, so that $C_m \gg C_i$, that is, the attack occurs when it is not profitable to start the IPS, which makes the defense probability low. In case the probability of defense is high, it means that the IPS has probably been launched because the losses to be attacked are greater than the cost of having the IPS in operation, that is, $C_m \gg C_i$ which indicates that the attack probability must be low.

In conclusion the probabilities of attack and defense are inversely proportional and the system will be in Nash's mixed equilibrium when $\sigma = \delta$.

Manshaei [3] discusses a two-player Bayesian game, a defense node and a malicious or regular one. The malicious node can choose between attacking and non-attacking, while the defense node can choose between monitoring and nonmonitoring. The security of the defender is quantifiable according to the goods it protects w, therefore, when there is a security breach the damage is represented by $-w$. The rewards matrix is presented below:

$$\begin{bmatrix} (1-\alpha)w - C_a, (2\alpha - 1)w - C_m & w - C_a, -w \\ 0, \beta w - C_m & 0,0 \end{bmatrix}$$

In this matrix the columns represent the behaviors of the defender (monitor and not monitor) and the rows attacker behaviors (attack and not attack), C_a and C_m are costs of attacking and monitoring, α and β are the detection rate and the false alarm rate of the IDS respectively and μ_0 the probability that a player is malicious.

Finally they show that when $\mu_0 < \dfrac{(1+\beta)w + Cm}{(2\alpha + \beta - 1)w}$ the game supports a strategy of pure balance (attack if it is malicious, do not attack if it is regular), do not monitor, μ_0 and when $\mu_0 > \dfrac{(1+\beta)w + Cm}{(2\alpha + \beta - 1)w}$ the game does not have a pure strategy.

4 Proposed Model

From the model described by Manshaei and establishing that the two players are intruder and defender, since the intruder is ready to carry out the attack because has done a vulnerability study and has planned the different strategies to follow in order to enter authorized to the network, the time when no attack represents a C_w cost (waiting cost) because the network can change and the investment mentioned above both of time and of resources can be lost. Therefore, the payment matrix is:

$$\begin{bmatrix} (1-\alpha)w - C_a, (2\alpha - 1)w - C_m & w - C_a, -w \\ -C_w, -\beta w - C_m & -C_w, 0 \end{bmatrix}$$

Depending on the strategy that the other actor takes and the respective payments they obtain, it is possible to determine if there is a Nash equilibrium.

When the defender does not monitor, the attacker has two possible strategies: Attack, with gain $w - C_a$; and do not attack, where he gets $-C_w$, therefore you will always choose to attack.

When the defender also monitors the attacker can choose between attacking and not attacking, assuming a detection rate greater than 90%, attacking would lose the cost of attacking, while not attacking would lose the cost of waiting.

Generally the deployment of an attack to take control of the network or the information it contains is more expensive than carrying out a recognition and learning of the network and its vulnerabilities, therefore the attacker will choose not to attack.

$$-C_w > -C_a$$

Now it is necessary to fix the behavior of the attacker and analyze the possible strategies that the defender will perform.

First, when the attacker decides to attack and assuming a detection rate greater than 90%.

$$(2\alpha - 1)w - C_m > -w$$

This means that the defender will choose to monitor.

Second, when the attacker decides not to attack

$$-\beta w - C_m < 0$$

Then, the defender will always choose not to monitor.

After analyzing the different strategies, it is clear that in neither scenario will both actors be satisfied with their reward. Which prevents that pure Nash equilibrium exists in the proposed game.

5   Model Evaluation

Because there is no point in the rewards matrix in which both the defender and the attacker feel comfortable with the situation, it is necessary to determine if the model is in mixed Nash equilibrium, for this the probability that the attacker attack σ and the probability that the defender defends δ.

The mixed strategy of the attacker is:

$$U_A = \left[(1 - \alpha)\omega - c_a\right]\delta\sigma + \left[\omega - c_a\right](1 - \delta)\sigma + \left(-c_\omega\right)\delta(1 - \sigma) + \left(-c_\omega\right)(1 - \delta)(1 - \sigma) \tag{13}$$

The mixed strategy of the defender is

$$U_I = \left[(2\alpha - 1)\omega - c_m\right]\delta\sigma + [-\omega](1 - \delta)\sigma + \left[\beta\omega - c_m\right]\delta(1 - \sigma) \tag{14}$$

Using the extreme value method to solve the strategy of the Nash mixed model, the equations are derived (16) and (17) regarding δ and σ respectively and are equal to zero.

$$\frac{\partial U_A}{\partial \sigma} = -\delta\alpha w + w - C_a + C_w = 0 \tag{15}$$

$$\frac{\partial U_I}{\partial \sigma} = 2\sigma\alpha w - \beta w - C_m + \sigma\beta_w = 0 \tag{16}$$

From the Eq. (18) its possible find δ:

$$\delta = \frac{w - C_a + C_w}{w\alpha} \tag{17}$$

From the Eq. (19) its possible find σ:

$$\sigma = \frac{\beta w + C_m}{2\alpha w + \beta_w} \tag{18}$$

To analyze the Nash equilibrium by mixed strategy, you can start assuming that the probability of attacking δ be high, for this to be $Cm >> 2aw$, this means that the attacker could attack comfortably when the goods that protects the defender are not so valuable to him, which is why I will not have activated the IPS. In the case where you come from protect are valuable the defense probability will increase and the probability of attack will decrease.

Therefore, it is found again that the probabilities of attack and defense are inversely proportional and the system will be in mixed Nash equilibrium when:

$$\delta = \sigma \tag{19}$$

It is known that

$$0 \leq P(a) \leq 1 \tag{20}$$

Then

$$\delta = \frac{\omega - c_a + c_\omega}{\omega\alpha} \leq 1 \tag{21}$$

$$\sigma = \frac{\beta\omega - c_m}{2\alpha\omega + \beta\omega} \leq 1 \tag{22}$$

From Eq. (23)

$$C_m \leq 2\alpha\omega \tag{23}$$

Equation 25 indicates that when the detection rate or value of the good to be protected decreases in the same way, the detection cost used to protect the good should decrease. From Eq. (24)

$$C_\omega \leq C_a - (1 - \alpha)\omega \tag{24}$$

Equation 26 indicates that the attacker's maximum wait cost is directly linked to the undetected rate of the intrusion prevention system, as that rate increases the maximum wait cost should decrease. From Eq. (21)

$$\omega = \frac{(-C_a(2\alpha + \beta) + C_m\alpha - C_\omega(2\alpha + \beta)}{\alpha(\beta - 2) - \beta} \tag{25}$$

6   Conclusions

The game theory model for intrusion prevention adds the waiting cost of the attacker and it is shown theoretically that it is a decisive factor in the prevention of an attack. This is because if the attacker spends several resources planning the attack, it is possible for the defender to make modifications in the network that prevent the initially planned attack or simply migrate the information desired by the attacker, in which case the attacker will have spent more resources than he is going to get when he succeeds in violating the security of the defender.

Describes the condition to be in Mixed Nash Equilibrium, it means that each player has the same probability in their options of behavior.

$$\omega = \frac{\left(-C_a(2\alpha + \beta) + C_m\alpha - C_\omega(2\alpha + \beta)\right)}{\alpha(\beta - 2) - \beta} \tag{26}$$

References

1. Kolias, C., Kambourakis, G., Stavrou, A., Gritzalis, S.: Intrusion detection in 802.11 networks: empirical evaluation of threats and a public dataset. IEEE Commun. Surv. Tutor. **18**(1), 184–208 (2016). https://doi.org/10.1109/COMST.2015.2402161
2. Huang, H., Hu, Y., Ja, Y., Ao, S.: A whole-process WiFi security perception software system. In: 2017 International Conference on Circuits, System and Simulation, ICCSS 2017, pp. 151–156 (2017). https://doi.org/10.1109/CIRSYSSIM.2017.8023201
3. Manshaei, M.H., Zhu, Q., Alpcan, T., Bacar, T., Hubaux, J.-P.: Game theory meets network security and privacy. ACM Comput. **45** (2013). https://doi.org/10.1145/2480741.2480742
4. Sharma, P.K., Moon, S.Y., Moon, D., Park, J.H.: DFA-AD: a distributed framework architecture for the detection of advanced persistent threats. Cluster Comput. **20**(1), 597609 (2017). https://doi.org/10.1007/s10586-016-0716-0
5. Wang, K., Du, M., Yang, D., Zhu, C., Shen, J., Zhang, Y.: Gametheory-based active defense for intrusion detection in cyber-physical embedded systems. ACM Trans. Embed. Comput. Syst. **16**(1), 121 (2016). https://doi.org/10.1145/2886100

Security for the Internet of Things and the Bluetooth Protocol

Rodrigo Alexander Fagua Arévalo[1]([✉]), Octavio José Salcedo Parra[1,2], and Juan Manuel Sánchez Céspedes[3]

[1] Department of Systems and Industrial Engineering, Faculty of Engineering, Universidad Nacional de Colombia, Bogotá D.C., Colombia
{rafaguaa,ojsalcedop}@unal.edu.co
[2] Faculty of Engineering, Intelligent Internet Research Group, Universidad Distrital "Francisco José de Caldas", Bogotá D.C., Colombia
osalcedo@udistrital.edu.co
[3] Faculty of Engineering, GIIRA Research Group, Universidad Distrital "Francisco José de Caldas", Bogotá D.C., Colombia
jmsanchezc@udistrital.edu.co

Abstract. This document analyzes the state of art regarding the security of the Internet of Things as well as some well-known vulnerability scenarios. Using the Blueborne exploit, we intend to take control of a mobile device, run malicious code on it and gain access to sensitive information.

Keywords: Internet of things · Wireless · WPA2 · KRAKCs · Blueborne Bluetooth · Encryption · Authentication

1 Introduction

The current trend in technology is to be connected all the time not only through computers and smartphones but it is also sought that all electronic devices that we use every day stay connected so they can be remotely controlled or enable browsing or recollecting data [1].

This type of technology implies great benefits such as the consolidation of smart cities, smart signaling to handle traffic, resource management, safety, emergency services and medical monitoring [2] with devices that can understand the environment and react to it.

Technological breakthrough often leads to massive implementation before it can be deemed completely safe and even if the security protocols are strictly followed for this type of remote connections they are not flawless and are still susceptible to be exploited by some individuals.

The security failures in IoT devices can go from losing information such as passwords and company private data up to the loss of control of devices. This is particularly preoccupying when it is related to medical care devices, vehicles or robots that interact with humans [3], which can affect the integrity of people.

© Springer Nature Switzerland AG 2018
T. K. Dang et al. (Eds.): FDSE 2018, LNCS 11251, pp. 74–79, 2018.
https://doi.org/10.1007/978-3-030-03192-3_7

1.1 Bluetooth

Originally presented in 1998, Bluetooth is now a wireless communication standard for short distances which is present in millions of devices, especially in the consumer electronics sector. It has received constant upgrades and one of its versions (BLE Bluetooth Low Energy) has gained popularity since it is low-consuming and can be applied to small sensors that are then connected to powerful devices. Bluetooth device-based networks have also been developed which allow the connection of several devices at the same time and using them to carry information from one device to another which would normally be out of physical communication range.

As it is the case with other technologies, for the implementation of Bluetooth the available documentation and protocols must be followed but they have become complex due to the way this technology has been developed. This has led to fragmented and unclear documentation which complicates the security work.

2   Background

Informatics security is a highly debated and studied subject since it moves large amounts of money every year and is researched thoroughly. However, the developments attained on the matter not always translate to IoT devices.

When security flaws are detected, the researcher in charge contacts the affected companies and asks them for a time frame to correct the flaws before publishing his research.

In 2015, Symantec which is a well-known informatics security company published a report [4] where the safety of 50 smart devices is put to the test. The devices were used at home and it was determined that none of them had strong passwords, mutual authentication or accounts protected against brute force attacks. Additionally, some of the apps used to control the devices did not use encryption to upload data onto the cloud.

Even when the manufacturers are informed of these vulnerabilities, they do not always take actions to correct them or simply ignore the warnings.

2.1   Security in Hospitals

Hospira is a company that produces Lifecare PCA3 and PCA5 pumps that serve to give medicine to patients in several hospitals all over the world. Researcher Billy Rios [5] found that the libraries were the minimum and maximum dosage limits can be easily accessed since no authentication is required. This means that the information can be accessed by anyone that can connect to the hospital's network or over the internet. Following the adequate procedure, the company was notified of this failure a year before publishing this article, but no measures were taken to correct it and, to make it even worse, there is evidence of the same failure in newer versions of the product.

2.2   Wireless Connections

A recent publication that attracted a lot the attention of the media is related to protocol 802.11 for wireless communications, specifically with the WPA2 security. Researcher Mathy Vanhoef discovered a security breach in the encryption protocol which means that any device with the protocol's correct implementation is vulnerable affecting all sorts of devices (Apple, Windows, Android, Linux, etc.).

The attack called KRACK (Key Reinstallation Attack) consists on taking advantage of a flaw in the 4-way handshake of the WPA2 protocol which is executed when a client wants to join a protected Wi-Fi network and confirms that both parties have the right credentials; In this process, the encryption password is also sent which will be used to send packages between the parties. This is where the attack can occur by producing a reinstallation of the encryption password meaning that the attacker can know it. The technical details can be found in the paper [6] presented in the conferences on Computer and Communications Society (CCS) and Black Hat Europe.

In summation, by using this vulnerability, sensitive information can be encrypted such as passwords, credit card numbers, e-mails, chat messages and more. Since the attack affects Wi-Fi connections, the attacker must be physically close to the victim and the connection point.

Big software companies have released update with patches that correct this vulnerability. Although most Android devices do not have the latest version of the system or new security updates (for over 2 years) since their fragmentation level is very high. The home routers do not receive these updates showing yet another defect of IoT security.

2.3   Propositions to Improve Security in IoT

Time synchronization protocol with improved security [7].

Time synchronization is used for industrial IoT applications (IEEE 802.15.4e) which allows low consumption and high reliability. If the nodes are under attack, the entire network is paralyzed. Yang et al. present two types of attacks. The first one is absolute slot number (ASN) where the nodes of the network receive incorrect ASN values which interfere with their synchronization to the rest of the network. The second type named time slot template attack involves the introduction into the network of a corrupted node which causes a shift in phase of the clock calculation for the legitimate nodes. Tests reveal that the network is severely affected by these attacks: Sec_ASN and the threshold filter (TOF) algorithm are improvements implemented to the protocol in order to enhance security and effectively respond to these types attacks.

2.4   Zero Knowledge Authentications [8]

The communication layer of the network is the most affected by security issues referring to the authentication process. A strong cryptographic solution is proposed named Zero Knowledge which is based on proving to the other party that there is knowledge without giving any information other than the veracity of this fact. It is based on two-way mathematical algorithms and has already been proved and compared with other authentication methods.

3   Methodology

To breach a mobile device's security and eventually take control over it, the Bluetooth vulnerability discovered by Armis Labs [9] will be used.

To attack the device, a laptop with a Linux operative system will be used and the targeted device will be a Motorola smartphone with Android 4.4.2. It is important to state that it is not necessary that the smartphone has the 'visible for all devices' feature activated but only the Bluetooth has to be activated.

First, the computer identifies all the Bluetooth devices that are active nearby. Then, the computer identifies the MAC address of the smartphone and the device's operative system, so the algorithm can be applied to take advantage of the specific vulnerabilities of Android.

L2CAP (Logical Link Control and Adaptation Protocol) is a protocol that establishes a logical link between two terminals so that an app can use it. In this case, an error related to this protocol is profited; a stack overflow occurs on the variable that stores the connection parameters and it can only store up to 64 bytes. A setting package that exceeds this size is intentionally sent to the stack, so the rest of the package is stored by overwriting the adjacent memory which corresponds to the system's core. The malicious code is introduced with this method.

Since this exchange is made before any authentication method (when the connection is being established), it is not affected by passwords or other security methods.

Once the code is introduced, some vulnerabilities can be used that allow the remote execution of the code such as the one affecting the BNEP (Bluetooth Network Encapsulation Protocol) making it possible to control the device. At this point, the smartphone's peripherals can be used such as the camera and microphone and files can be accessed and transferred via Bluetooth. Finally, the connection can be ended and there will be no record of it.

4   Design and Implementation

The PoC (Proof of Concept) [10] written in Python is executed from the Linux command window and was designed to exploit the vulnerabilities CVE-20170781 and CVE-2017-0785. It can also evade the ASLR (Address Space Layout Randomization) technique in charge of protecting the memory against stack overflows. For its operation, it is necessary to open specific ports in the computer that is attacking (1233–1235).

With the help of radars [11], the parameters corresponding to the smartphone model are obtained. They will be modified by code since the test was designed for Pixel and Nexus smartphones. Furthermore, the computer's IP is necessary to perform attack since it is modified in the code. Some details specific to the Android version (6.0) and related to the Bluetooth adapter are also necessary. This code only works with versions 6.0 and 7.0 of 64-bit Android.

5 Results

There were several problems in the implementation of the code since it is not easy to obtain the exact values to be replaced. In the case of Bluetooth values specific to the smartphone, these change every time the process is restarted.

The smartphone was detected when it was in the 'non-visible for other devices' mode which is a security problem per se.

The connection to the smartphone was possible even if only some data was obtained (such as the Android version and some memory addresses) and no malware was executed.

Taking as a reference the video shown in Armis Labs [12], the process can be deemed unsuccessful since no important information was extracted from the device.

6 Conclusions

Taking into account that more security errors were reported than those proposed for improving security, it can be said that the IoT is still at an early stage of development. The lack of response from manufacturers to security problems is alarming. This is because older devices are not updated with security patches and IoT is not always a priority in the newer systems.

IoT security needs more attention as it is being implemented massively. There are no regulations that require a demonstration of your security before implementation. In addition, the public is not aware that the devices they use are not completely safe.

The Blueborne and KRACK cases show that following the established protocols does not guarantee flawless security. Due to the high amount of information that is previously required to attack a spe-cific device (the code does not operate globally), the limited connection range (10 to 15 m) and the fact that the smartphone must be unblocked (PIN, fingerprint, pattern, etc.), Blueborne is not as big of a security issue as it was initially stated but it still must be solved. There are a number security patches for the different platforms available on the market.

References

1. Höller, J., Tsiatsis, V., Mulligan, C., Karnouskos, S., Avesand, S., Boyle, D.: From Machine-to-Machine to the Internet of Things: Introduction to a New Age of Intelligence, Amsterdam. Elsevier, The Netherlands (2014)
2. Yang, G., et al.: A health-IoT platform based on the integration of intelligent packaging unobtrusive bio-sensor and intelligent medicine box. IEEE Trans. Ind. Informat. **10**(4), 2180–2191 (2014)
3. Wilkin, R.: hackers-are-nowtargeting-car-washes, New York Post. https://nypost.com/2017/08/01/hackersare-now-targeting-car-washes/amp/

4. Barcena, M.B., Wueest, C.: Security Response, Symantec (2015)
5. Zetter, K.: Hacker Can Send Fatal Dose to Hospital Drug Pumps. Wired. https://www.wired.com/2015/06/hackerscan-send-fatal-doses-hospital-drug-pumps/
6. Vanhoef, M., Piessens, F.: Key Reinstallation Attacks: Forcing Nonce Reuse in WPA2, November 2017. https://www.krackattacks.com/#paper
7. Yang, W., Wan, Y., Wang, Q.: Enhanced secure time synchronisation protocol for IEEE802.15.4e-based industrial Internet of Things. Institution of Engineering and Technology IET 2017, vol. 11, no. 6, p. 369–376. IEEE xplore (2017)
8. Beydemir, A., Sogukpinar, I.: Lightweight zero knowledge authentication for Internet of things. In: 2017 International Conference on Computer Science and Engineering (UBMK), Antalya, pp. 360–365 (2017)
9. Armis Labs. Blueborne – Technical White Paper 2017. Disponible en. http://go.armis.com/blueborne-technicalpaper
10. Repositorio github con el código. https://github.com/ArmisSecurity/blueborneradare.org/r/
11. Reverse Engineering Framework. https://www.radare.org/r/
12. Armis Labs. https://www.armis.com/armis-labs/

Authentication and Access Control

A Light-Weight Tightening Authentication Scheme for the Objects' Encounters in the Meetings

Kim Khanh Tran[1,2]($\boxtimes$), Minh Khue Pham[1]($\boxtimes$), and Tran Khanh Dang[1]

[1] Faculty of Computer Science and Engineering, HCMC University of Technology,
VNUHCM, Ho Chi Minh City, Vietnam
khuebcc@gmail.com, khanh@hcmut.edu.vn
[2] Faculty of Information Technology, Can Tho University of Technology,
Can Tho city, Vietnam
ttkkhanh@ctuet.edu.vn

Abstract. Wireless sensor networks consist of a large number of distributed sensor nodes so that potential risks are becoming more and more unpredictable. In order to prevent a new object from joining, many previous research works applied the initial authentication process in the wireless sensor network and the wireless Internet of Things (IoT) network generally. However, the majority of the former articles only focused on a central authority (CA) or a key distribution center (KDC) which increased the computation cost and energy consumption for the specific cases in IoT. Hence, in this article, we address these issues through an advanced authentication mechanism, including key-based management and rating-based authentication. The scheme reduces costs between an object and its peers effectively. We refer to the mobility dataset from CRAWDAD collected at the University Politehnica of Bucharest and rebuild it into a new larger random dataset. Our protocol uses the new dataset as an algorithm's input. It enables the protocol to handle authentication rigorously for unknown devices into the secure zone. The proposed scheme helps to increase flexibility in difficult contexts, resource-constrained conditions.

Keywords: Authentication · Wireless sensor network
Moving items · Trust · Reputation

1 Introduction

With the technological advancements of the sensors rapidly, Wireless Sensor Networks (WSNs) has become the main technology for IoT. WSN are mobile ad-hoc networks in which sensors have limited resources and communication capabilities. They contain a large number of spatially distributed devices which are the low cost, ease of deployment, and versatility. Especially, limited battery energy and storage space is the tightest resource constraint on WSNs. These

© Springer Nature Switzerland AG 2018
T. K. Dang et al. (Eds.): FDSE 2018, LNCS 11251, pp. 83–102, 2018.
https://doi.org/10.1007/978-3-030-03192-3_8

constraints make authentication of nodes more difficult. To easily experiment and emphasis on constrained objects, WSN is the main model of our scenario which represents the IoT environment. Earlier research works [12,14,16] have proposed an authentication protocol based on a Credential Service Provider (CSP) or a trusted third party (TTP) to grant, distribute, and manage the key in an area or the whole system. These protocols need high overhead of exponential computations in resource-constrained environments. In the past, the issue of mobility and applying mobility devices to mutual authentication proposed in wireless sensors networks [4,5], yet such protocol had a definite goal in mind: "eventually all" nodes should establish a security association with each other. Friendship or neighborship is known somebody you know which is a factor to help confirm a person [15] and reuse the closest objects discovery [13]. Our research focuses on the mutual authentication of individual nodes with the ability to move through areas on WSN. In addition, our proposed scheme employs the strength of symmetric-key encryption and human factors into a trustworthy interaction to reduce the cost of communication as well as energy.

The rest of the paper is organized as follows. In the next section, we present related work at Sect. 2 and Sect. 3 discusses about the application scenario, assumptions and offers the security goals. In Sect. 4, we describe the detail of the light-weight authentication scheme. Section 5 divides two pieces including Subsect. 5.1 shows the description of our simulation and CRAWDAD's dataset we referenced to rebuild a new dataset, Subsect. 5.2 is an analysis of result. Security analysis will be mentioned in Subsect. 6.2. Finally, we conclude the paper with a summary of our results and discuss opportunities for further research in Sect. 7.

2   Related Works

The evaluation of trust evidence was solved as a path on a direct graph, whereas nodes as objects and edges as relations between two objects. There was no pre-established infrastructure. Their idea is [18] also based on the direct trust relations of intermediate nodes. It takes the weight of the edges to calculate trust values and establishes an interaction later. However, at each step of computation, the source node computes many times so the implementation may be complicated. The conventional view of security based on cryptography is only not adequate to the novel misbehaviors or colluding attacks in sensor networks. They proposed a reputation-based framework [11] for sensor networks where nodes maintain a reputation for other nodes to evaluate their trust through metrics, represent past behavior of other nodes and be used as an inherent aspect for predicting their future behavior. They used Bayesian formulation for the algorithm steps of reputation representing, updates, integration and trust evolution. A trust/reputation model [9] proposed a method which filters bad nodes through its behavior and distributes workloads by using multiple agents to measure the trust of the system. A scenario is a person who needs to evaluate another person by the value of trust and parameters which are similar to the idea of our evaluation. They defined many types of trust including direct trust, indirect trust, recent trust and expected trust and built a strong, rigorous model. Because of many functions and complex computing, they are not suitable for small devices. Moreover, [6] created a temporary trust relationship to improve the authentication process which is also relatively close to our idea. Because they were the vehicular ad-hoc networks, the requirements were also appropriate to low the cost of computing in the smart networks. Once a valid vehicle is successfully validated, the protocol will become a valid vehicle for another vehicle which is a temporary police car. The protocol ensured that it was small, anonymous, location privacy security, data can not be changed (upon the hash function). Nevertheless, they did not care about the security of information exchange.

Amin et al. [2] built a novel architecture suitable for WSNs. They designed a user authentication protocol and session key agreement for accessing data of the sensor nodes through multi-gateway. They managed smart cards and passwords to register and get access to a system. We do not deny the strength of this work but only offers a solution in the event that their gateway nodes are no longer usable. Although their solution is secure and addresses the cost of energy, storage and time, the component involved in the authentication process is between the real user who comes with the smart device and the gateway to get real-time data from the sensor nodes. Similarly, [1] created a novel remote user authentication scheme using wireless sensor networks (for agriculture monitoring). The proposed scheme involves four types of participants: User/agriculture Professional U_i, a Base Station BS, a Gateway Node GWN_j, and Sensor Node SN_j which user prefer to get some environment data. Instead of broadcasting to multiple gateways in multiple areas as [2], they go through two authentication steps at BS and GWN. Thus, we prefer to provide an additional solution that alleviates the communication cost when the scheme focuses on the role of CA, BS or KDC and still ensure flexible, available and tightened all the traffic moving in and out of the secure zone.

3   Problem Solution and Proposed Approach

3.1   Scenario

The specific context in wireless network is a motivation to improve and analyze the protocol properly. The characteristic that we exploit is a ability to detect objects nearby and must be the active objects in a specific zone. Currently, the *IoT* wireless network is widely available, and we are going to cover a very basic example that will be covered throughout the article. We will analyze the movement of a human/object and the authentication when they meet in a zone/area, a meeting room through their relationships. They will use a smart object to represent themselves such as a smartphone, a smartpen, etc. Because the meeting is secure, we will scrutinize they do enjoy the room and hidden their personal privacy. Besides, it is crucial to identify whether this newcomer is trustworthy or not. We will not use any intermediate helper center (CA, KDC, etc.) since it will spend a lot of time and is an inconvenience for small objects, portable and low battery. Our objects can communicate with each other through wireless technologies such as Bluetooth, ZigBee, wifi-direct, etc. However, we do not assume that they are necessarily connected to the internet and can use the internet as an indirect connection mechanism. In terms of security, we make some assumptions that consider proximity and authentication of mobile objects, new challenging assumptions are not familiar to traditional approaches to authenticate in the network. We start the list below with the assumptions:

1. We do not assume that all honest nodes of the network are friends that eventually need to be connected.
2. To reduce the frequent mess, we assume that the objects move and join in the meeting rooms which are authorized. The movement speed of the nodes will not be considered.
3. Our protocol will be operated based on the information included after the network system is fully constituted at the beginning of time (bootstrapping). Therefore, if any new object wants to get involved in the network, it should be established some materials to qualify to join the zone: zone access token, initialize list of friends based on the zone they are assigned. The adversary will attack the objects that have some interactions or leave out the safe zone.
4. To disturb the traffic flow, we assume that the system can dynamically change the limit of the number of requests sent to a node. Depending on the speed of the wireless channel, the number of requests will be changed that be able to distinguish this is the spam or not.
5. The guide which is detected by the newcomer is a good object.

3.2   Security Requirements and Proposed Solutions

1. **Confidentiality and Privacy:** The message is only disclosed to authorized objects (entities, users, devices) and when they need to access the services. The private data, keys, and must cover security credentials from unauthorized objects. Our key exchange model supports encrypt all messages and identifies the device's friends through their shared materials which is applied to each node without any trusted parties. In the authentication process, each device has just a unique ID to represent itself and is possible to detect nearby devices.
2. **Integrity:** The attackers will use any technique to falsify the original target which would signify the receiver misunderstood. We developed a key exchange scheme to complement the voting process. An object gets an own ID and an encrypted message that contains a shared-key associated with a specific sequence number. An attacker will not be able to know two people he is aiming which stage they are in or how many times they communicated. It will initially be locked with a private key, then with a special key depending on the order communication of the two people.
3. **Authentication:** They are the major issues to exchange information or access a zone's services and the results that the protocol we want to be. Not everyone can enter or be able to access certain documents in the group. A newcomer device uses its former secure interactions which have the corresponding material to verify. The authentication mechanism for a device is

not only concerned with the safety and correctness, but also the promptness, availability and energy savings. We observed that the authentication mechanisms focused their trust on the central distribution, that is a reason why they both made the cost of authentication increase and could do much to the energy. With the above impulses, we relied on human relationships in real life to solve problems of energy saving on trust-based. A real scenario implemented which was the management of secure meetings. Based on the results of our implementation, our protocol limited the number of members entering the zone quickly and it saved a lot of energy for moving subjects with the assumptions we have addressed.

4. **Availability:** The system is not based on mobile network access or internet, the third-party to perform authentication. The IoT systems are required to be robust to provide services for accessing anytime. As long as it does in accordance with context assumptions, our scheme can work every time with new requests to join a specific zone. First, the format of the message is correct. Then, some adjacent neighbors are detected enough to organize a short voting process (the parameters are limited to show the suitability and availability of the protocol, the results are presented in the experiment section). In addition, we specify the number of messages from the outside that is only one message at the beginning of the authentication phase. Thus, the protocol avoids affection from DOS attack for the power of the receiver and congestion. Because when a newcomer is searching and wanting to join the zone by sending the request, the two objects must be near each other and send the message using short-range wireless technologies to save energy as well as have the high probability of receiving.

4   Protocol Details

We build the idea of the organization and operation of the meeting room and staffs could use a smart device to identify and join a secure zone. However, we use the trust relationship of nodes (all the neighbors at the same level), a reputation-based method for each small device. These devices do not rely entirely on the trusted third parties to alleviates the cost of communication, the processing time and then ready to allow nodes' authentication anytime when authentication procedures are happening.

4.1   A Light-Weight Tightening Authentication Scheme

Figure 1a describes the overview and operation of our protocol. And Fig. 1b, we introduce three main roles that will be involved: a newcomer (Newton), an intermediate guide (Ginny) and some helpers (Harries). Firstly, Newton is an unknown person, moves close to the zone's bound and will choose one of the adjacent objects by a discovery mechanism. The selected object as the intermediate guide will reach a request to make a secure association and accept Newton to take part in this zone by his previous relationships and his permissions. From

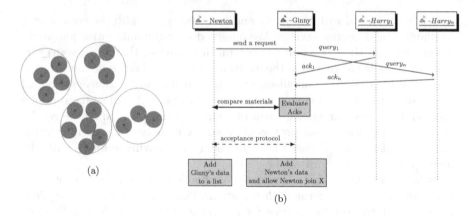

Fig. 1. Overview and the enhanced authentication process for object's encounters in the meetings

our assumptions Sect. 3.1, this protocol will be utilized after the system has organized completely: all honest nodes meet and have a secure interaction, distribute the zone's permission to all honest nodes. At this time, each node will contain:

- ID_O: a unique ID of each object O.
- (S, P): a set of private keys and public keys.
- $ls\_fr(O)$: list of nodes that O has an interaction which will have a key-list(the list of sequence shared keys that O received a message for the last time) and reputation scores, respectively.
- $ls\_zone(O)$ (the zones that O is allowed to move to - assigned early): list of zone's ID.

We will describe the steps of the mechanism below and use the specific name of the person to represent the respective components.

Moving and Joining: After the bootstrap phase is finished, the objects will have enough information to join the given zone. Newton moves to zone X and sends a request to enter zone X to Ginny who is detected the nearest one. At this point, Newton will run his check algorithm that satisfies the conditions: a zone's permission and must be Ginny's friend or be a friend of the nodes who is in zone X. [has_permission(X) and is_friend(Ginny)] or [has_permission(X) and is_friend(some of all zone's member)]. We distinguish two cases in this step:

- *Are you Guide's friend?:* From the is_friend (Ginny) function, will we consider in her friends' list whether or not Newton's ID exists? From the very beginning, Newton had discovered Ginny and sent a request to the zone. Because Newton and Ginny established a relationship, he would also enclose the shared key as a secret key between the two objects with the message to interact. From there, Ginny can easily check the message content and evaluate whether to enter or not.

- *Are you neighbours' friend?:* If Newton does not know Ginny, Newton will only send his permissions and ID to her. After that, Ginny explore the closest objects around her and send the messages which expect some helps to verify. Our protocol regulates not only the number of node in-out to the zone but also the choice of helpers. Hence, it will reach the maximum number of nodes based on her existing energy, and the number must be odd to avoid the number of ACKs being equal to the number of nodes who do not know the guide. There will be two situations: find the appropriate number of helpers which Ginny sees the nearby h nodes and sends them h messages with the same content; there is too little or no more helper around which Newton will not enter into zone X.

Checking and Response: When the helper receives an asking from Ginny, they execute the is_friend (Newton) function to detect a previous interaction with Newton. To reduce the communication cost, Helpers will not need to send back Ginny if they have a negative ack. Here, Harry will represent the helper to illustrate the task, [Newton is my friend]: Harry will send the format including Newton's ID, Harry's ID, Newton-Harry's shared key (K_{ABi}), $Snum$, roll call sound for Newton. In this message, K_{ABi} and $Snum$ are as the proofs to prove they were friends and these materials will be detailed in Sect. 4.2.

Evaluation the Acknowledgements (ACKs): The received ACKs from the Harry (ies), Ginny will filter these ACKs by their reputation. Because Harry(ies) are all Ginny's friends with her in zone X, she will have the respective reputation scores2, thenceforward, we may have the averaged Newton's reputation from Harry(ies).

Before evaluating the value of $T(G, C)$, Ginny will ask Newton to compare shared materials which will prove actually this association was created. And then, we set up a scoring system and a multiplier for the reputation score. The scoring system is created by committing points based on behaviors and the numbers of members over time, we need to choose a limited score that is large enough for the reputation of each node. To make it easy to identify trustworthiness, we classify three different levels of reputation: 0: too bad [0], 1: Fine $[1, \frac{N}{2}-1]$, 2: Good $[\frac{N}{2}, N]$. Based on that the multiplier corresponds to raise or lower the trust value of the answer from the helper.

$$Total\_trust(H, C) = \sum_{i=1}^{h} \frac{M(G, H_i) * R(H_i, C)}{m} \tag{1}$$

H is a list of helpers, $H = \{Harry_1, Harry_2, ..., Harry_h\}$.
C is the newcomer object.
G is the intermediate guide.
h is the number of neighborhoods who G detected.
$M(G, H_i)$ is a value of H_i's multiplier by G.
$R(H_i, C)$ is a value of C's reputation by H_i.
m is the sum of $M(G, H_i)$.

Because we do not use trusted third parties, we use the secure voting mechanisms such as human-to-human relationships. We are interested in two types of value: $Total\_ACKs$ and $Total\_ToDeliver$. (2) will consider the number of friends H positive comments about C. Then, we proposed an algorithm based on (1) to build a combined evaluation result from Harry(ies). Accepting individual scores from each node will produce maybe intuitive and sentiment, so we consider the number of responses received as the votes for Newton to balance the final results (if there are the disparities in scores). The value of reputation ranges from 0 to N, where although '0' is computed in (2), it is as an empty vote when we evaluate the number of ACKs. Besides, we add a threshold value for trust issue is $\Delta_{reputation}$ which we compare to the value of $T(G, C)$ which is an trust value average of newcomer by guide. If $T(G, C) \geq \Delta_{reputation}$, a new association is established, both of Newton and Ginny will store their information and he can take part in zone X.

$$NA(H, C) = \frac{Total\_ACKs}{Total\_ToDeliver} \tag{2}$$

$$T(G, C) = \frac{NA(H, C) + Total\_trust(H, C)/N}{2} \tag{3}$$

$T(G, C)$: is an trust value average of newcomer by guide.
$NA(H, C)$: The ratio of the positive number of H about C over total number of queries has been sent.
$Total\_ACKs$: the total number of acknowledgements received.
$Total\_ToDeliver$: is the total number of messages to be delivered. ($0 <= Total\_ACKs <= Total\_ToDeliver$).
N is a limited score in our scoring system.

While the system is operating, the algorithm computes the reputation points on each node that always be activated. In order to be able to build algorithms flexibility, we will not spell out any specific behavior in this article. We will raise or reduce the value of reputation (two nodes are friends of each other) against the communication status between two nodes. We divide it into four cases including sending-receiving (in a friendship), evaluation, suspicion, and attack. The former two categories, if it's successful, reputation will be plus 1 and the latter two categories that the node will subtract 1 for the opponent based on their behavior.

4.2 Key Exchange Model

Key exchange scheme is a mechanism by which two objects that communicate over an adversarially controlled network can produce a secret shared material. Key exchange protocols are essential for enabling the use of shared-key cryptography to protect transmitted data during a authentication process. In this article, we propose a key exchange mechanism to match the constraint requirements of participants. Our technique can not only save time but also ensure that forwarding data are secure. As we have studied the symmetric and asymmetric encryption, both of them can support encryption processes for resource-constrained devices, but there are drawbacks that make them flawed. So that, we combine their advantages to increase security and processing time. We will describe the condensed mechanism of exchanging messages which are communicated by two objects.

As illustrated by the example on Fig. 2, we will use the same object's name. The scheme consists of two stages: first-meet and meet-again. In the first phase, Alice uses Bob's public key to encrypt a packet which is included:

- ID_A: determines who is sending mess for Bob.
- $Snum_B$: a sequence number of Bob. At the first time, $Snum_B = 0$, after Alice has sent a message to Bob, $Snum_B$ will increase the value by one when Bob did decrypt Alice's message successfully to determine the correct order of messages, also to avoid the attack which sends fake messages to Bob.
- K_{ABi}: a key to decrypt for the following message that Alice will send to Bob (with i is a version number). This key will be attached to the original message.
- E_M: an encrypted message which combine an message and K_{ABi}. $E_K(M)$ is a function to encrypt a message M with a key named K.
- $De_K(E_M)$: a function to decrypt an encrypted message E_M with a key named K.

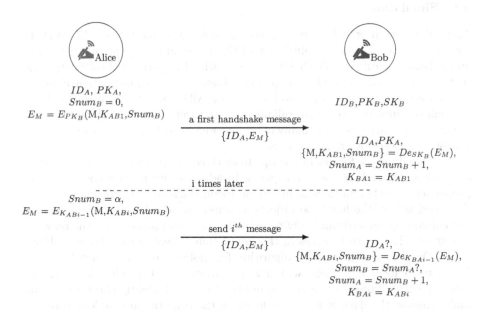

Fig. 2. Key-exchange scheme (KES)

This scheme will perform in a secure chanel. Alice has an original message (M), creates a shared-key (K_{ABi}) and an initial sequence meeting number ($SnumB$) and afterwards Bob has received an encrypted message (E_M) from Alice, he will use his private key to decrypt EM with a function as $De_K E_M$. Subsequently, Bob will get an original message (M), K_{ABi} which decrypts the next exchange message and $SnumB$ which Alice sent Bob to check for the exchange later. At that time, Alice also stores K_{ABi}, $SnumB$ in her private memory. We use the asymmetric cryptography once at the initial exchange to help make the scheme's security safer.

We assume that when Bob and Alice meet again after a period of time, she sends a message to Bob at $i^t h$ (with $i \geq 1$). Because Bob has had the shared key (K_{ABi-1}) from the previous communication with Alice to decrypt the current message, Alice will use the key she gave Bob at the $i - 1^t h$ to encrypt the ith message and build a new shared-key to send to him with the above same information. After Bob has received the message and decrypted it successfully, he will check $SnumB$ which represents the number of times they have interacted with each other. If valid, he stores the key and $Snum$ that Alice has just sent, increases $Snum$ by 1. At the same time, Alice and Bob will keep the new shared-key to replace the old one, a new $Snum$ and surely ID_A is not changed.

5   Protocol Evaluation

5.1   Simulation

To validate our protocol we referenced the mobility dataset from CRAWDAD collected at the University Politehnica of Bucharest in the spring of 2012, using an application entitled HYCCUPS Tracer [7] with the purpose of collecting contextual data from Android smartphones. They gathered information about a device's encounters with other nodes by using AllJoyn. This software has been and will continue to be openly available for developers to download, and runs on popular platforms, especially many other lightweight real-time operating systems without the need for internet access.

We based on the format of a script from Hyccups database which is a list of contacts and friends of the participants. In addition, we remove and add some properties, objects and lines to fit our protocol, the script of encounters will be described below. We have 100 objects as sensor nodes. A list of friends and a current zone generated randomly to simulate the initial phase when the network was stable. There are 19 zones in this simulation. There are all 100 availability zones. Our experiment runs an algorithm for mobility devices in nearby areas. It creates a local movement within zones (from 1 to 19). The big result we reached 99999 connections. Since the period of time will greatly affect our results and formulas, the dataset should unfocus on the time the interactions connect. Assume that each line will appear at different times and gradually increase from top to bottom in our script. As we mentioned, our protocol has two phases: bootstrap, move and join.

Table 1. Energy cost

Operations	Energy cost (μJ)
Transmit 1 byte	5.76
Receive 1 byte	6.48
AES-128 128-bit encrypt	9

In the first phase, we set up information for an object (details are presented in section), friendships (taken from the dataset), zones, grant zone's permission to the objects. Our project has two sections: node and script data making and illustrating. In the first section, we create some information for 100 objects, including their own ID, friendList, zone (a list of zones where the object is allowed) and a current zone where they. Then the last one is all script which we need to run the whole process of illustration. Our protocol will work with two types of files: *script.json* (which contains the migration script of the whole system), *nodes.json* (information of an object, the file being simulated as a storage space of an object). Each line of *script.json* file represents a specific time of an encounter with the first element (the left ID) as Newton and the second one (the right ID) as Ginny. Both of IDs are the same line when Newton meets Ginny. Besides, we add Ginny's current zone attribute each line and lists nearby neighbors that Ginny has traced. The algorithm will start running after the end of the safe phase, each time Newton goes to zone X, detects the closest one to him and sends a request with a certificate to join the zone (if any) because we want to reduce the confusion, we make the zone's rule fixed. Ginny gets the signal, checks the zone certificate for real or not and their interaction. Each node not only has a specific grant but requires prior interaction for authentication.

5.2 Results

Based on the chart, this shows that our protocol is strict in accepting any node that joins the zone. The parameters that we set to find the appropriate threshold values dynamically. They are Δ_{max_friend} and $\Delta_{reputation}$. Firstly the number of friends in the same zone is important as it affects the energy of the Guide. Δ_{max_friend} is the maximum number of targets which the Guide is able to send a query. That will also be affected by which we've mentioned above. Then, a threshold of reputation is a comparable value to make the final decision. We set up the parameter which is an arithmetic progression with common difference of 1 with the following conditions: $3 \leq \Delta_{max\_friend} < 100$, we have 100 active objects in 19 zones. However, an object that does not have enough energy to detect all objects. In addition, $\Delta_{reputation}$ ($50 \leq \Delta_{reputation} < 100$) is the threshold of reputation after The Guide receives assistances from nearby neighbors. As far as the best data we received, the result of successful authentication is approximately 22% of the total number of nodes which want to enter a specific zone. In addition, asymmetric encryption is only used once in the first time of interaction.

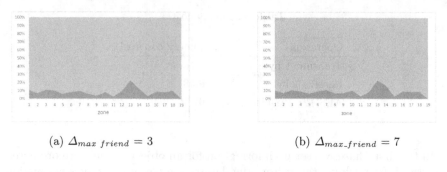

(a) $\Delta_{max\_friend} = 3$ (b) $\Delta_{max\_friend} = 7$

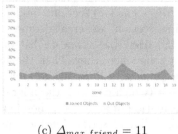

(c) $\Delta_{max\_friend} = 11$

Fig. 3. The number of incoming objects according some defined parameters with $\Delta_{reputation} = 80$

Table 2. The number of messages from our experiment

Scenario	Transmit (1 byte)	Receive (1 byte)	Queries (1 byte)	ACKs (1 byte)
1. Friends	1	1	0	0
2. Do not have mutual friends	1	0	7	0
3. Have some mutual friends but fail	1	0	7	7
4. Have some mutual friends and success	2	2	7	7

It indicates that the performance is improved and saves the energy in transmitting messages between nodes in the network. Also, we create a count function of messages when: requesting, sending queries, sending the result of own evaluation, successful authentication notification. Table 1 describes the values that we deduced from a research work [10].

Based on the estimated values, we evaluated the energy consumption of the proposed protocol and key exchange mechanism. In the Table 2, we show all worst-case scenarios which give the largest energy consumption. The percentage of the number of accepted members into the active zones we tested on multiple parameters (Fig. 3). We experiment on the change of the total cost of the our mechanism by the variation of Δ_{max_friend} from 3 to 11. The difference of final results (Fig. 4) is not negligible (from 1.5% to 2.5%). The comparison result

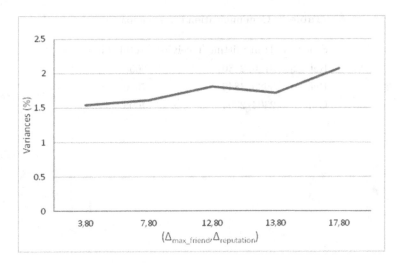

Fig. 4. The difference in the number of successful authentication of each zone with different parameter data.

Table 3. Computational cost comparison.

Schemes	Newcomer	The Guide + Helper	Total	Execution time (ms)
Ref. [2]	$8T_H$	$7T_H + 5T_H$	$20T_H$	0.008
Ref. [1]	$6T_H + 2T_S$	$(5T_H+2T_S)+(8T_H+3T_S)$ $+ (5T_H + 1T_S)$	$24T_H + 8T_S$	1.052
Our	$1T_H$	$(1T_H + 28T_S)$	$2T_H + 28T_S$	3.60

Note: T_H is the execution time for hash function; T_S is the execution time for symmetric-key (encryption or decryption).

of the authentication process with the precious works from average parameters ($\Delta_{max\_friend} = 7$).

Table 2 indicates scenarios which get the average numbers from our algorithm. From here, we can calculate the maximum cost each a sensor object, based on the above assumptions and validation process. In case 4, from the Table 2, Ginny (as a Guide) is a mediator that spends more execute time and energy than others, so we focus on Ginny's message flow. Ginny is supported to run on the TelosB platform (the configuration mentioned above in this section), she will receive a request (16 bytes), a list about Newton's friends (16 bytes). Besides, Ginny will send 7 queries to ask her neighbors who she could be found to support and be able to get back 7 ACKs. After that, If there is a validation result, Ginny will send a confirmation to Newton. From the number of messages and the value from the Table 1, we can deduce the maximum energy consumption of each sensor as a Guide that the protocol has gotten in all scenarios: $(7 * 16 + 2) * C_{transmit} + (7 * 16 + 2) * C_{receive} + 7 * C_{AES} * 4 = 1521\,\mu J$. In the Table 3, we calculate the computational cost for the whole process from the time of sending a request to join to the successful verification of a sensor

Table 4. Communication costs comparison.

Schemes	Transmitting/receiving	Total (bits)
Ref. [2]	3776/2880	6656
Ref. [1]	3680/3840	7520
Our	2304/2304	4608

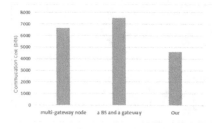

(a) Communication costs (bits)

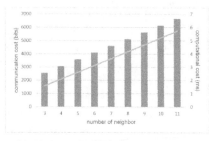

(b) The growth of communication and computational costs to changing Δ_{max_friend} and $\Delta_{reputation}$

Fig. 5. The comparison of the proposed authentication scheme

node with the related participants. According to the experimental results, the encryption time for symmetric encryption (AES-128bit) from [1] is 0.130 ms. Therefore, we compute that the average execution time is 3.6 ms. In this regard, our mechanism is slower than [1] about $\simeq 2$ ms–3 ms. Because we have to use multiple queries and send encrypted messages to neighbors rather than to rely on the base stations or central authorities or gateway nodes like [1,2] which gives our scheme more flexibility, central station is broken without issue. In addition, we also compared the communication cost for all objects in an authentication process with the two works above. For calculating the communication cost, we consider the length of IDs, cipher messages, requests, confirm messages, and so on. Table 4 shows the comparison of the total communication cost with [1,2] (see Fig. 5 for more detail). In general, though our mechanism can not optimize for both computational and communication issues, we reduce the number of requests and components involved in the authentication process greatly while the computational cost we consume is only slightly slower.

6   Security Analysis

6.1   Security Model and Proof

BAN-logic is a beneficial tool to formally analyzing an authentication scheme. Inspired by the analyzing methods proposed by [3], we utilize this approach to formally evaluate our proposed scheme. We first briefly introduce the model including goals and assumptions, BAN-logic and use it to analyze our scheme in terms of security. We present our proof based on Burrows's BAN-logic model to formally evaluate the ability of an authentication scheme. According to BAN-logic, there are three objects in logic: principals, encryption keys, and statements (we will identify a message as a statement in logic). Therefore, they let symbols P and Q be principals, X and Y range overstatements, and K represent the cryptographic key. Below are notations used in BAN-logic model.

- $P \mid \equiv X$: It means that P believes that in the current run of the protocol that the formula X is true.
- $P \lhd X$: P who can read, receives a message including X.
- $P \mid \sim X$: P sent a message containing the statement.
- $P \Rightarrow X$: P has jurisdiction over X.
- $\#(X)$: a statement X is fresh. It has not been used before.
- $P \xleftrightarrow{K} Q$: The secret key K is only usable in the communication between P and Q.
- Encryption of X with key K is denoted in the standard way: $\{X\}_K$.
- K_{ABi-1}: is a secret key between A and B at the i-1 and is used to authenticate mutually at the i.
- $Snum_A$: a sequence number of A which represents the time of encounters with B. This number will be stored in A's archive.

We will list the main BAN-logic's rules that we will use to prove our protocol:

- Message meaning rule: $\dfrac{P|\equiv P \xleftrightarrow{K} Q, P \lhd \{X\}_K}{P|\equiv Q|\sim X}$
- Nonce Verification rule: $\dfrac{P|\equiv \#(X), P|\equiv (Q|\sim X)}{P|\equiv Q|\equiv X}$
- Jurisdiction rule: $\dfrac{P|\equiv Q \Rightarrow X, P|\equiv Q|\equiv X}{P|\equiv X}$
- Freshness rule: $\dfrac{P|\equiv \#(X)}{P|\equiv \#(X,Y)}$
- Elimination rule: $\dfrac{P|\equiv Q|\equiv (X,Y)}{P|\equiv Q|\equiv X}$

In our protocol, there are four main goals we need to achieve. We use symbols A and B to represent two objects in the KES as Alice and Bob, respectively (see Fig. 2).

- G_1: $A|\equiv A \xleftrightarrow{K_{ABi-1}} B$
- G_2: $B|\equiv A|\equiv A \xleftrightarrow{K_{ABi-1}} B$
- G_3: $B|\equiv A \xleftrightarrow{K_{ABi-1}} B$
- G_4: $A|\equiv B \xleftrightarrow{Snum_B} A$
- G_5: $B|\equiv A|\equiv B \xleftrightarrow{Snum_B} A$
- G_6: $A|\equiv B|\equiv B \xleftrightarrow{Snum} A$

We use BAN-logic to demonstrate the key exchange scheme between two objects that have created an interaction and shared the secure proof or the two objects need to prove their through an intermediary. At each transaction, an object sends an message X to an other object at the same level and A and B reciprocate their roles so that we only prove towards sending the request and the verify the request. In particular, A is the sender and B is the receiver.

$$X = <A \xleftrightarrow{K} B, B \xleftarrow{Snum} A, M, ID>$$

1. Assumptions concerning the initial state are written as
   - A_1: $B| \equiv B \xleftarrow{K_{ABi-1}} A$
   - A_2: $B| \equiv A \Rightarrow A \xleftarrow{K_{ABi-1}} B$
   - A_3: $B| \equiv \#ID$
   - A_4: $A| \equiv \#M$
   - A_5: $A| \equiv B \Rightarrow B \xleftarrow{Snum} A$
2. Analyzing our proposed scheme based on the BAN-logic rules and assumptions about authenticating among the mobility objects follow the above main goals.
   - Because A and B have created K_{ABi-1} after they establish a secure encounter, it means that A believes that A has shared the secret shared-key K_{ABi-1} with B; therefore we achieve G_1.
   - In accordance with A_1 and the message X, applying the message-meaning rule to derive:

$$\frac{B| \equiv B \xleftarrow{K_{ABi-1}} A, B \lhd \{X\}_K}{B| \equiv A| \sim X} \tag{4}$$

   - According to A_3, applying the freshness rule to derive:

$$\frac{B| \equiv \#(ID)}{B| \equiv \#(X)} \tag{5}$$

   - In accordance with 4 and 5, we apply the nonce-verification rule to derive:

$$\frac{B| \equiv \#(X), B| \equiv A| \sim X}{B| \equiv A| \equiv X} \tag{6}$$

   - In accordance with 6, we apply elimination rule to derive:

$$\frac{B| \equiv A| \equiv X}{B| \equiv A| \equiv A \xleftarrow{K_{ABi-1}} B}(G_2) \tag{7}$$

$$\frac{B| \equiv A| \equiv X}{B| \equiv A| \equiv A \xleftarrow{Snum} B}(G_5) \tag{8}$$

   - According to G_2 and A_2, applying jurisdiction rule to evolve:

$$\frac{B| \equiv A \Rightarrow A \xleftarrow{K_{ABi-1}} B, B| \equiv A| \equiv A \xleftarrow{K_{ABi-1}} B}{B| \equiv A \xleftarrow{K_{ABi-1}} B}(G_3) \tag{9}$$

- The combination of the G_1 and message-meaning rule, we can infer:

$$\frac{A| \equiv A \xleftarrow{K_{ABi-1}} B, A \lhd \{X\}_K}{A| \equiv B| \sim X} \tag{10}$$

- According to A_4, applying the freshness rule to derive:

$$\frac{A| \equiv \#(M)}{A| \equiv \#(X)} \tag{11}$$

- In accordance with 10 and 11, we apply nonce-verification rule to deduce:

$$\frac{A| \equiv \#(X), A| \equiv B| \sim X}{A| \equiv B| \equiv X} \tag{12}$$

- From 11, we combine with elimination rule to derive:

$$\frac{A| \equiv B| \equiv X}{A| \equiv B| \equiv B \xleftarrow{Snum} A}(G_6) \tag{13}$$

- In accordance with G_6 and A_5, applying the jurisdiction rule to infer:

$$\frac{A| \equiv B \Rightarrow B \xleftarrow{Snum} A, A| \equiv B| \equiv B \xleftarrow{Snum} A}{A| \equiv B \xleftarrow{Snum} A}(G_4) \tag{14}$$

3. In accordance with G_1, G_2, G_3, G_4, G_5 and G_6, we can suppose that B believes that A believes they have a good shared-key K_{ABi-1} in their connection. Additionally, A (B) also believes that B (A) believes $Snum$ is a good share-material between them.

6.2 Additional Discussion

We assume that the information of token is completely stolen by some reasons. In this subsection, we will discussion the security of our authentication scheme. Based on many known attacks, our scheme could detect, against and limit the risk of creating the lowest possible corruption so that our protocol meets the security goals mentioned in the section. We also suppose that the communication chanel between the newcommer and the guide is open, anybody can eavesdrop and capture the message packages transmitted in this channel.

- *Impersonation and Modification Attack:* When a device moves into an unsafe environment, unfortunately, it is attacked and masquerades some unprotected information inside its memory to make false positives or may join in unauthorized zones. There are two cases:

  ◇ When an outsider forges a relationship with a guide: they have made an interaction with each other. In our special key exchange model, they will distribute the following keys by the number of times their messages are sent which includes the sequence number and the shared key (K_{ABi}, i is the communication time). K_{ABi} stored in encrypted and can not be revealed even though a fake outsider acknowledges the sequence number.

⋄ When an outsider never encounters an X's Guide: this is also the goal to authenticate an object to the secure zone. At this point, we will rely on the reputation score and the number of responses to make the final decision on this object. However, Ginny will simultaneously detect proximity friends and send a list contains these friends to Newton. Depending on the available energy that Newton will send one of the evidence of interaction with someone in the friend's list. Ginny receives and sends to the selected helper, especially the helper that Newton sends $E_M = (K_{Newton,helpers}||snum = \alpha)$ (Ginny can not read this content, will also be added to Ginny's original message to the helper. If one of K and $snum$ has problems then the helper will automatically remove the reputation to 0.

- *Man-in-the-Middle:* Authentication occurs on an insecure channel between the outsider and the insider. Although the two objects are real, it may appear the eavesdropper who stands in the middle to tap or catch confidential information. Moreover, they can replace another message. The exchange between the two objects will be applied the key exchange model, the query format is $<ID|E_M>$, E_M is encrypted from the original message, the corresponding sequence number. E_M is encrypted initially with an asymmetric key then it will be encrypted using a symmetric key. Although the asymmetric lock is extremely complex and secure, it takes a lot of time to open up while weak objects cannot complete the exchange process. Therefore, we use both types of cryptography with numbers in order to save time and safer.

- *Replay Attack:* The intruder sends to the target the same message which has already used in the victim's communication. The attacker either eavesdropped a message between two objects before or knew the message format from his previous communication with one of the sides. The transmission message contains shared-keys to use for mutual authentications. However, it is encrypted with the previous shared-key which only two objects can understand.

7   Conclusion

Several *IoT* scenarios include aspects of mobile objects that face several encounters at different secure zones. These locations can be a meeting room, a building, an area or a city based on the network is established. For encounters with strangers, the objects meet various security challenges and the flexibility of authentication to participate in secured zones with its resource constraints.

In this article, we introduced a new approach to mutual authentication that includes a reputation-based evaluation and a key exchange mechanism to lower the costs effectively between the two objects. We described the term of reputation based on friendships as a trustworthy value. At the same time, we proposed the key exchange mechanism rely on the number of previous communication to replenish for the process, each sequence the cryptographic key will be different. Additionally, the scheme selected the relevant modern cryptosystems to improve the effectiveness of information security. Our approach solved the problem of previous works that depends on a CA trust model, the execution time, the cost of communication solutions and the limit the number of passing devices. We devised a simulation based on the dataset of CRAWDAD combined with our additional dataset which created a pretty big and messy script for more realistic testing. Our formulas are just average operations in which the dividend is not large. Thus, the execution time is quite small for the whole large script. Besides, we compared the results of two previous secure authentication protocols that determined our algorithm is lower communication cost than.

To develop a large scenario like IoT, the number of objects which displace to other authorized areas quickly, the other security issues such as privacy for objects are still the challenges. Future work will create a large set of data that works on the modified protocol. In addition, the association of the authentication process and the different kinds of privacy protection ([8,19]) are possibilities to support to extend the security of our protocol, especially with resource-constrained IoT devices. The dynamic scenario which we proposed will touch the access request since authentication process begins getting the wanderer's joining request, we apply a new approach [17] to our scheme whereby the system will greater flexibility, availability while ensuring security for circumstances can happen in wireless IoT network system.

Acknowledgments. This research is funded by Vietnam National University HoChiMinh city (VNU-HCM) under grant number B2018-20-08. We also thank D-STAR LAB members for their meaningful help during this manuscript preparation.

References

1. Ali, R., Pal, A.K., Kumari, S., Karuppiah, M., Conti, M.: A secure user authentication and key-agreement scheme using wireless sensor networks for agriculture monitoring. Future Gener. Comput. Syst. **84**, 200–215 (2018)
2. Amin, R., Biswas, G.: A secure light weight scheme for user authentication and key agreement in multi-gateway based wireless sensor networks. Ad Hoc Netw. **36**, 58–80 (2016)
3. Amin, R., Kumar, N., Biswas, G., Iqbal, R., Chang, V.: A light weight authentication protocol for IoT-enabled devices in distributed cloud computing environment. Future Gener. Comput. Syst. **78**, 1005–1019 (2018)
4. Čapkun, S., Hubaux, J.P., Buttyán, L.: Mobility helps security in ad hoc networks. In: Proceedings of the 4th ACM International Symposium on Mobile Ad Hoc Networking & Computing, pp. 46–56. ACM (2003)
5. Capkun, S., Hubaux, J.P., Buttyan, L.: Mobility helps peer-to-peer security. IEEE Trans. Mobile Comput. **5**(1), 43–51 (2006)
6. Chuang, M.C., Lee, J.F.: TEAM: trust-extended authentication mechanism for vehicular ad hoc networks. IEEE Syst. J. **8**(3), 749–758 (2014)
7. Ciobanu, R.I., Dobre, C.: CRAWDAD dataset upb/hyccups (v. 2016-10-17), October 2016. https://doi.org/10.15783/C7TG7K. https://crawdad.org/upb/hyccups/20161017/2012. Traceset: 2012
8. Dang, T.K.: Ensuring correctness, completeness, and freshness for outsourced tree-indexed data. Inf. Resour. Manag. J. **21**, 59–76 (2008)

9. Das, A., Islam, M.M.: Securedtrust: a dynamic trust computation model for secured communication in multiagent systems. IEEE Trans. Depend. Secur. Comput. **9**(2), 261–274 (2012)
10. De Meulenaer, G., Gosset, F., Standaert, F.X., Pereira, O.: On the energy cost of communication and cryptography in wireless sensor networks. In: IEEE International Conference on Wireless and Mobile Computing Networking and Communications, WIMOB 2008, pp. 580–585. IEEE (2008)
11. Ganeriwal, S., Balzano, L.K., Srivastava, M.B.: Reputation-based framework for high integrity sensor networks. ACM Trans. Sens. Netw. (TOSN) **4**(3), 15 (2008)
12. Khemissa, H., Tandjaoui, D., Bouzefrane, S.: An ultra-lightweight authentication scheme for heterogeneous wireless sensor networks in the context of Internet of Things. In: Bouzefrane, S., Banerjee, S., Sailhan, F., Boumerdassi, S., Renault, E. (eds.) MSPN 2017. LNCS, vol. 10566, pp. 49–62. Springer, Cham (2017). https://doi.org/10.1007/978-3-319-67807-8_4
13. Michalevsky, Y., Nath, S., Liu, J.: MASHaBLE: mobile applications of secret handshakes over bluetooth LE. In: Proceedings of the 22nd Annual International Conference on Mobile Computing and Networking, pp. 387–400. ACM (2016)
14. Moon, A.H., Iqbal, U., Bhat, G.M.: Implementation of node authentication for wsn using hash chains. Procedia Comput. Sci. **89**, 90–98 (2016)
15. Pointcheval, D., Zimmer, S.: Multi-factor authenticated key exchange. In: Bellovin, S.M., Gennaro, R., Keromytis, A., Yung, M. (eds.) ACNS 2008. LNCS, vol. 5037, pp. 277–295. Springer, Heidelberg (2008). https://doi.org/10.1007/978-3-540-68914-0_17
16. Porambage, P., Schmitt, C., Kumar, P., Gurtov, A., Ylianttila, M.: PAuthKey: a pervasive authentication protocol and key establishment scheme for wireless sensor networks in distributed IoT applications. Int. J. Distrib. Sens. Netw. **10**(7), 357430 (2014)
17. Son, H.X., Dang, T.K., Massacci, F.: REW-SMT: a new approach for rewriting XACML request with dynamic big data security policies. In: Wang, G., Atiquzzaman, M., Yan, Z., Choo, K.-K.R. (eds.) SpaCCS 2017. LNCS, vol. 10656, pp. 501–515. Springer, Cham (2017). https://doi.org/10.1007/978-3-319-72389-1_40
18. Theodorakopoulos, G., Baras, J.S.: On trust models and trust evaluation metrics for ad hoc networks. IEEE J. Sel. Areas Commun. **24**(2), 318–328 (2006)
19. Thi, Q.N.T., Si, T.T., Dang, T.K.: Fine grained attribute based access control model for privacy protection. In: Dang, T.K., Wagner, R., Küng, J., Thoai, N., Takizawa, M., Neuhold, E. (eds.) FDSE 2016. LNCS, vol. 10018, pp. 305–316. Springer, Cham (2016). https://doi.org/10.1007/978-3-319-48057-2_21

A Privacy Preserving Authentication Scheme in the Intelligent Transportation Systems

Cuong Nguyen Hai Vinh, Anh Truong$^{(\boxtimes)}$, and Tai Tran Huu

HoChiMinh City University of Technology – VNU-HCM, Ho Chi Minh City, Vietnam
anhtt@hcmut.edu.vn

Abstract. In Intelligent Transportation System (ITS) applications, such as traffic monitoring and road safety, privacy-preservation is one of the most important issues. The vehicles in the system can communicate by transmitting messages to another vehicle or servers. In order to detect and discard messages from attackers, the receivers need to authenticate the message that normally requires the identity of the sender. This leads to the privacy violation of the sender because the user's identity is sent in clear text in the transmitted messages, and thus it can be linked to the vehicle's sensitive information. To overcome this issue, pseudonyms are provided to vehicles and servers and then the receivers request a trusted authority (TA) to authenticate the sender with its pseudonym. However, it faces the congestion challenge since the number of exchanged messages in a real ITS is extremely large. In this paper, we review the state-of-the-art solutions to the privacy preserving in authentication and then propose an authentication scheme to preserve senders' privacy in ITS. In this scheme, the vehicles and servers need to contact TA only one time to get secret information and then, based on their secret information, they can generate pseudonyms for the authentication at the receiver side. Some scenarios are also provided for the security analysis of the schemes.

Keywords: Authentication · Intelligent transport systems · Privacy
Autonomous authentication

1 Introduction

The continuous evolution and development of the entire transport industry are constantly demanding endless and rapid advancement in vehicle performance and efficiency. This crucial and imperative need in our transportation system is not only important but also extremely essential for the present and future of road network, vehicle and user sustenance. Improvement in road and vehicle transport technology has continued to redefine the current expectations and subsequently future prospects of sustainable transport and traffic management. Intelligent Transportation System (ITS), which is part of the Internet of Things, includes vehicle-to-vehicle (V2V) and vehicle-to-infrastructure (V2I) technology and incorporates both wireless and wire line communications-based information and electronics technologies are growing fast. In such systems, the vehicles can communicate with each other and with smart city infrastructure [2]. To do this, the vehicles will

© Springer Nature Switzerland AG 2018
T. K. Dang et al. (Eds.): FDSE 2018, LNCS 11251, pp. 103–123, 2018.
https://doi.org/10.1007/978-3-030-03192-3_9

be equipped with on-board communication units (OBUs) that allow on-demand vehicle-to-vehicle (V2V) communications. Besides, roadside units (RSUs) have been developed to allow vehicles to communicate with infrastructure (V2I). This new paradigm has tremendous growth potentials for promoting socio-economic development [3]. For instances, the messages are transferred among vehicles and between the vehicle and infrastructure may include information such as vehicle movements, traffic light status, and priority vehicles notification or collision avoidance [4, 5].

However, these systems face great challenges of privacy-preservations. Since the nature of information transmitted, message authentication is one of the most important security requirements that should be satisfied so that misbehaviors from the sender of a message can be detected [1]. To do this, the sender of the message in general needs to send their identity for authentication to prove that he or she is a valid user of the system. Clearly, exposing the identity of a user may lead to the privacy violation since it allows eavesdroppers to track people's activities (travel routes, timetables, destinations, etc.) by linking the location information of transmitted messages to the transmitter's real identity. Therefore, it is to provide a mechanism to support conditional privacy. Pseudonyms should be provided to vehicles and servers (traffic monitoring service ...), the messages should be validated by receivers (RSUs or OBUs), and a trusted authority (TA), acting as a third party, should be able to retrieve the vehicle's real identity in case of misbehavior. However, this might cause network congestion or be infeasible in some situations due to the lack or scarcity of deployed infrastructure.

The main idea of this scheme is that at the initialization phase, the TA sends all the necessary information to the vehicles and servers, this only happened once, after that the vehicles and servers can use that information to sign on the sending message with their signature. The signature then will be verified by the receivers of the message. Clearly, the authentication is performed at the receivers without the requirement of establishing contact with TA. This may overcome the congestion challenge as well as the case in which the vehicles cannot get an internet connection to contact TA for authentication.

The paper is structured as follows: Sect. 2 caters for the mathematical preliminaries to understand the privacy-preserving authentication scheme proposed; Sect. 3 introduces the system model on which the authentication scheme works; Sect. 4 reviews existing solutions to the privacy preserving in authentication; Sect. 5 considers a new authentication scheme proposed for ITS. Section 6 provides a thorough description of our autonomous authentication scheme for ITS; Sect. 7 discusses some anonymity revocation mechanisms. Section 8 illustrates some scenarios for the security analysis; and finally, Sect. 9 concludes this paper.

2   Preliminary Background

This section provides some mathematical background required to understand the construction of our proposed privacy-preserving authentication scheme. In addition, we provide the definition of the Collusion Attack Algorithm with k Traitors and n Examples ((k, n)-CAA) [23], and the Elliptic Curve Discrete Logarithm Problem (ECDLP), upon which the security proof of our scheme is based.

2.1 Bilinear Maps

We briefly review the necessary facts about bilinear maps and bilinear map groups with the following notation:

- G_1, G_2 are two (multiplicative) cyclic groups of prime order p.
- Security parameter K, which defines the length of p.
- e is a bilinear map e: G_1 x G_1 → G_2.

Let G_1 and G_2 be two groups as above. A bilinear map is a map e: G_1 x G_1 → G_2 with the following properties [24]:

- Bilinearity: $\forall \alpha, \beta \in Z_p^*$ and P, Q, R $\in G_1$, we have $e(\alpha P + \beta Q, R) = e(P, R)^\alpha e(Q, R)^\beta$ and $e(R, \alpha P + \beta Q) = e(R, P)^\alpha\ e(R, Q)^\beta$.
- Non-degeneracy: $e(G_1, G_1) \neq 1$.
- Complexity: the group action in G can be computed efficiently and there exists a group G1 and an efficiently computable bilinear map G_1 x G_1 → G_2 as above.

2.2 ECDLP (Elliptic Curve Discrete Logarithm Problem)

Let G is a cyclic group of points prime number p in an elliptic curve, assume the equation Q = nP, where P is a generator of G and $\forall Q \in G$. In some case, Q can be calculated easily but it is relatively very hard to determinate n for a given Q and P. This problem is called Elliptic Curve Discrete Logarithm Problem. The number n is called discrete logarithm of Q given base P.

For illustration consider the elliptic curve group G1: $x^2 = x^3 + x + 1$ with a generator P = 7 and the base P(2, 2). In this case, what is the discrete logarithm n of Q = (0, 6) to the base P(2, 2)? To solve the problem we can determine n such that Q = nP. In other words, we can say that we have to compute additive multiple of P until Q is found. Here, with P = (2, 2) and Q = (0, 6), then 3P = Q, so n = 3 is a solution to the discrete logarithm problem.

2.3 (k, n)-CAA Problem (Collusion Attack Algorithm)

Let G_1, G_2, and e be given as above, and the ECDLP is hard in both groups. With k traitors and n examples then we can define the Collusion Attack Algorithm ((k, n) - CAA) problem as follows [23]:

P, P_1 ... $P_n \in G_1$, and x, a_1 ... $a_k \in Z_p$, with k and n be integers. For some $aP_j \notin \{a_iP_j | 1 \leq i \leq k, 1 \leq j \leq n\}$, the (k, n)-CAA problem consists of obtaining P/(x + a) with set $\{xP, xP_j | 1 \leq j \leq n\}$ is given.

The (k, n) - CAA is considered to be hard in literature. That is the probability of success of any probabilistic, polynomial-time, 0/1 in solving (k, n)-CAA problem is negligible. A function F(y) said to be negligible if it is less than $1/y^l$ for every fixed $l > 0$ and sufficiently larger integer y.

2.4   Fiat-Shamir Heuristic

The Fiat–Shamir heuristic is a technique in cryptography for taking an interactive proof of knowledge and creating a digital signature based on it. This way, some fact (for example, knowledge of a certain number secret to the public) can be proven without revealing underlying information. The original interactive proof must have the property of being public-coin, for the method to work. In the random oracle model, The Fiat-Shamir heuristic is used to convert an interactive proof of knowledge into a digital signature. In a proof of knowledge a prover Alice (A) proves the knowledge of a secret to a verifier Bob (B):

1. A holds a secret x, which is the discrete logarithm of $y = g^x$. Where g is a generator of a cyclic group G of prime order q where the discrete logarithm problem is hard, and y is the A public value belonging to such group.
2. A selects randomly $v \in Z_q$ and computes $t = g^v$. A sends t to B.
3. B chooses a random value $c \in Z_q$ and sends it to A. This value is called nonce.
4. A computes $r = v - cx$ and sends r to B. Note that this step is only feasible if A knows x.
5. B can verify the correct construction of r by verifying that $t = g^r y^c$ holds.

In the interactive proof, there is a 3-way handshake, where A sends two messages to B and B sends a nonce to A. The Fiat-Shamir heuristic substitutes the step 3, the nonce random generation, with a random oracle, which is a hash function in practice. Hence, in the non-interactive proof, there is only one message sent from A to B which contains the proof that can be considered as a digital signature. The proof generation follows the steps:

1. A holds a secret x, and wants to prove that she knows x, which is the discrete logarithm of $y = g^x$.
2. A selects randomly $v \in Z_q$ and computes $t = g^v$.
3. A uses a Hash function to generate $c = H(g, y, t)$.
4. A computes $r = v - cx$.
5. A sends the proof (t, r) to B. Note that B also knows g and y since these values are public, hence B can obtain $c = H(g, y, t)$.
6. B checks whether $t = g^r y^c$ holds.

2.5   Hash Chains

The successive application of cryptographic hash function is to convert a hash chain to a data stream that yields multiple random values. From the original data stream can obtain the set of random values. We can define $Hi(\mu)$, which is a hash chain of length i, and as the operation of hashing $i - 1$ times the output of $H(\mu)$, i.e.

$$H^i(\mu) = H^{i-1}(H(\mu)) = H^{i-2}(H(H(\mu))) = \ldots = H(H(H(\ldots H(\mu)))).$$

Where H be a hash function H: $G \rightarrow G$, and μ be a data stream belonging to a group G. Here, the aim of using hash chains is that in the random oracle model, the i values obtained are random, however, in order to recover the set of i values, it only need the value μ.

Example:

A server stores $H^{\{1000\}}$(password) which is provided by the user. Now, if the user wants to authenticate, he will supply $H^{\{999\}}$(password) to the server. Then, the server computes $H(H^{\{999\}}$(password)) = $H^{\{1000\}}$(password) and verifies if this matches the hash chain stored. After that, server will store $H^{\{999\}}$ (password) for the next time the user wishes to authenticate.

2.6 RSA (Rivest, Shamir and Adleman) Algorithm

RSA is a public key cryptographic algorithm [24], Also known as asymmetric key encryption algorithm. As the names suggest, anyone can be given information about the public key, whereas the private key must be kept secret. Anyone can use the public key to encrypt a message, but only someone with knowledge of the private key can hope to decrypt the message in a reasonable amount of time. The power and security of the RSA cryptosystem are based on the fact that the factoring problem is "hard." That is, it is believed that the full decryption of an RSA ciphertext is infeasible because no efficient algorithm currently exists for factoring large numbers. The keys used for the RSA algorithm are generated as follows:

- Two distinct prime numbers p and q are chosen. In order for the system to be secure, the integers p and q should be chosen at random and should be of similar bit-length. To find large primes, the numbers can be chosen at random and, using one of several fast probabilistic methods.
- The product n will be computed by n = pq and used as the modulus for both the public and private keys. Its length, usually expressed in bits, is the length of the key. Then Euler's totient function is used to compute eu (n) = eu(p) eu(q) = (p − 1)(q − 1).
- The public key exponent e is an integer that 1 < e < eu (n) and gcd (e, eu (n)) = 1 (that is, e and eu (n) are coprime).
- The private key exponent d is determined as $d \equiv e^{-1}$ mod eu(n). That is, d is the multiplicative inverse of e (modulo eu (n)). This is often computed using the extended Euclidean algorithm.

The public key is formed by the pair (n, e), where n is called the modulus and e is called the public (or encryption) exponent. The private key is formed by the pair (n, d), where d is called the private (or decryption) exponent. It is imperative that the decryption exponent d is kept secret. In addition, the numbers p, q, and eu (n) must also be kept private because they can be used to calculate d. Once the keys are determined, secure messages can now be sent.

An example:

Alice would like to create a public and private key to use for her secure internet transactions. In order to create these keys, she chooses p = 101 and q = 113. Next, she computes n = pq = 11413 as well as the totient eu(n) = (p − 1)(q − 1) = 11200. In order to find the public key exponent, Alice must choose a number 1 < e < 11200 which is also coprime to 11200. For this example, Alice will choose e = 3533. Finally, Alice must compute $d \equiv e^{-1}$(mod eu(n)) = 6597. Alice publishes the public key pair (n = 11413, e = 3533) while keeping p, q, and d private.

Now suppose that Bob would like to send the message m = 9726 to Alice. Bob would compute $c \equiv 9726^{3533}$ (mod 11413) = 5761, which he would then send to Alice. After receiving the ciphertext c, Alice can decode the message using her private key $m \equiv 5761^{6597}$ (mod 11413) = 9726.

3   System Model

The proposed scheme is applied to traffic monitoring and road safety applications running over an ITS, as shown in Fig. 1. In the system, vehicles can communicate with other vehicles (V2V) and with the network infrastructure (V2I). Communications between vehicles and network infrastructure can be supported with direct point-to-point links between vehicles or relayed by RSU. RSUs are fundamental in high-density areas in order to reduce the number of uplink connections, which could produce a network bottleneck in current cellular networks [22].

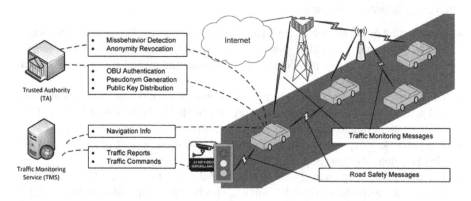

Fig. 1. System model for traffic monitoring and road safety applications

Messages (e.g. road safety messages) between vehicles are transmitted directly by V2V communication for road safety applications, whereas V2I communication enables information exchange (e.g. traffic monitoring messages, navigation info, traffic reports) between vehicles and traffic monitoring servers. In addition, credential generation, credential distribution, misbehavior detection, and anonymity revocation are TA' responsibility. The TA can link the issued credentials to the SN by the onboard tamper-resistant serial number (SN), which is attached to the OBUs. The message cannot be

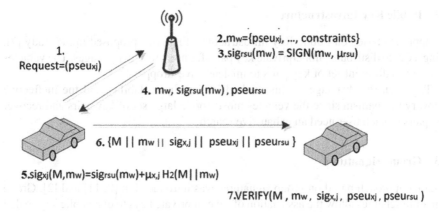

Fig. 2. A step by step portray the proposed privacy-preserving authentication scheme.

linked to the transmitter's real identity by the receiver or any eavesdropper since transmitted messages are signed with the private key associated with the vehicle's active pseudonym and only the TA can link the OBU's SN to the pseudonym.

4 Existing Solutions to Preserve Privacy in Authentication Scheme

Many studies have previously suggested ensuring the privacy of a user such as anonymous credentials [6], Public Key Infrastructure [10], group signatures group signatures, cooperation, and pseudo-identities... In this section, we review some mechanism that focuses on the privacy-preserving authentication.

4.1 Anonymous Credentials

There have been a number of studies that have provided a number of validation mechanisms for data transmission while maintaining conditional privacy in VANET, such as the work [6]. In this scheme, certificates are distributed by the certification authority (CA) to the vehicles in the system and these devices use certificates to sign anonymous messages. The CA monitors the real identity information, serial number (SN) and vehicle related certificates to restore the user's identity in case of abnormal behavior.

The advantage of this approach is that it relies on digital signatures, no need to preliminary handshake to establish the key between the vehicles at the time of transmitting an authenticated message. However, the scalability is the weakness of this system, as the vehicles required to contact CAs whenever they want to be granted a new certificate, each certificate has a short lifetime so they will expire very quickly. In addition, in case of detection of abnormal behavior, the CA must perform a comprehensive search on the certification file to identify the pair (Certificate, SN), which reduces the efficiency in the receiver mechanism.

4.2   Public Key Infrastructure

An approach based on Public Key Infrastructure (PKI) was proposed in the study [7], using a CA to distribute public/private key pairs for users to sign messages. The vehicles may have a different set of key pairs to mislead eavesdroppers.

The main disadvantage of this method is also the scalability and the ineffective recovery mechanism since the vehicles must store a large set of key pairs and recover key pairs of each user need an exhaustive search.

4.3   Group Signatures

Another approach based on group signatures was introduced in [8, 11] and [2]. Group signature schemes allow the association of several private keys to one public key so that the recipient can verify the received signature and associate the signature with a group of vehicles, hence the recipient will not be able to identify the exact sender.

This approach also guarantees the ability to secure the conditional privacy as the CA can still track the identity of the sender. Some problems of this approach are the computational complexity [12] when combining multiple private keys and the difficulty in detecting misbehavior due to the signature used by the whole group and the permanent contact with CA which will lead to the network congestion.

4.4   Cooperation

The Cooperation scheme is an approach based on clustering. In this scheme, all the vehicles are treated as nodes, which combine into groups, with a leader in every group. The leaders are responsible for collecting the packets of the team members and forwards them. Team leaders sacrifice their privacy to protect the identity of the rest of the group.

This strategy can be supported by dynamic grouping algorithms [13]. However, it raised some other security issues, particularly in the leader group. The fact that the team leader is not protected and the possibility of a broken car being selected as a leader can endanger the privacy of all team members.

4.5   Pseudo – Identities

Compared with other models, the model based on Pseudo – Identities does not face security threats in clustering methods and they are far more effective than signature-based models. Here, the cars are given a pseudonym and secret key by TA, which are used to sign the message. Then, cars can send a message with signature and pseudonym. The receiver of the packets can verify the correctness of pseudonym and the signature with only public parameters from the TA. The Vehicles only need to contact TA and retrieve a set of pseudonyms as described in [9, 14–17], these pseudonyms are periodically shifted to avoid possible trace back to their true origin. There are as well existing mechanisms for efficient revocation such as the one proposed in [18] and others suitable strategies to switch the active pseudonym that prevent linkability between different pseudonyms of the same users such as [19–21].

The model based on Pseudo – Identities can ensure that the sender's personal information is kept confidential, eliminating misbehavior in a timely manner. However, the vehicle must be provided with a big set of pseudonyms to avoid traceability and contact periodically with the server to get the set of pseudonyms, which can cause network congestion.

5   Autonomous Privacy-Preserving Authentication Scheme in ITS

An authentication scheme for the privacy preserving in ITS was proposed in [1]. This is a privacy-preserving authentication scheme which provides a self-generation of a number of pseudonyms from a simple credential, an off-line operational mode in case of TA unavailability. In this scheme, the TA issues a credential to the vehicles, and the vehicles are able to self-generate a number of pseudonyms. In the following section, we will summarize the scheme as well as the advantages and disadvantages of this scheme.

The scheme in includes the following 5 algorithms: System Initialization, Generation of credentials, Pseudonym self-generation, Message signing, and Signature verification.

5.1   System Initialization

At the initialization phase, the TA generates a set of public parameters and a secret master key. The TA also updates periodically a time-variant restriction key. This key changes over time in order to control pseudonym self-generation on the vehicle side. The time is divided into slots and this division is known by all vehicles requiring only loose time synchronization. A function $slot_i = T\ (time)$, where the input is the current time and the output the $slot_i$, is defined for all vehicles to obtain the current time-slot. The TA also keeps a registry of the set of vehicles, each vehicle is identified by its OBU's serial number and the user's identity. Vehicles need to provide this information when they authenticate towards the TA.

5.2   Generation of Credentials

The TA generates and issues a credential for every vehicle and every RSU. These credentials will be used by vehicles and RSUs to generate pseudonyms and sign messages. The cases for vehicles and RSUs are slightly different. Vehicles require a multiple credentials, allowing the generation of several pseudonyms per time-slot, hence it will be referred as multi-credential. These pseudonyms should be unlinkable between them and unlinkable with the user's real identity. On the other hand, the RSUs only require one pseudonym that is linked to the RSU's GPS coordinates.

For the vehicle multi-credential generation, the vehicle V_j with a unique and valid OBU's serial number SN_j, should contact the TA, authenticate itself by means of any well-known user authentication mechanism, and establish a secure channel with the TA. Then, the TA provides V_j with a multi-credential, denoted as $MCre_j$, through the secure channel.

It is worth noting that not only a secure channel between the vehicles and the TA is required since the vehicle's secret key is transferred from the TA to the vehicle. But also required confidentiality. Long-term public keys can be used to provide such a secure channel, providing authentication and data confidentiality, between the vehicles and the TA. This secure channel is needed only once to issue the multi-credential.

5.3 Pseudonym Self-generation

Each vehicle and RSU self-generate pseudonyms respectively. A vehicle can generate up to n pseudonyms per time-slot with the values obtained from TA. The vehicle and the RSU self-generate pseudonyms by computing the $Q_i = H_1$ (T (time)) then the set of n pseudonyms.

When sending a signed message, the RSU attaches the pseudonym and the coordinates used to generate such pseudonym. Hence, the receivers of the packet can validate if the pseudonym was generated with the received coordinates and that such coordinates match with their own calculated location information.

5.4 Message Signing

A vehicle V_j holds a pseudonym $pseu_{x,j}$ which is associated with a secret value $S_{ux,j}$. can sign a message by packing a packet $PcktV_j = \left(M \middle\| sig_{x,j} \middle\| pseu_{x,j} \right)$ including the message, the signature and the pseudonym then Vj transmits that one. An RSU can sign a   message   and   includes   the   coordinates   in   the   packet $Pckt_{rsuj} = \left(M \middle\| sig_{rsuj} \middle\| pseu_{rsuj} \middle\| coord \right)$ for the pseudonym verification following the same steps as the vehicles.

5.5 Signature Verification

The receiver (i.e. a vehicle or a traffic monitoring server) of a message can verify the correctness of the signature in order to ensure authentication and data integrity. The receiver obtains the $sig_{x,j} = (T, c, z_1, z_2)$ and validates the signature if the equality $c' = c$ holds with $c' = H_2 \left(M \| T \| R'_{G1} \| R' \| pseu_{x,j} \| T(time) \right)$.

Additionally, the receiver can check whether received coordinates and whether the pseudonym was computed are correct in case the transmitter is an RSU, the receiver validates the signature if $pseu = Q_i H_3$ ($gps\_coordinates$) holds.

5.6 Advantages and Disadvantages of the Autonomous Privacy-Preserving
Authentication Scheme

The privacy-preserving authentication scheme provides message authentication and data integrity as well as conditional privacy, pseudonym unlinkability and forward unlinkability with an efficient anonymity revocation mechanisms. In addition, the scheme is characterized by two main advantages: (i) it allows an autonomous operational mode for cases when the trusted authority is not available through RSU and (ii) it minimizes the communication overhead since the vehicles are not required to establish a point-to-point secure link with the trusted authority to renew the pseudonyms.

However, this solution still has some disadvantages such as the authentication of the sender's signature as well as some time constraints still need to periodically connect to the TA meaning at some point after system initialization a connection to TA is required. In places where there is the lack or scarcity of deployed infrastructure, these actions cannot be made.

6 New Autonomous Authentication Scheme to the Privacy Preserving in Its

We adopt the solution provided in [1, 23] and propose a privacy-preserving authentication scheme enabling self-generation of a number of pseudonyms from a simple credential, an off-line operational mode in case of TA unavailability. This section provides a new authentication scheme to overcome the disadvantages from the previous work for privacy-preservation in ITS discussed in Sect. 5. The biggest difference of this solution in comparison with the previous one is to remove time constraints required for the verification of the messages transmitted between vehicles or RSUs.

The proposed privacy-preserving scheme is constituted of a TA, a number of RSUs denoted as rsu_1, rsu_2..., and the vehicles denoted as V_1, V_2... The vehicles are equipped with OBUs, which are identified with unique serial numbers SN_1 SN_2... registered in the TA together with the users' real identity. The TA stores the V_j tuples $V_j = \{SN_j, ID_j\}$ and the RSUs are identified by their geographical position $coordr_{suj}$. The TA is in charge of generating the public parameters and the credentials for eligible vehicles and RSUs. Once the vehicle is provided with a credential, it can self-generate pseudonyms, which allow the vehicle to sign messages privately. The receivers of those messages, using the TA public parameters, are able to validate the signature of the message and authenticate the transmitter of the message if and only if the transmitter's pseudonym was generated from a valid credential. The TA is also in charge of detecting misbehavior and, eventually, triggering the anonymity revocation mechanism. Revocation can be done in by revoking the vehicle's pseudonyms or credentials. The former provides an efficient manner to revoke temporarily the users' privileges while preserving the forward unlinkability feature, whereas the latter revokes the vehicle's anonymity permanently.

The main advantage of the proposed privacy-preserving authentication scheme is its autonomy since the vehicles do not need to contact the TA again, after receiving the credential, in order to obtain new pseudonyms. In our scheme, the vehicles generate a number of pseudonyms on their own from the same credential, hence the vehicles only need to contact the TA once.

The proposed privacy-preserving authentication scheme is a pseudonym-based signature scheme composed of the following 5 algorithms: System Initialization, Generation of credentials, Pseudonym self-generation, Message signing, and Signature verification.

6.1 System Initialization

At the initialization phase, a set of public parameters and a secret master key are generated by the TA. In addition, The TA also keeps a registry of the set of vehicles $V = \{V_1, V_2...\}$ which is identified by its OBU's serial number (SN_j) and the user's identity (IDj). This information $(V_i = \{SNj, ID_j\})$ is required when vehicles first connect to the TA to be authenticated (Fig. 3).

TA

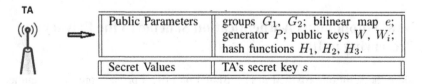

Public Parameters	groups G_1, G_2; bilinear map e; generator P; public keys W, W_i; hash functions H_1, H_2, H_3.
Secret Values	TA's secret key s

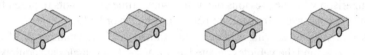

Fig. 3. A set of public parameters and a secret master key are generated by the TA

The TA performs the following steps in order to generate public and secret values:

1. The TA picks at random a prime number p of K bits, where K is a given security parameter. Then the TA selects two cyclic groups of order p, G_1 and G_2, in which the ECDLP is hard. The TA also selects P as a generator of G_1.
2. The TA selects a bilinear map e such that e: $G_1 \times G_2 \to G_2$ following the properties defined in Sect. 2.
3. The TA picks two cryptographic hash functions H_1, H_2: $\{0, 1\}^* \to G_1$.
4. The TA picks a third hash function $H_{3:}$ $\{0, 1\}^* \to Z_p$, where Z_p is a multiplicative group of order p.
5. The TA chooses at random a secret s $\in_R Z_p$
6. The TA computes the permanent public key W = sP

At the end of System Initialization phase the following public parameters and secret values are generated:

- Public parameters: groups G1, G2, bilinear map e generator P, hash functions H1, H2, H3, and H4.
- Secret key: s.

6.2   Generation of Credentials

TA generates and issues a credential for every vehicle and every RSU. These credentials will be used by vehicles and RSUs to generate pseudonyms and sign messages. The cases for vehicles and RSUs are slightly different. Vehicles require multiple credentials, allowing the generation of several pseudonyms per time-slot, hence it will be referred as multi-credential... On the other hand, the RSUs only require one pseudonym, which is linked to the RSU's GPS coordinates.

It is worth noting that a secure channel between the vehicles and the TA is required since the vehicle's secret key is transferred from the TA to the vehicle. Long-term public keys can be used to provide such a secure channel, providing authentication and data confidentiality, between the vehicles and the TA. This secure channel is needed only once to issue the multi-credential.

1. The TA picks a random value $\mu_{1,j} \in_R Z_p$, and then obtains the set $\mu_j = \{\mu_{1,j}, ..., \mu_{n,j}\}$ with a hash chain: $\mu_{2,j} = H_3(\mu_{1,j})$, ..., $\mu_{n,j} = H_3(\mu_{n-1,j})$., i.e. $\mu_{i,j} = H_3^{i-1}(\mu_{1,j})$.
2. The TA uses its secret value s to compute the vehicle's secret keys $Su_{x,j} = P/(s + \mu_{x,j})$. Note that there is one secret key per value $\mu_{x,j}$.
3. The TA sends to V_j the set $MCre_j = (\mu_{1,j}, Su_{1,j}, ..., Su_{n,j})$. The vehicle V_j can verify the correctness of the credential by checking if $e(\mu_{x,j}P + W, Su_{x,j}) = e(P, P)$ holds for every value $\mu_{x,j}$.

To construct a credential for an RSU_{rsuj}, the TA calculates $\mu_{rsuj} = H3$ (gps_coordinates), where gps_coordinates is obtained from the RSU location information. Afterward, the TA transmits to the RSU, over a secure channel, the credential $Cre_{rsuj} = (\mu_{rsui}, Su_{rsuj})$, where $Su_{rsuj} = P/(s + \mu_{rsuj})$.

The secret values regarding the credential material for both vehicles and RSUs are $MCre_j = (\mu_{1,j}, Su_{1,j}, ..., Su_{n,j})$, Su_{rsuj} and the public values are RSU gps_coordinates, μ_{rsuj} (Fig. 4).

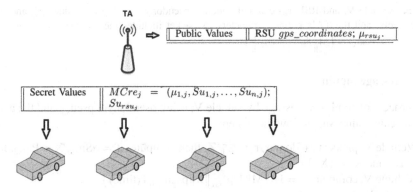

Fig. 4. TA generates and issues a credential for every vehicle and every RSU. These credentials will be used by vehicles and RSUs to generate pseudonyms and sign messages.

6.3 Pseudonym Self-generation

Every vehicle V_j and RSU rsu_j can self-generate pseudonyms using the values $\mu_{x,j}$, and μ_{rsuj} respectively, and the TA's public parameters. A vehicle V_j can generate up to n pseudonyms per time-slot {$pseu_{1,j}$, ..., $pseu_{n,j}$} with the values {$\mu_{1,j}$, ..., $\mu_{n,j}$}.

The vehicle V_j and the RSU rsu_j have to perform the following steps to generate each pseudonym.

1. The vehicle V_j and the RSU rsu_j acquire the time slot by computing $Q_i = H_1$ (T (time)).
2. The vehicle V_j obtains each pseudonym by computing $pseu_{x,j} = \mu_{x,j}Q_i$ and create a set of n pseudonyms {$pseu_{1,j}$, ..., $pseu_{n,j}$}. The RSU rsu_j also obtains its own pseudonyms $pseu_{rsuj} = \mu r_{suj}Q_i$.

The values $\mu_{x,j}$, and V_j of a vehicle are the secret parameters, otherwise, the pseudonyms of a vehicle V_j could be obtained and linked to each other. However, the μ_{rsuj} value of the RSU rsu_j is the public parameters, since it is computed with the GPS coordinates (Fig. 5).

Secret Values	$\mu_{x,j}$;
Public Values	$pseu_{x,j}$; RSU $gps\_coordinates$; μ_{rsu_j}; $pseu_{rsu_j} = \mu_{rsu_j}Q_i$.

Pckt$_{vj}$ = (M‖sig$_{x,j}$‖pseu$_{x,j}$)

Fig. 5. Vehicle V_j and RSU r_{suj} can self-generate pseudonyms using the values $\mu_{x,j}$ and μ_{rsuj} respectively, and the TA's public parameters. A packet includes the message, signature and pseudonym: $\text{Pckt}_{V_j} = \left(M\middle\|sig_{x,j}\middle\|pseu_{x,j}\middle\|\right)$.

6.4 Message Signing

The process of signing a message by vehicle V_j using pseudonym $pseu_{x,j}$, and the associated secret value $Su_{x,j}$ follow those step:

1. Vehicle V_j picks at random $\alpha, r, r' \in_R Z_p$ then computes $T = \alpha Su_{x,j}$, $X = P \mu_{x,j}$, $R_{G1} = rQ_i$ and $R = e(X, T)^{1/\alpha}$.
2. Vehicle V_j computes $c = H_2\left(M\|T\|R_{G1}\|R\|pseu_{x,j}\|T(time)\right)$.
3. V_j calculates $z_1 = c\alpha + r'$ and $z_2 = c\mu_{x,j} + r$.
4. The signature of the message M with the pseudonym $pseu_{x,j}$ is composed by the tuple $sig_{x,j} = (T, c, z_1, z_2)$.

The vehicle V_j can transmit a packet including the message, signature and pseudonym: $Pckt_{Vj} = \left(M\|sig_{x,j}\|pseu_{x,j}\right)$. An RSU follows the same steps to sign a message and includes the coordinates in the packet for pseudonym verification $Pckt_{rsuj} = \left(M\|sig_{rsuj}\|pseu_{rsuj}\|coord\right)$ (Fig. 6).

$$Pckt_{Vj} = (M\|sig_{x,j}\|pseu_{x,j})$$
$$Pckt_{rsuj} = (M\|sig_{rsuj}\|pseu_{rsuj}\|coord).$$

Fig. 6. A Vehicle's packet includes the signature and pseudonym $Packt_{Vj} = \left(M\|sig_{x,j}\|pseu_{x,j}\|\right)$ and an RSU includes the coordinates in the packet $Psckt_{rsuj} = \left(M\|sig_{rsuj}\|pseu_{rsuj}\|coord\right)$

6.5 Signature Verification

The receiver (i.e. a vehicle or a traffic monitoring server) of a message can verify the correctness of the signature in order to ensure authentication and data integrity. The receiver obtains the $sig_{x,j} = (T, c, z1, z2)$ and performs the following steps:

1. It computes the time variant parameters T (time) = $slot_i$ and $Qi = H_1$ (T (time)).
2. Then it computes $R'_{G1} = z_2 Q_i - cpseu_{x,j}$ and $R' = e(P, P)/e(W, T)$.
3. The receiver calculates $c' = H2\left(M\|T\|R_{G'1}\|R'\|pseu_{x,j}\|T \text{ (time)}\right)$.
4. The receiver validates the signature if the equality $c' = c$ holds.

Additionally, if the transmitter is an RSU, then the receiver can check whether these coordinates are correct and whether the pseudonym was calculated with the received coordinates, this is if pseu = $Q_i H_3$(gps_coordinates) holds. However, before the signature verification is performed the receiver must check that the pseudonym used by the transmitter is not included in the most up to date RL received from the TA. If the pseudonym is included in the RL, then the packet is discarded without proceeding with the verification of the signature. Figure 2 describes the signature verification process (Fig. 7).

7 Anonymity Revocation Process

Despite its benefits, anonymity can also act against the proper utilization of the vehicular networking information systems since malicious entities can transmit misleading information while keeping their identities hidden. For traffic monitoring applications, misleading information can lead to inefficient traffic route calculation by navigation systems, and for road safety applications, it can trigger false alarms in real time, which can be especially harmful and difficult to detect. The capability of the vehicles to change pseudonyms is an advantage for malicious entities, since they can change their identity once they are detected by other vehicles. Therefore, it is of paramount importance to provide efficient mechanisms for anonymity revocation to enable the exclusion of malicious entities from the traffic monitoring and road safety systems. This is accomplished by publishing the pseudonyms of the malicious entities in the RL. However, one of the main challenges in the anonymity revocation process for vehicular networks is the distribution mechanism of the RL, which is a delay sensitive task. The RL must be transmitted with no delay to prevent vehicles from authenticating misleading received messages from malicious or malfunctioning vehicles (Fig. 8).

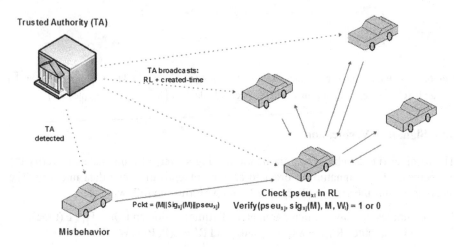

Fig. 8. Anonymity revocation process using RL

In this section, we discuss two efficient, fast and low complex revocation mechanisms that can be supported by our privacy-preserving authentication scheme to achieve exclusion of malicious entities.

$$c = H_2(M\|T\|R_{G1}\|R\|pseu_{x,j}\|T(time))$$
$$sig_{x,j} = (T, c, z_1, z_2).$$

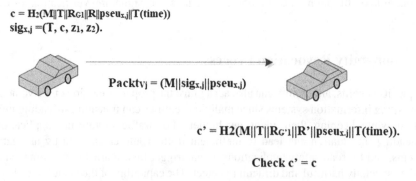

$$Packtv_j = (M\|sig_{x,j}\|pseu_{x,j})$$

$$c' = H2(M\|T\|R_{G'1}\|R'\|pseu_{x,j}\|T(time)).$$

$$Check\ c' = c$$

Fig. 7. Signature verification process.

A. *Time-slot based anonymity revocation mechanism*

The TA can consider revoking the pseudonyms of the misbehaving vehicle for some given time-slots. In this case, the TA includes in the RL the set of n pseudonyms of the vehicle V_j for the active slot $\{pseu_{1,j} \ldots pseu_{n,j}\} = \{\mu_{1,j}Q_i, \ldots., \mu_{n,j}Q_i\}$ and transmits the RL to the other vehicles and RSUs. The transmitted RL only includes the revoked pseudonyms of the current time-slot, which limits the size of the RL to n pseudonyms per revoked vehicle. This is important since decreasing the list size also reduces the transmission delay. Additionally, this mechanism supports autonomy for a number of time-slots, since the TA can publish in advance the revoked pseudonyms for future time-slots. In such case, the restriction keys of future time-slots $(W_i, W_{i+1},\ldots.)$ are published in advance, hence the RL size is increased. However, this is compensated by the fact of transmitting the RL only once at the beginning of the first slot.

B. *Permanent anonymity revocation mechanism*

The TA can revoke the user's anonymity privileges permanently when the vehicle's misbehavior is systematic or the misleading information transmitted is sufficiently harmful to be considered as an attack. This is accomplished by including the value $\mu_{1,j}$ of a misbehaving vehicle V_j in the RL and publishing this list. Note that with this value any other vehicle or traffic monitoring server can compute the set of pseudonyms of the vehicle V_j at any time-slot. It is worth noting that publishing the list of pseudonyms, according to the time-slot based revocation mechanism, ensures forward unlinkability whereas publishing the value $\mu_{1,j}$ provides vehicles and traffic monitoring servers with the ability to track and discard past messages. The permanent revocation of anonymity is highly efficient since it revokes all pseudonyms for all time-slots by including only one value in the RL. Moreover, the computation of the revoked pseudonyms from the revoked value $\mu_{1,j}$ is low complex, since it relays on a hash chain and a multiplication $pseu_{x,j} = \mu_{x,j}Q_i$, where $Q_i = H_1(T(time))$.

C. *Delegation mechanism to avoid RL verification*

A mechanism proposed in study [1] that allows the avoidance of RL verification when there is RSU coverage. The mechanism was used to overpass RL verification can and could enhance the performance of delay-sensitive communications such as vehicle collision avoidance or priority vehicles signaling in terms of latency. In this mechanism, every vehicle obtains delegated rights from an RSU for one of the pseudonyms in the set that is reserved for delay sensitive services. The RL is kept in the RSUs and vehicles contact the RSUs to obtain a warrant stating that they are not included in the RL. The vehicle afterward sign and transmit messages including this warrant. Receivers can trust the RSU that signed the warrant and overpass the RL verification. An RSU *rsuj* can state that a pseudonym $pseu_{x,j}$ of a vehicle V_j is not in the RL by delegating the RSU signature on a V_j pseudonym $pseu_{x,j}$. An RSU *rsu_j* with value μrsu_j and pseudonym $pseu_{rsuj} = \mu_{rsuj}\, Q_i$ can delegate a vehicle V_j with pseudonym $pseu_{x,j} = \mu_{x,j}Q_i$.

It is worth noting that getting a delegation for more than one pseudonym would require a secure channel between vehicles and RSUs to provide confidentiality. Otherwise, the request would include several pseudonyms in clear text, making these pseudonyms linkable.

8 Security Analysis

In this section, we provide the security proof of the proposed pseudonym-based signature scheme.

Case 1: Choose any value of $pseu_{x,j}$; $\mu_{x,j}$; $Su_{x,j}$ provided by TA to test this scheme. This is to prove that for any values of $pseu_{x,j}$ generated from $\mu_{x,j}$; $Su_{x,j}$ provided by TA, the scheme always validates the signature signed by using $pseu_{x,j}$.

Message Signing

- First, Vehicle 1 as a prover choose $\mu_{1,1}$ and compute $pseu_{1,1} = \mu_{1,1}Q_i$, $Su_{1,1} = P1/((s + \mu_{1,1}))$.
- Then, Vehicle V_1 picks at random α, r, r' $\in$ R Z_p and computes $T = \alpha Su_{1,1}$, $X = P \mu_{1,1}$, $R_{G1} = rQ_i$, $R = e(T, W + X)^{-\alpha}$.
- Vehicle V_1 computes $c = H_2\big(M\|T\|R_{G1}\|R\|pseu_{1,j}\|T(time)\big)$,
- V_1 calculates $z_1 = c\alpha + r'$ and $z_2 = c\mu_{1,j} + r$.
- V_1 sign the message M with the signature $sig_{1,1} = (T, c, z_1, z_2)$.
- Finally, V_1 Send the packet to the receiver: $PcktV1 = \big(M\|sig_{1,1}\|pseu_{1,1}\big)$.

Signature Verification

We have vehicle 2 as a receiver, after it receives a packet from an unknown sender, proceeds to validate the signature by following these step:

- V_2 computes the time variant parameters $T(time) = slot_i$ and $Qi = H_1(T(time))$.
- Then it computes $R'_{G1} = z_2Q_i - cpseu_{1,1}$ and $R' = e(P, P)$.
- V_2 calculates $c' = H2\big(M\|T\|R_{G'1}\|R'\|pseu_{1,1}\|T(time)\big)$.
- V2 proceeds to validates the signature by checking if the equality $c' = c$ holds which mean $R' = R$.

If R was constructed follow Sect. 6.4 then $R = e(T, W + X)^{-\alpha}$. Replace $T = \alpha Su_{1,1}$ and $X = P \mu_{1,1}$, we have $R = e (\alpha Su_{1,1}, P\mu_{1,1})^{-\alpha}$. Follow the bilinear properties in Sect. 2, R can be computes as $R = e (\alpha Su_{1,1}, P\mu_{1,1})^{-\alpha} = e(P,P) = R'$.

With this we can make the conclusion: any value of $pseu_{x,j}$; $\mu_{x,j}$; $Su_{x,j}$ generated by a validated sender can be checked by any receiver without any connection to the TA.

Case 2: An attacker will guess values $\mu_{x,j}$ or/and s (secret key of TA) and use them to verify the correctness of a new signature. This is to prove that no values of $\mu_{x,j}$ and s except the real ones provided by TA can be used to validates the message again without the presence of TA.

Message Signing

- Vehicle 1 as an attacker hasn't contacted with TA to receive it's credential so it has to pick some random values for s, $_{1,1}$. Then, vehicle 1 combine them with the public values to compute $pseu_{1,1} = \mu_{1,1}Q_i$, $Su_{1,1} = p/((s + \mu_{1,1}))$.

- Follow the message signing process, vehicle 1 proceed to sign the message with his signature created by s, $\mu_{1,1}$. As a result, the values of R, c are computed as $R = e(X, T)^{1/\alpha}$ and $c = H_2\left(M\|T\|R_{G1}\|R\|pseu_{1,1}\|T(time)\right)$,
- In the end, vehicle 1 produce a new signature $sig_{1,1} = (T, c, z_1, z_2)$ and a packet $PcktV_1 = \left(M\|sig_{1,1}\|pseu_{1,1}\right)$.

Signature Verification

- Vehicle 2 as the recipient after receiving the signature $sig_{1,1} = (T, c, z_1, z_2)$ from vehicle 1 proceeds to verify the authenticity of this signature by following these steps.
- Vehicle V2 calculates the time variable $T(time) = slot_i$ and $Q_i = H_1 (T (time))$ and then calculates $R_{G1} = z_2Q_i - cpseu_{1,1}$ and $R' = e(P, P)/e(W, T)$.
- V2 the proceed to compute $c = H2\left(M\|T\| R_{G1}\|R\| pseu_{1,1}\|T(time)\right)$.
- The signature is verified by checking whether $c' = c$ hold.

Vehicle V2 uses its parameters and the value received from V1 to calculate $R' = e(P, P)/e (W, T)$ then compare with R' taken from the signature of V_1. However, in this case, since R is not constructed based on a valid certificate, the equality $R' = R$ will not hold.

The reason for this is because vehicle 2 uses its own W value to check the parameters received from the vehicle 1 and not completely depend on the values in the signature for validation:

$$e\left(W + X, Su_{1,1}\right) = e (sP + P\mu'_{1,1}, P/(s' + \mu'_{1,1}))$$

We can reach a conclusion: an unauthorized vehicle cannot use a random value to validate the signature.

9   Conclusions

This paper reviews some of the existing solutions for privacy preserving in authentication and then proposes an authentication model to preserve user's privacy in ITS. The main idea of this model is that the vehicles and server only need to contact TA once to obtain the secret information. After contacting TA, all devices in the system can use their signature to sign messages based on this secret information. The signature will then be verified by the receiver of the message. Clearly, the authentication is done at the receiver end and does not require a connection to TA. This has overcome the congestion challenge, as well as the lack of internet connection to contact TA.

In addition, some efficient anonymity revocation mechanisms can be deployed in the proposed privacy-preserving authentication scheme. However, we did not provide any new anonymity revocation mechanisms. In the future, the RL problem will be carefully studied and also consider more scenarios to prove the security of the proposed scheme.

Acknowledgements. This research is funded by Vietnam National University Ho Chi Minh City (VNU-HCM) under grant number C2017-20-17.

References

1. Sucasas, V., Mantas, G., Saghezchi, F.B., Radwan, A., Rodriguez, J.: An autonomous privacy preserving authentication scheme for intelligent transportation systems. Comput. Secur. **60**, 193–205 (2016)
2. Saad, M.N.M., Laouiti, A., Qayyum, A.: Vehicular Ad-hoc Networks for Smart Cities. Springer, Singapore (2014). https://doi.org/10.1007/978-981-287-158-9
3. Olariu, S., Weigle, M.C.: Vehicular Networks: From Theory to Practice, 1st edn. Chapman & Hall/CRC, Boca Raton (2009)
4. Juan, Z., Jianping, W., McDonald, M.: Socio-economic impact assessment of intelligent transport systems. Tsinghua Sci. Technol. **11**(3), 339–350 (2006)
5. Bekiaris, E., Nakanishi, Y.J.: Economic Impacts of Intelligent Transportation Systems. Elsevier, Amsterdam (2004)
6. Raya, M., Hubaux, J.-P.: The security of vehicular ad-hoc networks. In: Proceedings of the 3rd ACM Workshop on Security of Ad-Hoc and Sensor Networks, SASN 2005, pp. 11–21. ACM, New York (2005)
7. Liu, X., Fang, Z., Shi, L.: Securing vehicular ad hoc networks. In: 2nd International Conference on Pervasive Computing and Applications, ICPCA 2007, pp. 424–429, July 2007
8. Lin, X., Sun, X., Ho, P.-H., Shen, X.: GSIS: a secure and privacy-preserving protocol for vehicular communications. IEEE Trans. Veh. Technol. **56**(6), 3442–3456 (2007)
9. Lu, R., Lin, X., Zhu, H., Ho, P.-H., Shen, X.: ECPP: efficient conditional privacy preservation protocol for secure vehicular communications. In: The 27th Conference on Computer Communications, INFOCOM 2008. IEEE, April 2008
10. Calandriello, G., Papadimitratos, P., Hubaux, J.-P., Lioy, A.: Efficient and robust pseudonymous authentication in VANET. In: Proceedings of the Fourth ACM International Workshop on Vehicular Ad Hoc Networks, VANET 2007, pp. 19–28. ACM, New York (2007)
11. Sampigethaya, K., Li, M., Huang, L., Poovendran, R.: AMOEBA: robust location privacy scheme for VANET. IEEE J. Sel. Areas Commun. **25**(8), 1569–1589 (2007)
12. Lu, R., Lin, X., Shi, Z., Shen, X.: A lightweight conditional privacy-preservation protocol for vehicular traffic-monitoring systems. Intelligent Systems, IEEE **28**(3), 62–65 (2013)
13. Sucasas, V., Radwan, A., Marques, H., Rodriguez, J., Vahid, S., Tafazolli, R.: A cognitive approach for stable cooperative-group formation in mobile environments. In: 2014 IEEE International Conference on Communications (ICC), pp. 3241–3245, June 2014
14. Ma, Z., Kargl, F., Weber, M.: Pseudonym-on-demand: a new pseudonym refill strategy for vehicular communications. In: IEEE 68th Vehicular Technology Conference, VTC 2008-Fall, pp. 1–5, September 2008
15. Huang, J.-L., Yeh, L.-Y., Chien, Hung-Yu.: ABAKA: an anonymous batch authenticated and key agreement scheme for value added services in vehicular ad hoc networks. IEEE Trans. Veh. Technol. **60**(1), 248–262 (2011)
16. Huang, D., Misra, S., Verma, M., Xue, G.: PACP: an efficient pseudonymous authentication-based conditional privacy protocol for VANETs. IEEE Trans. Intell. Transp. Syst. **12**(3), 736–746 (2011)
17. Sun, Y., Lu, R., Lin, X., Shen, X., Jinshu, S.: An efficient pseudonymous authentication scheme with strong privacy preservation for vehicular communications. IEEE Trans. Veh. Technol. **59**(7), 3589–3603 (2010)
18. Haas, J.J., Hu, Y.-C., Laberteaux, K.P.: Design and analysis of a lightweight certificate revocation mechanism for VANET. In: Proceedings of the 6th ACM International Workshop on VehiculAr InterNETworking, VANET 2009, pp. 89–98. ACM, New York (2009)

19. Huang, L., Matsuura, K., Yamane, H., Sezaki, K.: Enhancing wireless location privacy using silent period. In: IEEE Wireless Communications and Networking Conference, vol. 2, pp. 1187–1192, March 2005
20. Ying, B., Makrakis, D., Mouftah, H.T.: Dynamic mix-zone for location privacy in vehicular networks. IEEE Commun. Lett. **17**(8), 1524–1527 (2013)
21. Kido, H., Yanagisawa, Y., Satoh, T.: An anonymous communication technique using dummies for location-based services. In: Proceedings of International Conference on Pervasive Services, ICPS 2005, pp. 88–97, July 2005
22. Araniti, G., Campolo, C., Condoluci, M., Iera, A., Molinaro, A.: Lte for vehicular networking: a survey. IEEE Commun. Mag. **51**(5), 148–157 (2013)
23. Zhang, Y., Chen, J.-L.: A delegation solution for universal identity management in SOA. IEEE Trans. Serv. Comput. **4**(1), 70–81 (2011)
24. Zhou, X., Tang, X.: Research and implementation of RSA algorithm for encryption and decryption. IEEE, September 2011

Big Data Analytics and Applications

Higher Performance IPPC$^+$ Tree for Parallel Incremental Frequent Itemsets Mining

Van Quoc Phuong Huynh$^{(\boxtimes)}$ and Josef Küng

Institute for Application Oriented Knowledge Processing (FAW),
Faculty of Engineering and Natural Sciences (TNF), Johannes Kepler University
(JKU), Linz, Austria
{vqphuynh, jkueng}@faw.jku.at

Abstract. IPPC tree provides incremental properties and high performance for mining frequent itemsets through shared-memory parallel algorithm IFIN$^+$. However, in the case of datasets comprising a large number of distinguishing items but just a small percentage of frequent items, IPPC tree becomes to lose its advantage in running time and memory for the tree construction. With a motivation of reducing the execution time for the tree building, in this paper, we propose an improved version for IPPC tree, called IPPC$^+$, to increase the performance of the tree construction. We conducted extensive experiments on both synthetic and real datasets to evaluate IPPC$^+$ tree against IPPC tree. Besides, the IFIN$^+$ with the new tree is also compared to the well-known algorithm FP-Growth and the other two state-of-the-art ones, FIN and PrePost$^+$. The experimental results show that the construction time of IPPC$^+$ tree is improved remarkably compared to that of IPPC tree; and IFIN$^+$ is the most efficient algorithm, especially in the case of mining at different support thresholds within the same running session.

Keywords: Incremental · Parallel · Frequent itemsets mining · Data mining
Big data · IPPC · IPPC$^+$ · IFIN · IFIN$^+$

1 Introduction and Related Works

Frequent itemsets mining can be briefly described as follows. Given a dataset of n transactions $D = \{T_1, T_2, \ldots, T_n\}$, the dataset contains a set of m distinct items $I = \{i_1, i_2, \ldots, i_m\}$, $T_i \subseteq I$. A k-itemset, IS, is a set of k items ($1 \leq k \leq m$). Each itemset IS possesses an attribute, *support*, which is the number of transactions containing IS. The problem is featured by a support threshold ε which is the percent of transactions in the whole dataset D. An itemset IS is called frequent itemset iff $IS.support \geq \varepsilon * n$. The problem is to discover all frequent itemsets existing in D.

This problem was started up by Agrawal & Srikant with algorithm Apriori [1]. This algorithm generates candidate $(k + 1)$-itemsets from frequent k-itemsets at the $(k + 1)^{\text{th}}$ pass and then scans dataset to check whether a candidate $(k + 1)$-itemsets is a frequent one. Many previous works were inspired by this algorithm. Algorithm Partition [7] aims at reducing I/O cost by dividing a dataset into non-overlapping and memory-fitting partitions which are sequentially scanned in two phases. In the first phase, local

© Springer Nature Switzerland AG 2018
T. K. Dang et al. (Eds.): FDSE 2018, LNCS 11251, pp. 127–144, 2018.
https://doi.org/10.1007/978-3-030-03192-3_10

candidate itemsets are generated for each partition, and then they are checked in the second one. DCP [8] enhances Apriori by incorporating two dataset pruning techniques introduced in DHP [9] and using direct counting method for storing candidate itemsets and counting their support. In general, Apriori-like methods suffer from two draw-backs: a deluge of generated candidate itemsets and/or I/O overhead caused by repeatedly scanning dataset.

Two other approaches, which are more efficient than Apriori-like methods, are also proposed to solve the problem: (1) frequent pattern growth adopting divide-and-conquer with FP tree structure and FP-Growth [2], and (2) vertical data format strategy in Eclat [10]. FP-Growth and algorithms based on it such as [11, 12] are efficient solutions since unlike Apriori, they avoid many times of scanning dataset and generation-and-test. However, they become less efficient when datasets are sparse. While algorithms based on FP-Growth and Apriori use a horizontal data format; Eclat and some other algorithms [7, 13, 14] apply vertical data format, in which each item is associated with a set of transaction identifiers, Tids, containing the item. This approach avoids scanning dataset repeatedly, but a huge memory overhead is expensed for sets of Tids when dataset becomes large and/or dense. Recently, two remarkably efficient algorithms are introduced: FIN [3] with POC tree and PrePost⁺ [4] with PPC tree. These two structures are prefix trees and similar to FP tree, but the two mining algo-rithms use additional data structures, called Nodeset and N-list respectively, to sig-nificantly improve mining speed.

Discovering frequent itemsets in a large dataset is an important problem in data mining. In Big Data era, this mining model, as well as other ones, has been being challenged by very large volume and high velocity of datasets which are fast accu-mulated over time. As a solution in our previous work [15] to deal with this problem, we proposed a tree structure, named IPPC (Incremental Pre-Post-Order Coding), which supports incremental tree construction; and an algorithm for incrementally mining frequent itemsets, IFIN (Incremental Frequent Itemsets Nodesets). For one our next work [16], we introduced a shared-memory parallel version of IFIN, named IFIN⁺, to enhance performance by exhausting as much as possible the computational power of commodity processors which are equipped with many physical computational units.

Through experiments, algorithm IFIN⁺ has demonstrated its superior performance compared to the well-known algorithm FP-Growth [2], and other two state-of-the-art ones FIN [3] and PrePost⁺ [4]. However, in case of datasets comprising a large number of distinct items but just a small percentage of frequent items for a certain support threshold, we investigates that IPPC tree becomes to lose its advantage in running time and memory for its construction compared to other trees such as FP, POC, and PPC of algorithms FP-Growth, FIN, and PrePost⁺. The reason is that these trees use just frequent items for their tree structures while the IPPC tree uses all items in datasets for its structure to be compensated with the abilities of incremental tree construction and mining. Table 1 reports the detail of construction time of the trees on a synthetic dataset and a real one, named Kosarak, at support threshold $\varepsilon = 0.1\%$ and 0.2% respectively. Note that, the construction of IPPC tree does not depend on the support thresholds.

Table 1. Tree construction time on datasets.

Tree construction time (Synthetic dataset of 1200k transactions, $\varepsilon = 0.1\%$)	200k	400k	600k	800k	1000k	1200k
IPPC tree	2.1 s	3 s	3.2 s	4.1 s	4.3 s	5.1 s
FP tree	3.4 s	5.7 s	8.7 s	10 s	12.9 s	16.2 s
POC/PPC trees	2.5 s	4.5 s	6.7 s	9.7 s	11.9 s	14.7 s

Tree construction time (Kosarak dataset of 990002 transactions, $\varepsilon = 0.2\%$)	200k	400k	600k	800k	1000k
IPPC tree	7.2 s	10.4 s	11.4 s	13.9 s	14.7 s
FP tree	3.3 s	6.2 s	9.2 s	12.6 s	15.3 s
POC/PPC trees	2.1 s	3.7 s	5.1 s	6.6 s	7.5 s

The synthetic dataset at $\varepsilon = 0.1\%$ includes 843 frequent items, 90% of all 932 items. In this case, the construction time of IPPC tree is superior to that of the other trees, approximate a ratio 1:3 to the time of POC/PPC trees at full size of 1200k transactions. In contrast, for Kosarak dataset at $\varepsilon = 0.2\%$, there are just 568 frequent items, 1.37% in the total of 41270 items. IPPC tree becomes to lose its advantage since extra computational overhead for a very large proportion of infrequent items, and its construction time is twice the time of POC/PPC trees at the full size of Kosarak dataset. Hence, the aim of this paper is to reduce the running time of the IPPC tree construction. Through experiments on these two datasets, the IPPC⁺ tree, the new version of IPPC tree, achieves a remarkable improvement of the tree building performance compared to IPPC tree and contributes better running time for the mining algorithm IFIN⁺ as well. Besides, the IFIN⁺ with the new tree is also compared with the well-known algorithm FP-Growth and the other two state-of-the-art ones, FIN and PrePost⁺. The experimental results show that IFIN⁺ is the most efficient algorithm, especially in the cases of mining at different support thresholds within the same running session.

Table 2. Example transaction dataset

ID	Items in transactions
1	b, e, d, f, c
2	d, c, b, g, f, h
3	f, a, c
4	a, b, d, f, c, h
5	b, d, c

The rest of the paper is organized as follows. Section 2 will mention some essential concepts of IPPC tree structure and then introduce the improved solution. Section 3 presents experiments; and finally, conclusions are given in Sect. 4.

2   Improved Solution

In this section, we propose a solution to reduce the construction time for the IPPC tree. More detail about the IPPC tree can be found in [16]. For convenience in reference and concise content, just the fundamental concepts and pseudo code of IPPC tree are presented; and based on that the IPPC$^+$ tree will be introduced.

2.1   IPPC Tree

IPPC tree is a compact and information-lossless structure of the whole items of all transactions in a given dataset. Its construction needs only one data scanning and does not require a given support threshold. Local order of items in a path of nodes from the root to a leaf is flexible and can be changed to improve compression. Each node in the tree is identified by a pair of codes: *pre-order* and *post*-order. With these characteristics, a built IPPC tree from a dataset D can be mined at different support thresholds and reused to build up a new IPPC tree corresponding to a new dataset $D' = D + \Delta D$.

To demonstrate the concept of IPPC tree building process, the Fig. 1 records transaction by transaction in Table 2 inserted into an empty IPPC tree. Initially, the tree has only the root node, and transaction $1(b, e, d, f, c)$ is inserted as it is in Fig. 1(a). The Fig. 1(b) is of the tree after transaction 2 (d, c, b, g, f, h) is added. The item b in transaction 2 is merged with node b in the tree. Although transaction 2 does not contain item e, but its common items $d, f,$ and c can be merged with the corresponding nodes. Item d is found common, so it is merged with node d after node d is swapped[1] with node e to guarantee the Property 2. Similarly, items f and c are merged with node f and

---

[1] Swapping two nodes is simply exchanging one's item name to that of the other.

c respectively; and the remaining items *g* and *h* are inserted as a child branch of node *c*. In Fig. 1(c), transaction 3 (*f*, *a*, *c*) is processed. Common item *f* is found that can be merged with node *f*, so node *f* is swapped with node *b*. Item *c* is also a common one, but it is not able to be merged with node *c* as node *d* does not satisfy the Descendant Swapping condition with node *c*. Then the items *a* and *c* are added as a branch from node *f*. When transaction 4 (*a*, *b*, *d*, *f*, *c*, *h*) is added in Fig. 1(d), common items *f*, *d*, *b*, and *c* are merged straightforwardly with corresponding nodes *f*, *d*, *b*, and *c*. The remaining items *a* and *h* are then inserted into the subtree having root node *c*. The item *h* is found common with node *h* in the second branch. Node *h* and item *h*, therefore, are merged together after node *h* is swapped with node *g*. The last item *a* is then inserted as a new child branch from node *h*. Insertion of transaction 5 (*b*, *d*, *c*) is depicted in Fig. 1 (e). All items in transaction 5 are common, but they cannot be merged with nodes *b*, *d*, and *c* as node *f* does not guarantee the Child Swapping condition. Thus, transaction 5 is added as a new child branch of the root node.

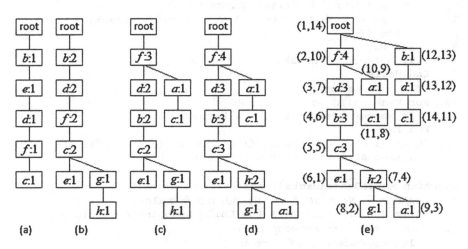

Fig. 1. An illustration for constructing an IPPC tree on example transaction dataset

After the dataset has been processed, each node in the IPPC tree is assigned a pair of sequent numbers (*pre-order, post-order*) by scanning the tree with pre-order and post-order traversals through procedure **AssignPrePostOrder**. For an example, node (4, 6) is identified by *pre-order* = 4 and *post-order* = 6, and it registers item *b* with *support* = 3.

Algorithm 1: BuildIPPCTree
Input: Dataset *D*, root node *R*
Output: An IPPC tree with root *R*, item list *L*
1. **For Each** transaction *T* ∈ *D*
2.   Update items and their supports in *L* from items in *T*;
3.   **InsertTransaction**(*T*, *R*);
4. **End For**
5. **AssignPrePostOrder**(*R*);

Procedure InsertTransaction(Transaction *T*, Node *R*)
1. *subNode* ← *R*; *notMerged*;
2. **While**(*T* ≠ ∅)
3.   *notMerged* ← **true**;
4.   **For Each** child node *N* of *subNode*
5.   **If**(*N*.*item-name* ∈ *T*)
6.   *notMerged* ← **false**; *N*.*support*++;
7.   *subNode* ← *N*; *T* ← (*T* \ *N*.*item-name*); **break**;
8.   **End If**
9.   **End For**
10.   **If**(*notMerged*) **break**;
11. **End While**
12. **If**(*T* = ∅) **Return**;
13. **For Each** child node *N* of *subNode*
14.   **If**(**MergeDescendants**(*T*, *N*)) **Return**;
15. **End For**
16. Insert *T* as a new branch from *subNode* (added nodes are initialized at 1 for their supports);

Function MergeDescendants(Transaction *T*, Node *N*)
1. *subNode* ← N; *mrgNode* ← N; *merged* ← **false**;
2. **While**(*subNode* satisfies the **Child Swapping** condition)
3.   *descendant* ← *subNode*.*child*;
4.   **If**(*descendant*.*item-name* ∈ *T*)
5.   *T* ← (*T* \ *descendant*.*item-name*); *merged* ← **true**;
6.   Exchange item names of *mrgNode* and *descendant*;
7.   *mrgNode*.*support*++; *mrgNode* ← *mrgNode*.*child*;
8.   **End If**
9.   *subNode* ← *descendant*;
10. **End While**
11. **If**(*merged*) Insert *T* as a new branch from *mrgNode*.*parent* (added nodes are initialized at 1 for their supports);
12. **Return** *merged*;

Procedure AssignPrePostOrder (Node *R*)
    // *PreOrder* and *PostOrder* are initialized at 1.
1. *R*.*pre-order* ← *PreOrder*; *PreOrder*++;
2. **For Each** child node *N* of *R* **Do AssignPrePostCode**(*N*);
3. *R*.*post-order* ← *PostOrder*; *PostOrder*++;

2.2 IPPC$^+$ Tree

The IPPC tree comprises the whole items of all transactions of a given dataset, regardless of the support threshold. Therefore, in case of a dataset possessing a large number of distinct items, the number of child nodes of every single node in the tree becomes larger. The more distinguishing items, the bigger the number of child nodes of the tree nodes will be. Before an item in a transaction is merged with a tree node or inserted into the tree as a new node, a sequence search for a child node (of the current parent node) having the same item name as the item will be done. Consequently, these factors have caused more computational overhead to insert a new transaction into the current IPPC tree.

To improve the performance of the transaction insertion, we replace the sequence search with the binary search based on the item name. The replacement requires maintaining the item name based order of child nodes of every single node in IPPC tree. The for-loop (**InsertTransaction** procedure, from lines 4 to 9) is changed from sequence traversal on lists of child nodes to sequence traversal on items of transactions. Since the cardinalities of child node lists are very much larger than the numbers of items in transactions, this changing obviously improves the performance. The more difference between the two kinds of cardinality, the more performance is enhanced. However, the changing does not guarantee the items with higher support are accumulated into the IPPC tree before other items with lower support. This drives the reduction of compression for higher-support items; and therefore, the *nodesets* [15, 16] of these items become longer that may decrease the performance in mining process of algorithm IFIN$^+$. To avoid this issue, the major items (refer the definition below) in transactions are extracted and sorted in the descending order of their supports and then combined, as the first part, with the remaining items of the transactions.

Definition (Major Items): Major items are items which their supports are greater than or equal to a given threshold, called major threshold α. As usual, α is greater than the support thresholds ε.

Based on the above concepts, the pseudocode in Subsect. 2.2 is redesigned for the IPPC$^+$ tree construction as follows.

Algorithm 2: BuildIPPC⁺Tree

Input: Dataset D, root node R, major item list MI, threshold α
Output: An IPPC⁺ tree with root R, item list $\mathcal{L}$, MI

1. **If**($MI = \emptyset$) Sample on dataset D to achieve major item list MI based on the major threshold α;
2. **For Each** transaction $T \in D$
3.   Update items and their supports in $\mathcal{L}$ from items in T;
4.   Based on MI, major items in T are extracted and sorted in descending order of their supports, then combined as the first part with the rest of T;
5.   **InsertTransaction**(T, R);
6. **End For**
7. Update MI based on the item list $\mathcal{L}$;
8. **AssignPrePostOrder**(R);

Procedure InsertTransaction(Transaction T, Node R)
1. *parentNode* $\leftarrow R$;
2. **While**(($mergedNode$ = **Merge**(T, $parentNode$)) $\neq$ **null**)
3.   *parentNode* = *mergedNode*;
4. **End While**
5. **If**($T = \emptyset$) **Return**;
6. **For Each** child node N of *parentNode*
7.   **If**(**MergeDescendants**(T, N)) **Return**;
8. **End For**
9. Insert T as a new branch from *parentNode* (added nodes are initialized at 1 for their supports);

Function Merge(Transaction T, Node N)
1. *mergedNode* $\leftarrow$ **null**;
2. **For Each** item name I of T
3.   *mergedNode* $\leftarrow$ binary search for the child node in child list of N which its item name equals I;
4.   **If**(*mergedNode* $\neq$ **null**)
5.   *mergedNode.support*++;
6. $T \leftarrow (T \setminus I)$;
7.   **Return** *mergedNode*;
8.   **End If**
9. **End For**
10. **Return null**;

```
Function MergeDescendants(Transaction T, Node N)
1.    subNode ← N; mrgNode ← N; merged ← false;
2.    While(subNode satisfies the Child Swapping condition)
3.        descendant ← subNode.child;
4.        If(descendant.item-name ∈ T)
5.            T ← (T \ descendant.item-name); merged ← true;
6.            Exchange item names of mrgNode and descendant;
7.            mrgNode.support++; mrgNode ← mrgNode.child;
8.        End If
9.        subNode ← descendant;
10.   End While
11.   If(merged)
12.       Correct the position of node N in the child node list of
          its parent node;
13.       Insert T as a new branch from mrgNode.parent (added nodes
          are initialized at 1 for their supports);
14.   End If
15.   Return merged;
```

The IPPC$^+$ tree construction needs a major item list to preprocess transactions (line 4 of **BuildIPPC$^+$Tree**) before the transactions are inserted into the tree. The IPPC$^+$ tree construction is not aware of the major item list of the initial dataset. Therefore, it can be achieved by sampling (or even a full scan) on the dataset. The major item list MI of a dataset D is used as acknowledge to incrementally build up the tree (corresponding to dataset D) with a new additional dataset ΔD, and then MI is updated based on the item list $\mathcal{L}$ of the dataset $D + \Delta D$. The procedure **AssignPrePostOrder** is the same as the one in Subsect. 2.1.

To insert a transaction into the tree, the function **Merge** is executed first to find and merged the first items of the transaction with the tree nodes, the binary search is employed in this process. If there are still items in the transaction, the function **MergeDescendants** is done to merge these items with descendant nodes. When at least a merger between an item and a descendant node happens in the function **MergeDescendants**, this means the item name of node N is changed that may cause to lose the right order of child node list of N's parent node. Therefore, an order correction is done at line 12. Finally, the remaining items in the transaction (if possible) is inserted as a new branch into the tree at lines 13 and 9 in function **MergeDescendants** and procedure **InsertTransaction** respectively.

After the tree construction has completed, a mining process with the algorithm IFIN + [16] on the built tree will be executed.

3 Experiments

All experiments were conducted on a 1.86 GHz Intel Core (MT) i3-4030U processor, and 4 GB memory computer with Window 8.1 operating system. To evaluate algorithms, we used the Market-Basket Synthetic Data Generator [5] based on the IBM

Quest to generate a synthetic dataset, and a real dataset named Kosarak [6], online news portal click-stream data. The properties of the datasets are shown in Table 3.

Table 3. The datasets' properties

	No. of transactions	Max length	Average length	No. of total distinct items	No. of frequent items at thresholds		
					0.001	0.002	0.006
Synthetic dataset	1200000	32	10	932	843	774	525
Kosarak	990002	2498	8.1	41270	1260	568	116

All the algorithms were implemented in Java. Experimental values of running time and used memory are the average values from three corresponding individual ones. In our previous works [15, 16], to guarantee available memory of 2 GB used for Java Heap, we set value "-Xmx2G" for _JAVA_OPTIONS, a Windows environment variable. However, we realize that this causes the Java garbage collector to run many times of full memory collection; and consequently, the total running time includes a significant percentage, approximate 45%, for the garbage collection. To avoid this in the current work, we set the value "-Xms2G -Xmx2G" for _JAVA_OPTIONS instead, and result in the running time for garbage collection is reduced to 6% which reflects more exactly the algorithms' performance.

For emulating scenarios of incremental mining, the synthetic dataset was divided into six equal parts, 200 thousand transactions for each one, and so on for Kosarak dataset with five parts in which the last one contains just 190002 transactions. The experiments start mining on the first part and then part by part from the second one is accumulated and mined. IFIN$^+$ can perform following three scenarios:

- **S1** (Incremental in Different Sessions): An IPPC/IPPC$^+$ tree corresponding with a dataset had been constructed, mined and stored in a running session. In the following sessions, the old tree is loaded and then built up with a new additional dataset.
- **S2** (Incremental in the Same Session): An IPPC/IPPC$^+$ tree corresponding with a dataset has been constructed and mined, and then it is built up with a new additional dataset in the same session.
- **S3** (Just Loading Tree): A stored IPPC/IPPC$^+$ tree in a previous session is loaded and mined in the following sessions.

Each execution scenario can be performed with different support thresholds in the same running session. The processor in our computer possesses two physical computational units, and we found that the performance achieved its best with two threads in parallel version IFIN$^+$. We set the major threshold $\alpha = 0.02$ for the construction of IPPC$^+$ tree. The experiments will be presented in three parts: comparisons between the two trees IPPC and IPPC$^+$ with the inclusion of IFIN$^+$, and algorithm IFIN$^+$ with the IPPC$^+$ tree against algorithms FP-Growth, FIN, and PrePost$^+$ on each of the two datasets.

3.1 Comparisons Between IPPC and IPPC⁺ Trees

In this subsection, we present comparisons between two versions of the algorithm IFIN⁺ mining on IPPC and IPPC⁺ trees with the synthetic and Kosarak datasets, based on the running time of the four partial processing phases.

Table 4 reports the running time in seconds of the execution phases of the two version of IFIN⁺ on the synthetic dataset in data accumulation steps from 200k to 1200k transactions. As shown in Table 3, the number of distinct items in this dataset is small that the efficiency improvement of the tree construction with binary search is not enough to compensate the computational overhead of extracting and sorting the major items in each transaction. This causes the tree construction time to increase, but the amounts are not considerable, 0.3 s in average. Almost there are no performance differences in the second phase, the Frequent 2-itemset Generation. The extracting and sorting the major items in each transaction causes appearances of these items to tend to be lesser and nearer to the root node, that makes the lengths of *nodesets* [15, 16] of major items reduce. This explains why the running time is decreased in the third and the fourth phases, 0.7 s and 0.2 s in average respectively. Consequently, the total efficiency is improved, around 1 s for steps of data accumulation from 600k to 1200k transactions.

Table 4. Running time of IFIN⁺ mining on the IPPC/IPPC⁺ trees with the synthetic dataset

Running time in tree construction phase (in S1 Scenario)						
	200k	400k	600k	800k	1000k	1200k
IPPC tree	2.1 s	3 s	3.2 s	4.1 s	4.3 s	5.1 s
IPPC⁺ tree	2.8 s	3.2 s	3.5 s	4.2 s	4.5 s	5.5 s
Running time in frequent 2-itemset generation phase ($\varepsilon = 0.001$)						
	200k	400k	600k	800k	1000k	1200k
IFIN⁺ (IPPC tree)	0.4 s	0.5 s	0.8 s	0.9 s	1 s	1.4 s
IFIN⁺ (IPPC⁺ tree)	0.3 s	0.5 s	0.8 s	0.9 s	1 s	1.4 s
Running time in nodeset generation phase ($\varepsilon = 0.001$)						
	200k	400k	600k	800k	1000k	1200k
IFIN⁺ (IPPC tree)	1.4 s	2.4 s	3.7 s	4.3 s	6 s	6.3 s
IFIN⁺ (IPPC⁺ tree)	1.1 s	2 s	2.8 s	3.8 s	4.8 s	5.4 s
Running time in discover frequent k-itemsets phase ($\varepsilon = 0.001$)						
	200k	400k	600k	800k	1000k	1200k
IFIN⁺ (IPPC tree)	1.6 s	2.3 s	3.1 s	4.6 s	5.0 s	6.1 s
IFIN⁺ (IPPC⁺ tree)	1.5 s	2.3 s	2.9 s	4 s	4.7 s	5.8 s
Total running time (in S3 Scenario, $\varepsilon = 0.001$)						
	200k	400k	600k	800k	1000k	1200k
IFIN⁺ (IPPC tree)	5.7 s	8.5 s	11.6 s	14.8 s	17 s	20 s
IFIN⁺ (IPPC⁺ tree)	6 s	8.4 s	10.7 s	13.8 s	15.8 s	19 s

Table 5 presents the running time of the four processing phases for the two version of IFIN$^+$ on Kosarak dataset in data accumulation steps from 200k to 990002 transactions. The Kosarak dataset includes a large number of distinct items as reported in Table 3. Therefore, the IPPC$^+$ construction performance based on the binary search is improved significantly, much far the computational overhead for extracting and sorting the major items in each transaction. As a result, the tree construction time is reduced considerably from 29% to 41% in the sequence of data accumulation steps. While the running time in the second phase increases, the running time in the third and the fourth ones reduce. In general, the performance of the last three phases between the two versions is not much difference, except the accumulation step of 1000k transactions. For the total effect, the running time is reduced approximately 25% for all steps of data accumulation of Kosarak dataset.

Table 5. Running time of IFIN$^+$ mining on the IPPC/IPPC$^+$ trees with Kosarak dataset

Running time in tree construction phase (in S1 Scenario)					
	200k	400k	600k	800k	1000k
IPPC tree	7.2 s	10.4 s	11.4 s	13.9 s	14.7 s
IPPC$^+$ tree	5.1 s	6.8 s	7.2 s	8.3 s	8.7 s

Running time in frequent 2-itemset generation phase ($\varepsilon = 0.002$)					
	200k	400k	600k	800k	1000k
IFIN$^+$ (IPPC tree)	0.6 s	1.1 s	1.7 s	2.3 s	2.5 s
IFIN$^+$ (IPPC$^+$ tree)	0.6 s	1.4 s	2 s	2.6 s	3.2 s

Running time in nodeset generation phase ($\varepsilon = 0.002$)					
	200k	400k	600k	800k	1000k
IFIN$^+$ (IPPC tree)	0.2 s	0.3 s	0.5 s	0.6 s	0.8 s
IFIN$^+$ (IPPC$^+$ tree)	0.2 s	0.3 s	0.4 s	0.5 s	0.7 s

Running time in discover frequent k-itemsets phase ($\varepsilon = 0.002$)					
	200k	400k	600k	800k	1000k
IFIN$^+$ (IPPC tree)	1 s	2.4 s	3.4 s	4.8 s	6.7 s
IFIN$^+$ (IPPC$^+$ tree)	1 s	2.2 s	3.3 s	4.5 s	5.4 s

Total running time (in S3 Scenario, $\varepsilon = 0.002$)					
	200k	400k	600k	800k	1000k
IFIN$^+$ (IPPC tree)	9.2 s	14.7 s	17.4 s	22.1 s	25.3 s
IFIN$^+$ (IPPC$^+$ tree)	7.1 s	11 s	13.2 s	16.5 s	18.6 s

In an overview of experiments on both datasets, the performance of IFIN$^+$ using IPPC$^+$ tree is improved compared to that of the version using IPPC tree. The larger the number of distinguishing items and transactions in a dataset; the more the running time is saved for the tree construction beside the minor efficient improvement in the mining process.

3.2   Comparisons with Other Algorithms on the Synthetic Dataset

In this subsection, we benchmark the running time and the peak consumed memory of IFIN⁺ using IPPC⁺ tree against that of the three algorithms FP-Growth, FIN, and PrePost⁺ on the synthetic dataset. In that, the algorithm IFIN⁺ experiments with all its possible execution scenarios S1, S2, and S3 as referred.

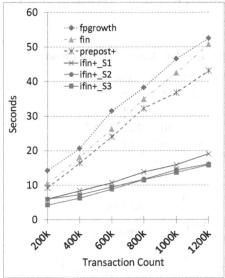
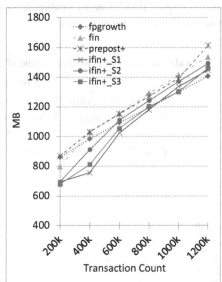

Fig. 2. Running time on the incremental synthetic datasets

Fig. 3. Peak memory on the incremental synthetic datasets

Figures 2 and 3 sequentially demonstrate the running time and the peak memory of the algorithms in steps of data accumulation at the support threshold $\varepsilon = 0.1\%$. For all algorithms, both running time and peak memory increase linearly when the dataset is accumulated. While FP-Growth is the slowest algorithm, it uses memory more efficient than FIN and PrePost⁺. Algorithm PrePost⁺ consumes the most memory, but it runs faster than FP-Growth and FIN. Algorithm IFIN⁺ is the most efficient for the running time. For data accumulation steps up to 600k transactions, IFIN⁺ uses memory more efficient than FP-Growth; and for remaining steps, FP-Growth takes this advantage, but not considerable. The slopes of the running time lines of IFIN⁺ are the same and lower than that of the three remaining algorithms. Hence, follow the data accumulation steps, the execution time of IFIN⁺ becomes more dominant, lesser than a haft, compared to the remaining algorithms'. Among the three execution scenarios of IFIN⁺, the running time of S2 and S3 is almost the same and better than S1's but not much difference.

Beside the high performance of mining phases in algorithm IFIN⁺, one more reason can be found out in Table 6 which reports the construction time of the four trees. Note that the IPPC⁺ tree construction does not depend on support threshold, but the other three trees. The POC, PPC trees of algorithms FIN and PrePost⁺ are almost the same, so their running time of the tree building is nearly equal. The IPPC⁺ tree construction of IFIN⁺ achieves the best performance, much better than the three algorithms'. Especially in scenario S3, the time ratios are approximately 1:7 and 1:6 compared to FP tree and PPC tree respectively. At the same dataset size, building tree in S3 is faster than that in S1, approximate 2.6 s in average; since the execution scenario S1 must build up the loaded tree with a new additional dataset of 200k transactions. This also reveals that constructing an IPPC⁺ tree by loading its stored data is much efficient than building the same tree from the same dataset.

Table 6. The tree construction time of the algorithms for the synthetic dataset

	200k	400k	600k	800k	1000k	1200k
IPPC⁺ tree (Scenario S1)	2.8	3.2	3.5	4.2	4.5	5.5
IPPC⁺ tree (Scenario S3)	0.5	0.7	1.2	1.6	2.1	2.2
FP tree ($\varepsilon = 0.001$)	3.4	5.7	8.7	10	12.9	16.2
POC/PPC ($\varepsilon = 0.001$)	2.5	4.5	6.7	9.7	11.9	14.7

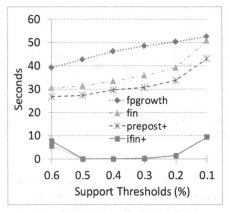 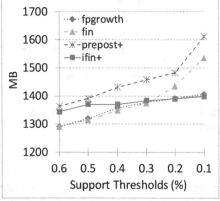

Fig. 4. Running time on the synthetic dataset at different support thresholds

Fig. 5. Peak memory on the synthetic dataset at different support thresholds

In Figs. 4 and 5, the running time and peak used memory are visualized for the algorithms mining on the synthetic dataset of 1.2 million transactions with different support thresholds ε. Start at $\varepsilon = 0.6\%$, IFIN⁺ can perform one of two scenarios S1 or S3 that their two running time values are shown in Fig. 4. For other ε values, IFIN⁺ just run its mining tasks since the built tree is completely reused. Furthermore, only a portion of its mining is performed. Consequently, with following values of $\varepsilon < 0.6\%$, the running time of IFIN⁺ takes an overwhelming dominance against that of the three algorithms. The memory used by IFIN⁺ increases slowly follow the steps of support thresholds, and approximates the memory used by FP-Growth for threshold values from 0.4 to 0.1. The algorithm FP-Growth uses memory more efficient than the two algorithms FIN and PrePost$^+$. However, its running time is considerably longer than that of FIN and PrePost$^+$. Algorithm PrePost$^+$ run faster than FIN and FP-Growth, but it uses the most memory.

3.3 Comparisons with Other Algorithms on Kosarak Dataset

Similar to the previous subsection, this one presents the running time and the used memory of the four algorithms for Kosarak dataset.

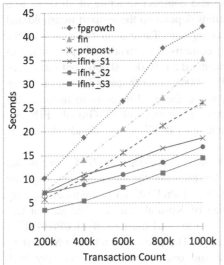
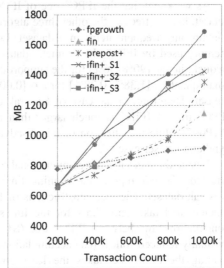

Fig. 6. Running time on the incremental Kosarak datasets

Fig. 7. Peak memory on the incremental Kosarak datasets

Figures 6 and 7 respectively visualize the algorithms' running time and peak used memory in data accumulation steps at the support threshold $\varepsilon = 0.2\%$. Like experiments on the synthetic dataset, for all algorithms, the running time and used memory increase linearly when the dataset is accumulated. Among the three algorithms FP-Growth, FIN and PrePost$^+$, the orders in the memory efficiency and the performance are similar to that in case of the synthetic dataset. FP-Growth still is the slowest algorithm, but it uses memory more efficient than PrePost$^+$ and FIN. While PrePost$^+$ runs remarkably faster than FP-Growth and FIN, it becomes to consume more memory than FP-Growth and FIN follow the data accumulation steps.

Shown in Fig. 6, the execution time of IFIN$^+$ is approximate that of PrePost$^+$ and FIN algorithms in the first two data accumulation steps; but for the following ones, IFIN$^+$ becomes to run faster than the others. In Fig. 7, IFIN$^+$ uses memory as good as the others at the dataset of 200k transactions. However, its consumed memory increases faster than the other algorithms' and is the most for larger sizes, approximate the PrePost$^+$'s memory at full size of Kosarak dataset in execution scenarios S1.

Table 7. Tree construction time of the algorithms for Kosarak dataset

	200k	400k	600k	800k	1000k
IPPC⁺ tree (Scenario S1)	5.1 s	6.8 s	7.2 s	8.3 s	8.7 s
IPPC⁺ tree (Scenario S3)	0.5 s	0.7 s	1.3 s	1.5 s	2.2 s
FP tree (ε = 0.002)	3.3 s	6.2 s	9.2 s	12.6 s	15.3 s
POC/PPC (ε = 0.002)	2.1 s	3.7 s	5.1 s	6.6 s	7.5 s

As we knew that the IPPC⁺ tree of IFIN⁺ is a compact structure of all items in a dataset, but the trees of the other three algorithms depend on the support threshold and contain only frequent items in a dataset. Looking into Table 3 for the reason of the memory used by IFIN⁺, the synthetic dataset comprises a considerable percentage of frequent items, [90%–56%] for the support threshold $\varepsilon \in [0.001–0.006]$; but just a very small quantity, [1.38%–0.28%] for $\varepsilon \in [0.002–0.006]$, is for frequent items in Kosarak dataset. Therefore, in the case of Kosarak dataset, the used memory to maintain the IPPC⁺ tree of IFIN⁺ is much larger than that of the trees of FP-Growth, FIN and PrePost⁺; while the affection of this disadvantage to IFIN⁺ is not considerable in the synthetic dataset case.

Beside the memory, the computational overhead to construct the IPPC⁺ tree is also affected. Table 7 reports the running time for building the trees of algorithms on Kosarak dataset. The tree constructions of FIN and PrePost⁺ on this dataset are very efficient and take only 7.5 s for the full size of Kosarak dataset. The gap in tree construction time between IPPC⁺ tree (S1 scenario) and the tree of FIN/PrePost⁺ is gradually reduced follow data accumulation steps. The tree construction in S3, just by loading the built tree, takes the least time and once again asserts its very high performance.

Figures 8 and 9 depict the running time and the peak memory of the algorithms mining on the full Kosarak dataset with different support thresholds ε. Start at $\varepsilon = 0.6\%$, IFIN⁺ can perform one of two scenarios S1 or S3 that their two running time values are shown in Fig. 8. For other ε values, IFIN⁺ just runs some portions of its mining tasks and reuses completely the built tree. Therefore, the same results as the corresponding experiments on the synthetic dataset in Fig. 4, the running time of IFIN⁺ takes an overwhelming advantage against that of the three remaining algorithms. IFIN⁺ consumes the most memory since it needs more memory to maintain IPPC⁺ tree. FP-Growth uses memory less efficient than the two algorithms FIN and PrePost⁺ for $\varepsilon > 0.3\%$. However, its consumed memory becomes lesser than other algorithms' for $\varepsilon \leq 0.3\%$. FP-Growth's running time is considerably longer than that of FIN and PrePost⁺. Algorithm PrePost⁺ runs faster than FIN, but this dominance of PrePost⁺ is not significant.

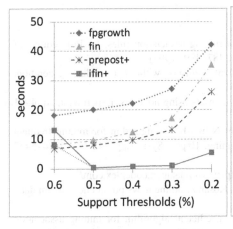

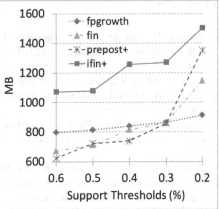

Fig. 8. Running time on Kosarak dataset at different support thresholds

Fig. 9. Peak memory on Kosarak dataset at different support thresholds

4 Conclusions

In this paper, we proposed an improved version of the IPPC tree, called IPPC$^+$, to enhance the performance of the tree construction. Beside the minor improvement in mining performance, the experiments show that with a datasets comprising a small number of distinct items, the tree construction performance of the two versions are approximate to each other; but in case of a dataset with a large number of distinguishing items, the tree construction performance of IPPC$^+$ tree is improved remarkably compared to that of IPPC tree. This contributes to significantly reducing the disadvantage in the tree construction phase of algorithm IFIN$^+$ compared to other algorithms such as FIN, PrePost$^+$ in cases of datasets with a huge number of distinct items but just a small percentage of frequent items.

Experiments also demonstrated that IFIN$^+$ is superior in performance compared to the three remaining algorithms, especially in mining circumstances when its incremental characters take effect. This provides IFIN$^+$ an efficient way to deal with the high-velocity property of Big Data and the data mining practices which often try with different threshold values.

In case of a dataset including a huge number of distinct items but just a small percentage of frequent items, IFIN$^+$ algorithm needs more memory than the other algorithms to retain its tree structure of all items in the dataset. However, when mining with small enough support thresholds, the gap in memory overhead between IFIN$^+$ and the other algorithms will be reduced; and when a dataset is more and more accumulated, all algorithms must face with the problem of memory scalability besides the running time. Therefore, as a potential approach, a distributed parallelization solution for IFIN$^+$ will be proposed to better confront with these problems of Big Data.

144 V. Q. P. Huynh and J. Küng

References

1. Agrawal, R., Srikant, R.: Fast algorithms for mining association rules. In: Proceedings of 20th International Conference on VLDB, pp. 487–499 (1994)
2. Han, J., Pei, J., Yin, Y.: Mining frequent itemsets without candidate generation. ACM SIGMOD Rec. **29**(2), 1–12 (2000)
3. Deng, Z.-H., Lv, S.-L.: Fast mining frequent itemsets using nodesets. Expert Syst. Appl. **41** (10), 4505–4512 (2014)
4. Deng, Z.-H., Lv, S.-L.: PrePost$^+$: an efficient N-lists-based algorithm for mining frequent itemsets via children-parent equivalence pruning. Expert Syst. Appl. **42**(13), 5424–5432 (2015)
5. Market-Basket Synthetic Data Generator. https://synthdatagen.codeplex.com/
6. Frequent Itemset Mining Dataset Repository: Kosarak, online news portal click-stream data. http://fimi.ua.ac.be/data/kosarak.dat.gz
7. Savasere, A., Omiecinski, E., Navathe, S.: An efficient algorithm for mining association rules in large databases. In: VLDB, pp. 432–443 (1995)
8. Perego, R., Orlando, S., Palmerini, P.: Enhancing the *Apriori* algorithm for frequent set counting. In: Kambayashi, Y., Winiwarter, W., Arikawa, M. (eds.) DaWaK 2001. LNCS, vol. 2114, pp. 71–82. Springer, Heidelberg (2001). https://doi.org/10.1007/3-540-44801-2_8
9. Park, J.S., Chen, M.S., Yu, P.S.: Using a hash-based method with transaction trimming and database scan reduction for mining association rules. IEEE Trans. Knowl. Data Eng. **9**(5), 813–825 (1997)
10. Zaki, M.J.: Scalable algorithms for association mining. IEEE Trans. Knowl. Data Eng. **12**(3), 372–390 (2000)
11. Grahne, G., Zhu, J.: Fast algorithms for frequent itemset mining using FP-trees. Trans. Knowl. Data Eng. **17**(10), 1347–1362 (2005)
12. Liu, G., Lu, H., Lou, W., Xu, Y., Yu, J.X.: Efficient mining of frequent itemsets using ascending frequency ordered prefix-tree. DMKD J. **9**(3), 249–274 (2004)
13. Shenoy, P., Haritsa, J.R., Sudarshan, S.: Turbo-charging vertical mining of large databases. In: 2000 SIGMOD, pp. 22–33 (2000)
14. Zaki, M.J., Gouda, K.: Fast vertical mining using diffsets. In: 9th SIGKDD, pp. 326–335 (2003)
15. Huynh, V.Q.P., Küng, J., Dang, T.K.: Incremental frequent itemsets mining with IPPC tree. In: Benslimane, D., Damiani, E., Grosky, W.I., Hameurlain, A., Sheth, A., Wagner, R.R. (eds.) DEXA 2017. LNCS, vol. 10438, pp. 463–477. Springer, Cham (2017). https://doi.org/10.1007/978-3-319-64468-4_35
16. Huynh, V.Q.P., Küng, J., Jäger, M., Dang, T.K.: IFIN$^+$: a parallel incremental frequent itemsets mining in shared-memory environment. In: Dang, T.K., Wagner, R., Küng, J., Thoai, N., Takizawa, M., Neuhold, E.J. (eds.) FDSE 2017. LNCS, vol. 10646, pp. 121–138. Springer, Cham (2017). https://doi.org/10.1007/978-3-319-70004-5_9

A Sample-Based Algorithm for Visual Assessment of Cluster Tendency (VAT) with Large Datasets

Le Hong Trang[✉], Pham Van Ngoan, and Nguyen Van Duc

Faculty of Computer Sicence and Engineering,
Ho Chi Minh City University of Technology, VNU-HCM,
268 Ly Thuong Kiet, District 10, Ho Chi Minh City, Vietnam
lhtrang@hcmut.edu.vn

Abstract. In this paper, a sampled-based version of the visual assessment of cluster tendency (VAT) algorithm [2] for large datasets is presented. The proposed algorithm consists of two main steps. We first propose a postprocessing task of the ProTraS algorithm [9] to obtain a sample of the dataset such that clusters in the sample are separated as much as possible while preserving the cluster structure of the whole dataset. The second one is to apply iVAT [5] on the sample to display the cluster tendency of the whole dataset. Algorithms are implemented. Numerical results are given and compared with siVAT to demonstrate the efficiency of our algorithm.

Keywords: Clustering · Cluster tendency · Sampling · VAT

1 Introduction

Clustering is the process of grouping a set of objects such that the objects in each group are similar to others than to objects in different groups. Clustering is a fundamental technique in data mining for exploring data and also called unsupervised learning. Many clustering algorithms and their variants have been proposed (see [3, 11, 12] for surveys on the clustering technique). Most of proposed techniques concentrate on how to separate objects into proper groups. Many algorithms, for example the family of k-means, require the number of clusters as an input. On the other hand, for other approaches, knowing an approximate number of clusters can help a clustering algorithm not only to speed up the process, but also to enhance its accuracy. Therefore, it is important to estimate a number of clusters before applying a suitable technique for the cluster analysis.

The visual assessment of cluster tendency (VAT) algorithm was introduced by Bezdek and Hathaway to determine whether cluster are presents in a given dataset [2]. The algorithm is to visualize cluster structures in relational matrices among objects of the dataset. VAT rearranges unlabeled objects so that similar ones will be located nearby, it then highlights the cluster structure of a dataset in

© Springer Nature Switzerland AG 2018
T. K. Dang et al. (Eds.): FDSE 2018, LNCS 11251, pp. 145–157, 2018.
https://doi.org/10.1007/978-3-030-03192-3_11

an intuitive image. VAT, thereby, helps to understand the essentially structural feature of the dataset without a priori knowledge. The VAT algorithm [2] takes a pairwise dissimilarity matrix of a dataset, denoted by D, as the input. Based on Prim's algorithm for finding the minimum spanning tree of a weighted graph [10], VAT determines a potential partition of the dataset using lengths of edges added to the tree, as in single-linkage clustering. A reordered matrix of D, denoted by D^*, is formed by the obtained partition. VAT finally visualizes D^* by a grayscale image $I(D^*)$, where each pixel of the image displays the scaled dissimilarity value of a pair of objects. The cluster tendency is indicated by the "dark blocks" along the diagonal of the image. The computational detail of VAT is described in Algorithm 1.

Algorithm 1. VAT [2]

Input: $D : n \times n$ pairwise dissimilarity matrix.
Output: D^*: reordered matrix.

$K = \{1, 2, \ldots, n\}, I = J = \emptyset$, and $P = (0, 0, \ldots, 0)$.
Select $(i, j) \in \arg\max_{p,q \in K} D_{pq}$.
Set $P(1) = i, I = \{i\}$, and $J = K \setminus \{i\}$.
for $r = 2, \ldots, n$ **do**
 Select $(i, j) \in \arg\min_{p \in I, q \in J} D_{pq}$.
 Set $P(r) = j$.
 $I \leftarrow I \cup \{j\}$ and $J \leftarrow J \setminus \{j\}$
end for
$D^*_{pq} = D_{P(p),P(q)}$, for all $1 \leq p, q \leq n$
return D^*.

A number of variants of the VAT algorithm have been introduced to deal with datasets of irregular structure and large size [4–8,13,14]. Wang et al. [14] proposed an improved VAT (iVAT) algorithm in which they applied a path-based distance in VAT. iVAT thus can produce clearly the images of datasets with highly complex cluster structures. This algorithm is then revised by Havens et al. [5] by reducing the computational complexity of the iVAT distance transformation. A disadvantage of both VAT is size limitation. Hathaway et al. [4] introduced the scalable VAT (sVAT) to solve the problem of large datasets. sVAT find a subset, called a sample, of the input. It then applies VAT to the sample and shows the corresponding reordered dissimilarity image. An extension of VAT can also be achieved by combining sVAT and iVAT, called siVAT, to overcome the problem of large size in iVAT. However, the method for sampling in sVAT need an overestimate of the true but unknown number of clusters. Furthermore, through it forms groups in the dataset, sample points is then chosen randomly in each group. The representativeness of the sample is thus not always ensured. Consequently, the VAT image of the sample does not reflect exactly the cluster tendency of the whole dataset.

Ros and Guillaume have just introduced a probabilistic traversing sampling algorithm called ProTraS [9]. The algorithm is based on farthest-first tracersal principle in which a representative is selected due to the highest probability of cost reduction. The sample consists of all representatives and preserves the cluster structure of datasets. In this paper, a sample-based VAT algorithm for large dataset is presented. We fist introduce a postprocessing task of the ProTraS algorithm for sampling a large dataset, and then apply iVAT to the resulting sample for visual assessment of cluster tendency. In particular, each point of a sample obtained by ProTraS is replaced by the center of set of patterns represented by the point. Thereby, we aims to move the sample inward to the inside of clusters. The VAT image of the new sample is thus sharper, while keeping the main structure of clusters. Therefore, the cluster tendency of a large dataset should be clearly displayed.

The rest of the paper is organized as follows. The next section briefly describes some related works including sVAT, iVAT, and ProTraS algorithms. Section 3 describes our algorithm. Numerical experiments are given in Sect. 4. Section 5 concludes our work.

2   Related Works

As mentioned in the previous section, the ordinary VAT algorithm suffers from the limitation of dataset size. sVAT [4] was introduced to overcome this limitation. This actually is a sample-based version of VAT for dealing with large datasets. Given a dataset, we suppose that c is true but unknown number of clusters. Let c, be an overestimate value of c', i.e., $c > c'$. sVAT determines c' distinguished objects in the dataset. These objects are tried to choose such that they can present in all clusters of the datasets. For each determined object, sVAT groups remaining objects with it, resulting c' groups in the dataset. In each group, a subsample size is computed and then select randomly a number of the subsample size of objects. The sample is then formed by combining subsamples of the groups. Finally, VAT is applied on the obtained sample. Algorithm 2 states steps of sVAT.

Wang et al. [14] proposed an improved VAT (iVAT) algorithm that deals with datasets including irregular clusters, where the Euclidean distance can not be applied as the measurement between two objects in the dataset. iVAT uses a path-based distance. We can consider D to be the representation of the weights of the edges of a fully connected graph. Let P_{ij} be the set of all paths connecting two vertices of the graph that corresponds to objects o_i and o_j in the dataset. For a path $p \in P_{ij}$, we denote by $p[h]$ and $|p|$ the index of the hth vertex and the number of vertices along p. Then, $D_{p[h]p[h+1]}$ is the weight of the hth edge along of p. The path-based distance is defined by

$$D'_{ij} = \min_{p \in P_{ij}} \max_{1 \le h \le |p|} D_{p[h]p[h+1]}. \tag{1}$$

By (1), we can understand that the length of p is specified by the maximum weight of edges of p. The distance between o_i and o_j is the minimum length

Algorithm 2. sVAT [4]

Input: $D : n \times n$ pairwise dissimilarity matrix, c': an overestimate of the (unknown) number of clusters, and n': sample size.
Output: $D_{n'}^*$: reordered dissimilarity matrix of the sample.

Select $m_1 = 1$.
Initialize $d = (d_1, d_2, \ldots, d_n) = (r_{1,1}, r_{1,2}, \ldots, r_{1,n})$
for $t = 2, \ldots, c'$ do
    $d = \left(\min\{d_1, r_{m_{t-1},1}\}, \min\{d_2, r_{m_{t-1},2}\}, \ldots, \min\{d_n, r_{m_{t-1},n}\} \right)$.
    Select $m_t \in \arg\max_{1 \le j \le n}\{d_j\}$.
end for
Initialize $S_1 = S_2 = \ldots = S_{c'} = \emptyset$.
for $= 1, \ldots, n$ do
    Select $k \in \arg\min_{r_{m_j,t}}$.
    $S_k \rightarrow S_k \cup \{t\}$.
end for
Initialize $S = \emptyset$.
for $t = 1, \ldots, c'$ do
    $n_t = \lceil \frac{n'|S_t|}{n} \rceil$.
    Choose randomly n_t indices in S_t.
    $S \rightarrow S \cup S_{n_t}$.
end for
Form D_N, the $n' \times n'$ submatrix of D corresponding to the row/column indices in S
Apply VAT to D_N to obtain D_N^*.
return D_N^*.

of all paths in P_{ij}. Computing D' in (1) might need to determine the shortest paths in the graph. There are n^2 pairs of vertices in the graph that we need to compute the shortest path connecting between them. The computational cost is thus expensive. Havens and Bezdek [5], in 2012, proposed an efficient formulation that reduces the computational complexity for (1) from $O(n^3)$ to $O(n^2)$. This revised version of iVAT is given by Algorithm 3.

Algorithm 3. iVAT [5]

Input: D^*: reordered matrix.
Output: $D'^* = [0]^{n \times n}$.

for $r = 2, \ldots, n$ do
    $j = \arg\min_{k=1,\ldots,r-1} D_{rk}^*$.
    $D_{rc}'^* = D_{rc}^*$ and $c = j$.
    $D_{rc}'^* = \max\{D_{rj}^*, D_{jc}'^*\}, c = 1, 2, \ldots, r-1, c \ne j$.
end for
$D_{rc}'^* = D_{cr}'^*$, since D'^* is symmetric.
return D'^*.

Our sampling method used in the next section uses the ProTraS algorithm given in [9]. The main idea of ProTraS is to select a representative point based on a probability of cost reduction. Given an $\epsilon > 0$, for each iteration of the algorithm, it adds a new representative into a group of the sample with highest probability of the cost reduction. When the cost comes below a threshold which depends on ϵ, the algorithm stops. The detail of the algorithm is given in Algorithm 4. Lines 4–7 of the algorithm find the nearest group for points that

---

Algorithm 4. ProTraS [9]

Input: $P = \{x_i\}$, for $i = 1, 2, \ldots, n$, a tolerance $\epsilon > 0$.
Output: A sample $S = \{y_j\}$ and $P(y_j)$, for $j = 1, 2, \ldots, s$.

Initialize a pattern $x_{init} \in P$.
$y_1 = x_{init}, P(y_1) = \{y_1\}, S = \{y_1\}$, and $s = 1$.
repeat
    for all $x_i \in P \setminus S$ **do**
        $y_k = \arg\min_{y_j \in S} d(x_l, y_j)$.
        $P(y_k) = P(y_k) \cup \{x_l\}$.
    end for
    $maxWD = cost = 0$.
    for all $y_k \in S$ **do**
        $x_{max}(y_k) = \arg\max_{x_i \in P(y_k)} d(x_i, y_k)$.
        $d_{max}(y_k) = d(x_{max}(y_k), y_k)$.
        $p_k = |P(y_k)| d_{max}(y_k)$.
        if $p_k > maxWD$ **then**
            $maxWD = p_k$.
            $y^* = y_k$.
        end if
        $cost = cost + p_k/n$.
    end for
    $x^* = x_{max}(y^*)$.
    $S = S \cup \{x^*\}$ and $s = s + 1$.
    $P(y^*) = \{x^*\}$.
until $cost < \epsilon$
return S and $P(y_j)$, for $j = 1, 2, \ldots, s$.

---

have not yet assigned to any group of the current sample. The point among them is determined to be new representative if it is farthest in its group and has also highest probability (Lines 9–18). This also means that the representative selected by ProTraS is the farthest-first traversal item.

3   Proposed Algorithm

As sVAT needs an overestimate c' the true number of clusters of a dataset, while the number is unknown, it is difficult to determine a proper value of c'. Even such a value of c', sampling technique in sVAT only covers on all clusters of the dataset if the clusters are is compact and separated (CS), i.e., each of the possible intra-cluster distances is strictly less than each of the possible intercluster distances [4]. However, a few datasets in practice have CS clusters. On the other hand, randomly choosing objects in groups S_t can not ensured the representative of sample, for $t = 1, 2, \ldots, c'$.

Since the ProTraS algorithm is based on the farthest-first traversal, it is high potential to cover all clusters of the dataset. Furthermore, an object is selected due to the highest probability of cost reduction which is defined by combining distance and density concepts. This means that selected objects in the sample are high representativeness. The termination condition of ProTraS depends only on the sampling cost, that is a given $\epsilon > 0$. It was shown in [9] that the sample obtained by ProTraS is a ϵ-coreset, a concept defined for geometric approximation of point set [1]. This allows us to generate samples in a consistent and manageable way. A sample by ProTraS can thus preserves the cluster structure of a dataset. However, this is not always useful for assessment of cluster tendency. If the value of ϵ is small, the number of objects in the sample is too. Due to the probability of cost reduction, the distribution of the objects should be uniform. In the case that the intercluster distances of the considering dataset is small, the distances between all pairs of objects in the sample are quite similar. It is the difficult to highlight the separation of clusters. Consequently, the grayscale image of the reorder dissimilarity matrix of the sample is not sharp distinction. The result of visual assessment of cluster tendency is thus not good.

In order to overcome the difficulty mentioned above, we propose a postprocessing task of ProTraS. We replace a representative in the sample obtained by ProTraS by the center of the group represented by it. Thereby, objects located at the boundary side of clusters will be replaced by interior ones of those. The new obtained sample thus should has separated clusters. This helps to improve the quality of VAT image. The detail of our algorithm is given by Algorithm 5. By ProTraS, for each representative $y_j \in S$, we have a set of patterns $P(y_j)$

Algorithm 5. Our algorithm

Input: $P = \{x_i\}$, for $i = 1, 2, \ldots, n$, a tolerance $\epsilon > 0$.
Output: A sample S and D'^*.

1: Call ProTraS for P and ϵ to obtain $S = \{y_j\}$ and $P(y_j)$.
2: $S' = \emptyset$.
3: **for all** $y_j \in S$ **do**
4:     $y_k^* = \arg\min_{y_k \in P(y_j)} \sum_{y_l \in P(y_j)} d(y_k, y_l)$.
5:     $S' = S' \cup \{y_k^*\}$.
6: **end for**
7: Form D^* the reordered matrix conrrespoding to S'.
8: Apply iVAT on D^* to obtain D'^* and produce $I(D'^*)$.
9: **return** S and D'^*.

---

represented by y_j. Then, the center of $P(y_k)$ can be determined by

$$y_k^* = \arg\min_{y_k \in P(y_j)} \sum_{y_l \in P(y_j)} d(y_k, y_l).$$

The algorithm then computes the reordered matrix D^* and applies iVAT on D^* for the visual assessment of cluster tendency.

4   Numerical Experiments

All algorithms are implemented in Matlab and run on Platform Mac OSX, Processor 2.5 GHz Core i5, 8 GB Memory. Datasets used for testing are from repositories such as SIPU[1] and Deric clustering[2] benchmarks. We tested also for some large simulated datasets.

We fist implemented the ProTraS algorithm and the proposed postprocessing task for sampling a number of large datasets. Table 1 shows results tested for 31 datsets, including both benchmark and simulated ones, for $\epsilon = 0.1$ and 0.2. As indicated in [9], the sample size mainly depends on the structure of data, but not the data size. Namely, the sample size should be larger as the structure of data is more complex. For example, though the data sizes of FLAME and JAIN are smaller compared with the others, the size ratio of their samples are larger (see Fig. 1).

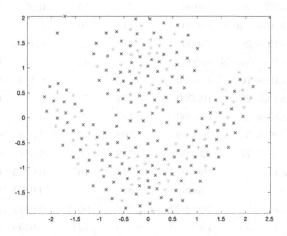

Fig. 1. The sample of FLAME dataset with $\epsilon = 0.1$, the size ratio is 69.17%.

---

1   cs.joensuu.fi/sipu.
2   github.com/deric/clustering-benchmark.

Table 1. Sample size with $\epsilon = 0.1$ and 0.2.

Ord.	Dataset	Data size (T)	Sample size (S)		Ratio S/T (%)	
			$\epsilon = 0.1$	$\epsilon = 0.2$	$\epsilon = 0.1$	$\epsilon = 0.2$
1	A.set 1	3000	261	97	8.7	3.23
2	A.set 2	5250	315	116	6	2.21
3	A.set 3	7500	341	119	4.55	1.59
4	FLAME	240	166	90	69.17	37.5
5	Birch-set 3	100000	424	153	0.424	0.153
6	JAIN	373	108	56	28.95	15.01
7	S.sets 1	5000	237	96	4.74	1.92
8	S.sets 2	5000	327	120	6.54	2.4
9	S.sets 3	5000	422	155	8.44	3.1
10	S.sets 4	5000	448	166	8.96	3.32
11	Dim sets 1	1351	17	10	1.26	0.74
12	Dim sets 2	2701	17	11	0.63	0.41
13	Dim sets 3	4051	20	8	0.49	0.2
14	Dim sets 4	5401	416	17	7.7	0.31
15	Dim sets 5	6751	379	19	5.61	0.28
16	data5k-CS	5000	44	17	0.88	0.34
17	data5k-NonCS	5000	264	95	5.28	1.9
18	data10k-CS	10000	25	10	0.25	0.1
19	data10k-NonCS	10000	114	40	1.14	0.4
20	data15k-CS	15000	61	22	0.41	0.145
21	data15k-NonCS	15000	111	44	0.74	0.293
22	data100k-10	100000	103	45	0.103	0.045
23	data100k-25	100000	191	73	0.191	0.073
24	data100k-27	100000	187	79	0.187	0.079
25	data200k-5	200000	108	44	0.054	0.022
26	data200k-17	200000	162	62	0.081	0.031
27	data1M	1000000	315	107	0.0315	0.0107
28	data1M-7	1000000	84	41	0.0084	0.0041
29	data1M-15	1000000	142	60	0.0142	0.006
30	data1M-55	1000000	355	131	0.0355	0.0131
31	data2M-77	2000000	457	159	0.023	0.008

We now demonstrate an example of our sampling for the VAT problem. Figure 2 displays results obtained by ProTraS tested for A.set 1 in the SIPU repository. The size of this dataset is 3000. For $\epsilon = 0.1$, the sample size is 261. The size should be reduced, i.e. 97, for $\epsilon = 0.2$. This is because of higher tolerance in ProTraS (see Algorithm 4). As shown in Fig. 2(b), in case of $\epsilon = 0.2$, the sample points are located at the boundary side of clusters. This does not well support for the cluster tendency assessment. Our method overcomes this difficulty by replacing each sample point by the center of group represented by the point. The obtained sample is displayed in Fig. 3(b). Such a sample would help to clearly visualize the cluster tendency as a sharp distinction between clusters in the sample.

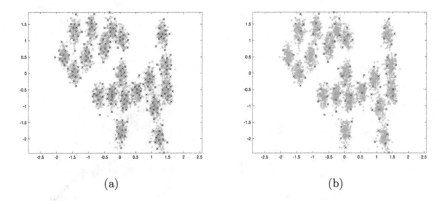

(a)   (b)

Fig. 2. The sample sets of A.set 1 with $\epsilon = 0.1$ (a) and $\epsilon = 0.2$ (b).

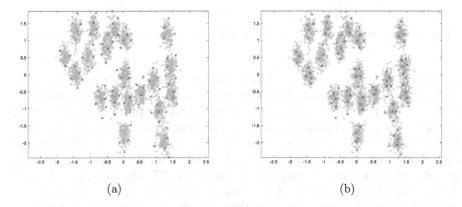

(a)   (b)

Fig. 3. The sample sets of A.set 1 obtained by ProTraS (a) and our method (b).

Figure 4 shows the VAT image of our algorithm (b) compared with that of siVAT (d) for A.set 1 dataset. Since the sample points are located toward the inside of clusters, the intra-cluster distances within the sample are increased. Hence, the VAT image is sharper. It is easier to determine potential clusters which are indicated by blocks along the diagonal in (b) than that in (c). We note also that A.set 1 dataset is not compacted and separated. This example

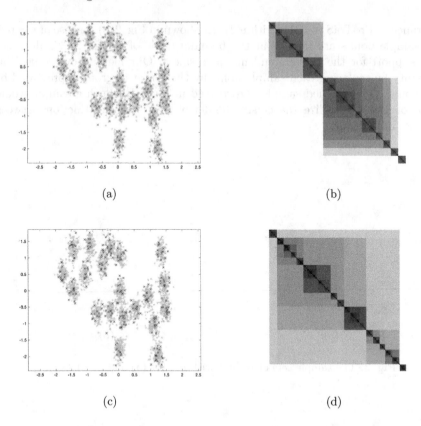

(a)   (b)

(c)   (d)

Fig. 4. The samples and VAT images: our method (a) and (b); siVAT (c) and (d).

thus demonstrate the efficiency of our algorithm. On the other hand, the sample is uniformly distributed. Therefore, the main structure of whole dataset is preserved (see Fig. 4(b)). This is also indicated when applying for datasets of irregular structure. As showed in Fig. 5(a), the number of clusters of the dataset is 9. The structure is complex, where the shapes of clusters are different and complicated. By our algorithm, the VAT image (Fig. 5(b)) not only presents potential clusters which is sharper than that by siVAT (Fig. 5(d)), but also well displays the cluster structure of the dataset. In siVAT, the sample points are randomly chosen in each group. It is thus difficult to always keep the structure of a dataset, specially with irregular structures.

Finally, we demonstrate the efficiency of our algorithm for large datasets. The last five rows in Table 1 report the details for very large datasets, with $\epsilon = 0.1$ and 0.2. The sample ratios are very small. It, however, can be used for a good visual assessment of cluster tendency. An example tested for data200k-5 dataset (25th row in Table 1) with the size of 200, 000 is displayed in Fig. 6. The value of ϵ is chosen to be 0.1. The size of obtained sample is 108 (the ratio is 0.054%), and

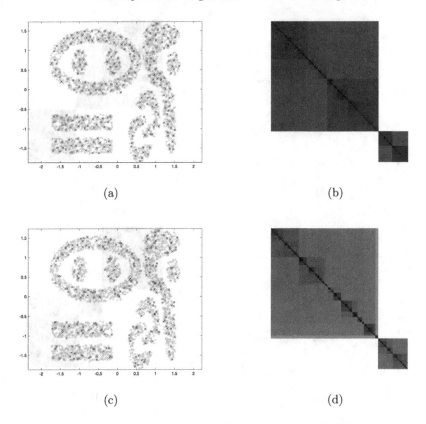

Fig. 5. The test for complex-9 dataset. (a) The sample and (b) VAT image by our method with $\epsilon = 0.1$. (c) The sample and (d) VAT image by siVAT.

the computing time is approximately 1.7 (s). We can see three clusters located on the top are overlap. In the case, the sample obtained by our algorithm are more distinct that that of siVAT. Therefore, the corresponding VAT image (Fig. 6(b)) is again sharper. The structural feature of the whole dataset also is presented more explicitly than that by siVAT (Fig. 6(d)). We note also that the sample size by siVAT in this case is more than twice of that by our algorithm. This means that, for large datasets, the improvement of our algorithm in both sample size and quality of VAT images is significant.

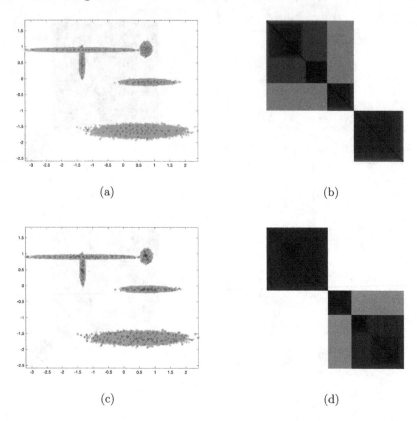

(a)   (b)

(c)   (d)

Fig. 6. The test for data200k-5 dataset ($n = 200,000$). (a) The sample and (b) VAT image by our method, $n' = 108$. (c) The sample and (d) VAT image by siVAT, $n' = 262$.

5   Concluding Remarks

The paper presented a sampled-based version of the visual assessment of cluster tendency (VAT) algorithm for large datasets. A postprocessing task of the Pro-TraS is introduced to obtain a sample of the dataset such that clusters in the sample are separated as much as possible while preserving the cluster structure of the whole dataset. Numerical experiments were carried out on a number of large datasets to show the efficiency of proposed algorithm. However, due to ProTraS the sampling in our algorithm is also based on farthest-first traversal. This means that, in the case of datasets with high noise or outlier, the algorithm might not be robust. We can overcome this challenge by removing from the sample some objects that their representativeness are low. This will be the topic in another work.

Acknowledgments. This research is funded by Ho Chi Minh City University of Technology – VNU-HCM under grant **number To-KHMT-2017-09**.

References

1. Agarwal, P.K., Har-Peled, S., Varadarajan, K.R.: Geometric approximation via coresets. Comb. Comput. Geom. **52**, 1–30 (2005)
2. Bezdek, J., Hathaway, R.: VAT: a tool for visual assessment of (cluster) tendency. In: Proceedings of International Joint Conference on Neural Networks (IJCNN), Honolulu, HI, USA, pp. 2225–2230 (2002)
3. Fahad, A., et al.: A survey of clustering algorithms for big data: taxonomy and empirical analysis. IEEE Trans. Emerg. Top. Comput. **2**(3), 267–279 (2014)
4. Hathaway, R., Bezdek, J., Huband, J.: Scalable visual asseessment of cluster tendency for large data sets. Pattern Recogn. **39**(7), 1315–1324 (2006)
5. Havens, T.C., Bezdek, J.C.: An efficient formulation of the improved visual assessment of cluster tendency (IVAT) algorithm. IEEE Trans. Knowl. Data Eng. **24**(5), 813–822 (2012)
6. Honda, K., Sako, T., Ubukata, S., Notsu, A.: Visual assessment of co-cluster structure through co-occurrence-sensitive ordering. In: Proceedings of Joint 17th World Congress of International Fuzzy Systems Association and 9th International Conference on Soft Computing and Intelligent Systems (IFSA-SCIS), Otsu, Japan, pp. 1–6 (2017)
7. Huband, J.M., Bezdek, J.C., Hathaway, R.J.: bigVAT: visual assessment of cluster tendency for large data sets. Pattern Recogn. **38**, 1875–1886 (2005)
8. Iredale, T. B., Erfani, S. M., Leckie, C.: An efficient visual assessment of cluster tendency tool for large-scale time series data sets. In: Proceedings of IEEE International Conference on Fuzzy Systems (FUZZ-IEEE), Naples, Italy (2017)
9. Ros, F., Guillaume, S.: ProTraS: a probabilistic traversing sampling algorithm. Expert Syst. Appl. **105**, 65–76. https://doi.org/10.1016/j.eswa.2018.03.052 (2018)
10. Prim, R.: Shortest connection networks and some generalisations. Bell Syst. Tech. J. **36**, 1389–1401 (1957)
11. Xu, D., Tian, Y.: A comprehensive survey of clustering algorithms. Ann. Data Sci. **2**(2), 165–193 (2015)
12. Xu, R., Wunsch, D.: Survey of clustering algorithms. IEEE Trans. Neural Netw. **16**(3), 645–678 (2005)
13. Wang, L., Geng, X., Bezdek, J., Leckie, C., Kotagiri, R.: SpecVAT: enhanced visual cluster analysis. In: Proceedings of the Eighth IEEE International Conference on Data Mining, Pisa, Italy, pp. 638–647 (2008)
14. Wang, L., Nguyen, U. T. V., Bezdek, J., Leckie, C., Ramamohanarao, K.: iVAT and aVAT: enhanced visual analysis for cluster tendency assessment. In: Proceedings of Pacific-Asia Conference on Knowledge Discovery and Data Mining (PAKDD), Hyderabad, India, pp. 16–27 (2010)

An Efficient Batch Similarity Processing with MapReduce

Trong Nhan Phan$^{(\boxtimes)}$ and Tran Khanh Dang

Faculty of Computer Science and Engineering, HCMC University of Technology,
VNU-HCM, Ho Chi Minh City, Vietnam
{nhanpt,khanh}@hcmut.edu.vn

Abstract. In this paper, we study an efficient way for batch similarity processing with MapReduce. With the inverted index as a backbone, we embed metadata inside the indexes to minimize redundant data so as to build lightweight indexes from the data sources. In addition, we propose a general query batch processing scheme that not only handles a single query but also deals with sets of query in an incremental manner. Moreover, we build the indexes in an ordered fashion so that we can perform quick pruning discarding unnecessary objects and supporting the performance of similarity search. Last but not least, we measure our proposed solution by conducting empirical experiments on real datasets. The results verify the efficiency of our method when we do similarity search with query batches, especially when both query sets and data sets are large.

Keywords: Similarity search · Batch processing
Lightweight indexing · MapReduce

1 Introduction

Similarity search is the principle operation not only in databases but also in inter-disciplinary fields of study such as machine learning, recommendation systems, biology, or data analytics. The main goal of similarity search is to find those object similar to the given pivot, know as the query. The similarity computing, unfortunately, suffers high overheads due to distance calculation pair-by-pair together with similarity metrics [16]. It is more inevitable especially when data grow bigger and have higher dimensionality. As a result, the challenge is how to speed up the similarity searching process to enjoy its benefits.

There are several approaches to achieve the goal by improving indexing [2,13], hashing [11], filtering [14], or processing in parallel [1], for example. Among them, the parallel processing approach in distributed environment calls much attention to researchers worldwide [5,7,8,12]. This trend keeps motivating both academia and industry due to large amount of data. Other approaches may fail to assure scalability when processing massive datasets. As a result, we catch the trend by

© Springer Nature Switzerland AG 2018
T. K. Dang et al. (Eds.): FDSE 2018, LNCS 11251, pp. 158–171, 2018.
https://doi.org/10.1007/978-3-030-03192-3_12

employ MapReduce, a large-scale processing paradigm [3], when enhancing the performance of similarity search.

Even though MapReduce helps us process enormous data, it would suffer heavy overheads when processing big unnecessary or irrelevant objects. The scenario becomes even worse when those big objects are involved in similarity search. In other words, those redundant objects are combined with every other object to evaluate their similarity pair-by-pair. Moreover, the MapReduce-based process is strictly bound by I/O costs, processing irrelevant or unnecessary data leads to extra penalty.

Meanwhile, those recent literatures only deal with a single similarity query. Consequently, when given a query batch, each query is processed one by one, which slows down the whole performance of batch processing. In fact, We observe that queries in the batch may share their search space. As a consequence, it would be better to search the same shared search space for all queries in the batch rather than looking for the search space for one query and then redo it several times for other queries.

In our work, we, therefore, take batch processing into account rather than single query processing. Furthermore, we propose our strategies to make our MapReduce-based solution more effective. Specifically, our main contributions are as follows:

- We present a query batch processing scheme that not only handles sets of query but also does similarity search in an incremental way.
- We introduce a simple but efficient method to support quick pruning in similarity search by sorted inverted indexes.
- We propose an indexing scheme with metadata so that we can diminish duplicate data and then build lightweight indexes.
- We perform our empirical experiments on real datasets. The results verify the efficiency of our proposed solution when it does similarity search with query batches.

The rest of paper is organized as follows. Section 2 presents our related work. Additionally, Sect. 3 introduces our background related to similarity search and MapReduce paradigm. In Sect. 4, we propose our solution for similarity search in general and that with query batches in particular. After that, we conduct our experiments in Sect. 5 before making our remarks in Sect. 6.

2   Related Work

Metwally and Faloutsos introduce a method for all-pair similarity joins of multisets and vectors [5]. Their method is composed of two main phases, each consumes two MapReduce jobs. While a MapReduce job is costly, the more MapReduce jobs the more costs. Additionally, their method does not consider duplicate data items during the execution of MapReduce job, which usually adds extra penalty. Besides, Tang et al. present their way, so-called HA-index, to speed up

Hamming-based distance computing for range queries [12]. In addition, redundancy during the computing is also eliminated. Their whole process consumes three phases with two MapReduce jobs. Nevertheless, the costs for data preprocessing and post-processing are excluded from the MapReduce jobs.

Gao et al. bring up efficient and scalable metric similarity joins with MapReduce [4]. They focus on the load balancing and how to avoid unnecessary object pairs with their filtering methods, including the range-object filtering, the double-pivot filtering, the pivot filtering, and the plane sweeping techniques, so that they can achieve better query performance. In the meantime, Phan et al. propose an efficient hybrid similarity search with MapReduce [8]. Their basic idea is to firstly cluster similar objects and secondly define upper and lower boundaries to shrink the search space before looking similar pairs. The method is then deploy in a hybrid MapReduce-based architecture that deals with challenges from big data. In addition, their empirical studies show that their method is efficient in terms of data processing and storage. Though their method works well with batch processing, each query is sequentially processed in a batch with regards to their search scheme while we have indexing strategy for query batches supporting quick similarity search.

Nguyen et al. build VP-tree algorithm on top of the MapReduce framework to achieve good performance, scalability, and fault tolerance for similarity search over the large datasets in the distributed environment [6]. Moreover, their method can reduce the number of data that need to scan during the search phase. In contrast, our approach is towards MapReduce-based scheme-driven algorithms that are independent of the underlying MapReduce framework. By doing this, we are able to gain the two main advantages as follows: (1) No internal or additional changes from the framework; and (2) Mutual support from both the top algorithms and the underlying framework.

3 Preliminaries

3.1 Similarity Search

Similarity search is the operation that looks for similar objects when implicitly or explicitly given a pivot. For example, a corpus, denoted as Ω, consists of a set of document objects D_p, which is formally represented as $\Omega = \{D_1, D_2, D_3, \ldots, D_n\}$. In addition, each document object D_p is composed of a set of words, which is shown as $D_p = \{word_1, word_2, word_3, \ldots, word_w\}$. When given a document object Q_j, the similarity search computes how much similar the pair (Q_j, D_p) is for every document objects D_p in the corpus.

To evaluate how similar a pair is, a similarity measure such as Euclidean distance, Cosine similarity, Hamming distance, or Jaccard coefficient is used to quantifies their similarity [16]. The similarity score is usually standardized into the interval [0, 1] in that the pair is more similar when its similarity score is close to 1 while that is less similar when its similarity score is near 0. Moreover, a similarity threshold, like 80% similarity is provided to filter those pairs whose similarity scores are greater or equal to 0.8.

In the meantime, we observe that modeling the content of a document as a set of words does not reflect much how really similar a pair is because the two same words in different objects do not bring the same meaning. To better improve a part of semantic similarity, a concept of K-shingles [9], known as any sub-string having the length K found in the document, is used instead. As a consequence, each document object D_p is composed of a set of K-shingles, which is shown as $D_p = \{S_1, S_2, S_3, \ldots, S_z\}$.

In general, a similarity search process consists of two main phases as follows:

1. Candidate generation phase. This is the phase where two objects are identified as a candidate pair.
2. Candidate verification phase. This is the phase where the pair is evaluated for its similarity score.

3.2   MapReduce

MapReduce is a parallel paradigm for large-scale processing [3]. The philosophy behind is to apply "divide-and-conquer" strategy to data. A large data is split into different smaller data chunks, which are then processed at various machines. The intermediate results generated by each machine are aggregated into the final result. In order to implement this strategy, a MapReduce job is composed of a Map task and Reduce task in that the former is specified by a Map function while the latter is determined by a Reduce function. When a MapReduce job is executed on a cluster of commodity machines, those machines assigned Map tasks called mappers whereas those assigned Reduce tasks known as reducers. A Map task emits intermediate key-value pairs while a Reduce task writes the final key-value pairs into the distributed file system. Moreover, there is a shuffle phase between the Map task and the Reduce task, which re-distributes data based on the output keys by the mappers.

Suppose that there are M mappers and R reducers, a single MapReduce job is described as follows:

1. Input data is loaded into the distributed file system and is then divided into partitions based on their data size.
2. Mappers read their data partitions, perform the Map function, and emit intermediate results in the form of key-value pair $[k_i, v_j]$. These intermediate key-value pairs are locally stored at mappers.
3. The shuffle process aggregates the intermediate key-value pairs $[k_i, [v_j]]$ into R data partitions, which is based on they key values.
4. Reducers retrieve the intermediate key-value pairs $[k_i, [v_j]]$ from R data partitions and perform the Reduce function. The final output is written back to the distributed file system.

4   Our Proposed Solution

4.1   Query Batch Processing Scheme

In this paper, we introduce our general similarity search scheme as illustrated in Fig. 1. Either data or query objects are indexed into either data or query pools, respectively. We employ inverted index as an index data structure. In addition, the indexes are organized in an ordered way to serve our quick pruning strategy later on. Moreover, we use one MapReduce job for indexing data objects. For instance, a set of data object $\{D_p\}$ is indexed into the data pool in the form of Sorted Inverted Index (SII), known as SII(D_p). Similarly, a set of query object $\{Q_j\}$ is indexed into the query pool in the form of Sorted Inverted Index, known as SII(Q_j). After the indexing phase, both data and queries are ready for similarity search. Later on, one MapReduce job computes the similarity (SIM) among queries against data and produces the final result in the form SIM(Q_j, D_p). In our work, we employ Jaccard coefficient, a well-known metric for fast set-based similarity [5,8,10,15], to derive the similarity score of a pair as described in the Eq. 1 below.

$$SIM(Q_j, D_p) = \frac{\| Q_j \cap D_p \|}{\| Q_j \cup D_p \|} \tag{1}$$

In that, $\| Q_j \cap D_p \|$ is the intersection cardinality between Q_j and D_p while $\| Q_j \cup D_p \|$ is the union cardinality between Q_j and D_p.

With the proposed scheme, we can perform similarity search in an incremental manner. The reason is that both data and query objects are available in the data and query pools in the form of SII. Hence, if there is new data or query objects, the pools will include them in the form of SII. Then, a MapReduce job for similarity search can be configured to compute a set of SII in the pools as required. Furthermore, the general scheme is applied not only to single query processing but also to query batches.

4.2   MapReduce-Based Similarity Search

Our MapReduce-based similarity search following the above scheme consists of two main phases: (1) Indexing; and (2) Searching. In the former phase, we will index data and query objects into the pool in the form of SII while doing the similarity search with Jaccard measure in the latter phase. Last but not least, we model our documents as bags of 4-shingles rather than sets of words [8,9].

Figure 2 shows an example of data indexing by a MapReduce job. Assume that we have three data documents $D_p = [D_1, D_2, D_3]$ with their corresponding shingle-based contents. The Map task is to emit intermediate key-value pairs in the form of $[SH_p, URL_p]$ in that SH_p is a shingle of a document D_p and URL_p is the path location of D_p in the distributed environment. It is worth noting that duplicate shingles from the same document are discarded because they do not contribute to the similarity scores with Jaccard measure. We, thus, filter them at this Map task to avoid additional overheads after that. The Reduce task then produces SII(D_p) in the form of $[SH_u, [URL_v @ S_v]]$. It is worth noting that we need to keep the size S of those documents so that we can derive their similarity scores later on. Likewise, Fig. 3 implies an example of query indexing by a MapReduce job. Assume that we have three query documents $Q_j = [Q_1, Q_2, Q_3]$ with their corresponding shingle-based contents. The Map task is to emit intermediate key-value pairs in the form of $[SH_j, URL_j]$ whereas the Reduce task produces SII(Q_j) in the form of $[SH_u, [URL_v @ S_v]]$.

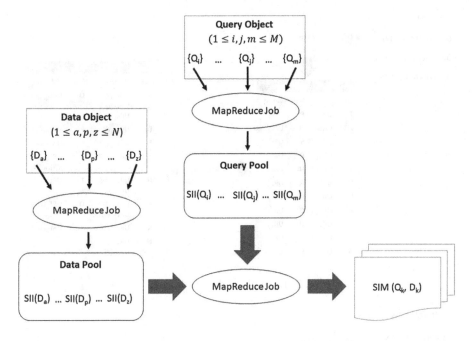

Fig. 1. Query batch processing scheme

The similarity search phase is done by one MapReduce job. Figure 4 illustrates the Map task with $SII(D_p)$ and $SII(Q_j)$. It compares key-by-key and emits the pair whenever they share the same shingles in the form of $[URL_j - URLp, (S_j + S_p)]$. To sooner discard unnecessary pairs, we apply our quick pruning strategy in the comparison to speed up the searching process. Due to the fact that we already organize $SII(D_p)$ and $SII(Q_j)$ in an ordered way, we can achieve the two following advantages for our quick pruning during the comparison:

1. We can stop the comparison half-way whenever $SH_p > SH_j$.
2. We can discard those shingles from the comparing set $SII(D_p)$ whenever $SH_p < SH_j$. By doing this way, we can reduce the size of the comparing set $SII(D_p)$ during the comparison.

Finally, the Reduce task aggregates the pairs emitted by the Map task and computes their similarity scores. As shown in Fig. 5, we will have 8 pairs with their corresponding similarity scores.

4.3   Lightweight Indexing

By observing, we see that there are duplicate values in the indexes. Whenever a pair shares the same shingle, the value of the form $URL_v @S_v$ emerges. In the distributed environment, a URL may be long due to its location path. Therefore, those repeated values make the size of indexes bigger. As a result, they add to

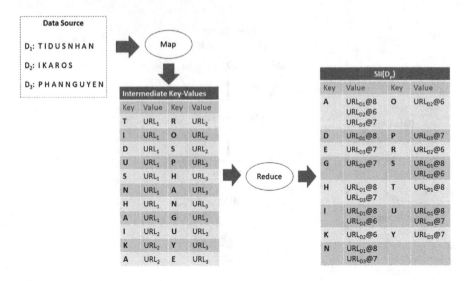

Fig. 2. Example of data indexing

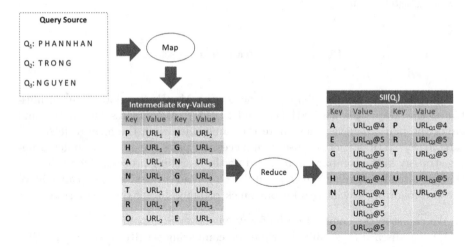

Fig. 3. Example of query indexing

the heavy cost of MapReduce-based processing since it is strictly bound by I/O cost. To minimize the redundancy and speed up the MapReduce-based searching process, we propose a metadata-based approach as follows.

1. We build a list L of document URLs as metadata in the form of $\{URL_v@S_v\}$. The metadata is put as header of each file produced by Reduce task.
2. For each value in the inverted index, we replace it with the index of L with regards to the corresponding document URL.

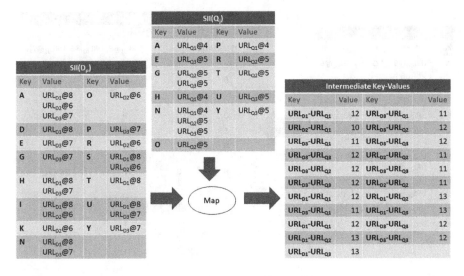

Fig. 4. Example of Map task

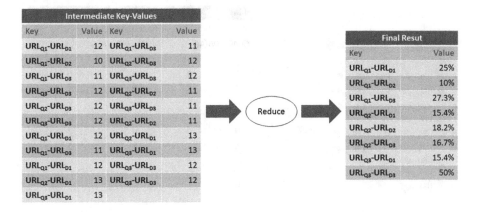

Fig. 5. Example of Reduce task

Figure 6 shows an example of metadata of data set while Fig. 7 gives an example of metadata of query set. For instance, we build the metadata from the data set as $\{URL_{D1}@8, URL_{D2}@6, URL_{D3}@7\}$. Consequently, the pair [A, [$URL_{D1}@8$, $URL_{D2}@6$, $URL_{D3}@7$]] is replaced by [A, [0, 1, 2]]. Meanwhile, we build the metadata from the query set as $\{URL_{Q1}@4, URL_{Q2}@5, URL_{Q3}@5\}$. Consequently, the pair [N, [$URL_{Q1}@4$, $URL_{Q2}@5$, $URL_{Q3}@5$]] is replaced by [N, [0, 1, 2]]. When the data input is large, the indexes with our metadata is much more lightweight than those without that. Our experiments in Sect. 5 show how much lightweight they are and how efficient they devote to speeding up the searching process.

SII(D_p)

Key	Value	Key	Value
A	$URL_{D1}@8$ $URL_{D2}@6$ $URL_{D3}@7$	O	$URL_{D2}@6$
D	$URL_1@8$	P	$URL_{D3}@7$
E	$URL_{D3}@7$	R	$URL_{D2}@6$
G	$URL_{D3}@7$	S	$URL_{D1}@8$ $URL_{D2}@6$
H	$URL_{D1}@8$ $URL_{D3}@7$	T	$URL_{D1}@8$
I	$URL_{D1}@8$ $URL_{D2}@6$	U	$URL_{D1}@8$ $URL_{D3}@7$
K	$URL_2@6$	Y	$URL_{D3}@7$
N	$URL_{D1}@8$ $URL_{D3}@7$		

SIIWMD(D_p)

MetaData Key		MetaData Value	
@METADATA@		$\{URL_{D1}@8, URL_{D2}@6, URL_{D3}@7\}$	
Key	Value	Key	Value
A	0 1 2	O	1
D	0	P	2
E	2	R	1
G	2	S	0 1
H	0 2	T	0
I	0 1	U	0 2
K	1	Y	2
N	0 2		

Fig. 6. Metadata of data set

SII(Q_i)

Key	Value	Key	Value
A	$URL_{Q1}@4$	P	$URL_{Q1}@4$
E	$URL_{Q3}@5$	R	$URL_{Q2}@5$
G	$URL_{Q2}@5$ $URL_{Q3}@5$	T	$URL_{Q2}@5$
H	$URL_{Q1}@4$	U	$URL_{Q3}@5$
N	$URL_{Q1}@4$ $URL_{Q2}@5$ $URL_{Q3}@5$	Y	$URL_{Q3}@5$
O	$URL_{Q2}@5$		

SIIWMD(Q_i)

MetaData Key		MetaData Value	
@METADATA@		$\{URL_{Q1}@4, URL_{Q2}@5, URL_{Q3}@5\}$	
Key	Value	Key	Value
A	0	P	0
E	2	R	1
G	1 2	T	1
H	0	U	2
N	0 1 2	Y	2
O	1		

Fig. 7. Metadata of query set

5   Empirical Experiments and Evaluations

5.1   Environmental Setting

We deploy Hadoop-based 2-node cluster on a PC, whose each node has 2.00 GB RAM and 50 GB HDD. The PC has Intel® Core™ i5-4460, 3.20 GHz CPU, 8.00 GB RAM, 500 GB HDD, and 64-bit operating system. Additionally, the Hadoop version is 2.7.3[1] run with its default settings. Nevertheless, we set the number of reducers to 4, which is based on 4 CPU cores.

---

[1] https://hadoop.apache.org/docs/r2.7.3/.

5.2   Dataset

We employ datasets retrieved from Gutenberg Project[2], an on-line data storage with over 56.000 free e-Books, for our experiments. The datasets are randomly chosen from the storage and organized into different data as well as query packages, which is illustrated in Table 1. The data type servers as the data input for similarity search while the query type is the pivot input to look for its similar documents from the data type. For the data type, we organize 5 different data packages as D5, D100, D300, D500, and D1K with 5, 100, 300, 500, and 1000 files, respectively. As randomly chosen, the file size ranges from 1 to 102 KB. In the meantime, the query type is organized into 4 different query packages as Q1, Q10, Q100, and Q1K with 1, 10, 100, and 1000 files, respectively. In addition, the query size range is 59 KB.

Table 1. Data organization

Type	Package no.	No. of files	Size range (KB)
Data	D5	5	40–102
	D100	100	15–59
	D300	300	1–32
	D500	500	1–32
	D1K	1000	1–59
Query	Q1	1	59
	Q10	10	59–59
	Q100	100	59–59
	Q1K	1000	59–59

5.3   Measurement

We compare the two different methods as follows:

- *Sorted Inverted Index (SII)*. This method follows our proposed similarity search scheme to build sorted inverted indexes for both data sources and query batches. In addition, the method performs query processing with our pruning strategy.
- *Sorted Inverted Index With MetaData (SIIWMD)*. This method follows our proposed similarity search scheme to build sorted inverted indexes with metadata organization. Additionally, the method performs query processing with our pruning strategy.

---

[2] http://www.gutenberg.org/.

5.4   Evaluation

In our first experiment, we measure the performance of the two comparing methods, known as SIIWMD and SII, for indexing query batches. Figure 8a shows query indexing time when the sorted inverted indexes are built for query batches. In general, the processing time of the two methods is not much different with Q10 and Q100. In fact, SII tends to have less query indexing time than SIIWMD when the number of queries sharply increases due to the fact that it does not suffer overheads for metadata organization. For example, the gap is around 4.55% with Q1K. Meanwhile, Fig. 8b indicates the query indexing size between the two comparing methods. Generally, SIIWMD generates sorted inverted indexes much lighter than SII. When the query batch size increases from Q10, Q100 to Q1K, SIIWMD saves nearly 12 times more data output than SII on the average. As a result, SIIWMD generates much more lightweight indexes than SII.

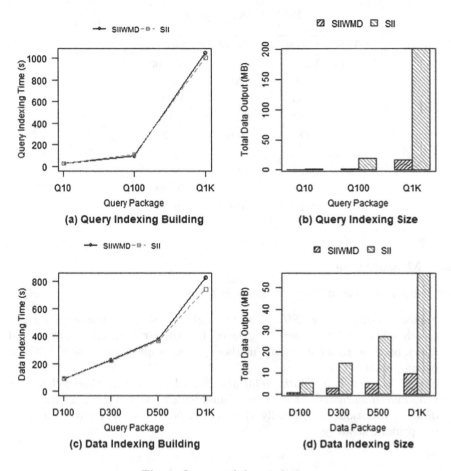

Fig. 8. Query and data indexing

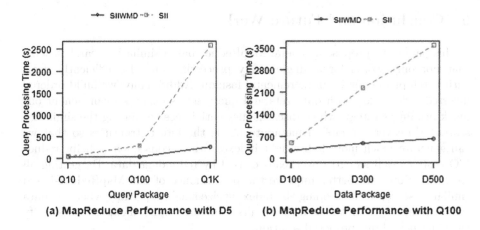

(a) MapReduce Performance with D5   (b) MapReduce Performance with Q100

Fig. 9. Query processing

To experience the indexing building with larger dataset, we do the same experiment for data packages. As illustrated in Fig. 8c, we observe that the indexing time between SIIWMD and SII is not much different when the dataset is small as with D100, D300, and D500. Nevertheless, when the dataset size is large as with D1K, the gap is around 10.37%. In the meantime, the result from Fig. 8d keeps enforcing the fact that SIIWMD saves much more data output when building indexes than SII. On the average, SIIWMD saves nearly 6 times more data output than SII.

In terms of query processing, we do our next experiments with different query and data packages. Figure 9a shows the MapReduce performance with D5. With the small number of queries such as Q10, the query processing time of SIIWMD is nearly 33% faster than that of SII. Nevertheless, when the query size rapidly increases, the performance gap between them is much bigger. More specifically, the performance gap sharply rises as 85.7% with Q100 whereas it is 90.28% with Q1K. Consequently, SIIWMD processes query batches 6 times much faster than SII does on the average. In parallel, Fig. 9b displays the MapReduce performance with Q100 when the dataset changes. Generally, SIIWMD gives much faster query processing time than SII does. When the dataset size is small as with D100, SII processes Q100 nearly 2 times slower than SIIWMD does. In addition, the query processing time of SII sharply rises when the dataset size becomes larger. More concretely, the performance gap sharply rises as 79% with D300 whereas it is 82.5% with D500. Furthermore, the query processing time of SIIWMD slightly increases while that of SII dramatically rises when the dataset size rapidly grows. As a consequence, SIIWMD processes query batches 4 times much faster than SII does on the average.

6   Conclusion and Future Work

In this paper, we propose a general MapReduce-based similarity search scheme that not only works for a single query processing but also efficiently deals with batch processing in an incremental fashion. Additionally, we build ordered inverted indexes for both data sets and query sets so that we can benefit our quick pruning strategy, which discards inessential accesses during the similarity search. Moreover, we embed metadata inside the index structures so that we can generate much more lightweight indexes, which consequently helps reduce I/O costs as well as extra overheads caused by redundant data. By getting all to work as one, we better improve the performance of our MapReduce-based similarity search while keeping the indexing size small, especially when the data set becomes larger. In the end, the results from our empirical experiments verify the efficiency of our proposed solution.

In our future work, we are going to apply our proposed method to the variants of similarity queries. Moreover, we plan to experience larger dataset size as well as the cluster size for large-scale similarity processing. Last but not least, we consider the load-balancing problem whose solution further improves the overall performance of MapReduce-based similarity search.

Acknowledgments. This research is funded by Ho Chi Minh City University of Technology-VNU-HCM, under the grant number T-KHMT-2017-50.

References

1. Alabduljalil, M.A., Tang, X., Yang, T.: Optimizing parallel algorithms for all pairs similarity search. In: Proceedings of the 6th ACM International Conference on Web Search and Data Mining, pp. 203–212 (2013)
2. Dang, T.K., Küng, J., Wagner, R.: The SH-tree: a super hybrid index structure for multidimensional data. In: Proceedings of the 12th International Conference on Database and Expert Systems Applications, pp. 340–349 (2001)
3. Dean, J., Ghemawat, S.: MapReduce: simplified data processing on large clusters. J. Commun. ACM **51**(1), 107–113 (2008)
4. Gao, Y., Yang, K., Chen, L., Zheng, B., Chen, G., Chen, C.: Metric similarity joins using MapReduce. IEEE Trans. Knowl. Data Eng. **29**(3), 656–669 (2017)
5. Metwally, A., Faloutsos, C.: V-SMART-join: a scalable mapreduce framework for all-pair similarity joins of multisets and vectors. Proc. VLDB Endow. **5**(8), 704–715 (2012)
6. Nguyen, D.T.-T., Yong, C.H., Pham, X.Q., Nguyen, H.Q., Loan, T.T.K., Huh, E.N.: An index scheme for similarity search on cloud computing using MapReduce over Docker container. In: Proceedings of the 10th International Conference on Ubiquitous Information Management and Communication, pp. 60:1–60:6 (2016)
7. Phan, T.N., Küng, J., Dang, T.K.: An adaptive similarity search in massive datasets. In: Hameurlain, A., Küng, J., Wagner, R., Dang, T.K., Thoai, N. (eds.) Transactions on Large-Scale Data- and Knowledge-Centered Systems XXIII. LNCS, vol. 9480, pp. 45–74. Springer, Heidelberg (2016). https://doi.org/10.1007/978-3-662-49175-1_3

8. Phan, T.N., Küng, J., Dang, T.K.: eHSim: an efficient hybrid similarity search with MapReduce. In: Proceedings of the 30th IEEE International Conference on Advanced Information Networking and Applications, pp. 422–429. IEEE Computer Society (2016)

9. Rajaraman, A., Ullman, J.D.: Finding similar items (Chap. 3). In: Mining of Massive Datasets, pp. 71–127. Cambridge University Press, Cambridge (2011)

10. Rong, C., Lu, W., Wang, X., Du, X., Chen, Y., Tung, A.K.H.: Efficient and scalable processing of string similarity join. IEEE Trans. Knowl. Data Eng. **25**(10), 2217–2230 (2013)

11. Satuluri, V., Parthasarathy, S.: Bayesian locality sensitive hashing for fast similarity search. Proc. VLDB Endow. **5**(5), 430–441 (2012)

12. Tang, M., Yu, Y., Aref, W.G., Malluhi, Q.M., Ouzzani, M.: Efficient processing of Hamming-distance-based similarity-search queries over MapReduce. In: Proceedings of 18th International Conference on Extending Database Technology, pp. 361–372 (2015)

13. Wang, J., Li, G., Deng, D., Zhang, Y., Feng, J.: Two birds with one stone: an efficient hierarchical framework for top-k and threshold-based string similarity search. In: 31st IEEE International Conference on Data Engineering, pp. 519–530 (2015)

14. Xiao, C., Wang, W., Lin, X., Yu, J.X., Wang, G.: Efficient similarity joins for near-duplicate detection. ACM Trans. Syst. **6**(3), 15:1–15:41 (2011)

15. Zadeh, R.B., Goel, A.: Dimension independent similarity computation. J. Mach. Learn. Res. **14**(1), 1605–1626 (2013)

16. Zezula, P., Amato, G., Dohnal, V., Batko, M.: Similarity Search - The Metric Space Approach. Advances in Database Systems, vol. 32, XVIII, pp. 1–220. Springer, Heidelberg (2006). https://doi.org/10.1007/0-387-29151-2. ISBN 0-387-29146-6

Vietnamese Paraphrase Identification Using Matching Duplicate Phrases and Similar Words

Hoang-Quoc Nguyen-Son[1,2(✉)], Nam-Phong Tran[1], Ngoc-Vien Pham[1], Minh-Triet Tran[1], and Isao Echizen[3]

[1] University of Science, Ho Chi Minh, Vietnam
{nshquoc,tmtriet}@fit.hcmus.edu.vn
[2] KDDI Research Inc., Saitama, Japan
[3] National Institute of Informatics, Tokyo, Japan
iechizen@nii.ac.jp

Abstract. Paraphrase identification is a core component for many significant tasks in natural language processing (e.g., text summarization, headline generation). A method suggested by Bach et al. for detecting Vietnamese paraphrase text using nine similarity metrics. The authors state that it is the first method for Vietnamese text. They evaluated the method on vnPara corpus with 3000 sentence pairs. However, this corpus is limited by collecting from few Vietnamese websites. Most other methods have focused on the English text. For instance, our previous method detected paraphrasing sentences by matching identical phrases and close words using Wordnet similarity. This method is unsuitable for Vietnamese due to the restriction of Wordnet corpora and morphological words in Vietnamese. Therefore, we extend the method to identify the paraphrase by proposing a $SimVN$ metric which measures the similarity of two Vietnamese words. We evaluated the proposed method on the vnPara corpus. The result shows that the method achieves better accuracy (97.78%) comparing with the state-of-the-art method (accuracy = 89.10%). The proposed method then creates a high diversity paraphrase corpus with 3134 sentence pairs in eight main topics from the top fifteen popular Vietnamese news websites.

Keywords: Paraphrase identification · Wordnet · Similarity
Matching · Duplicate phrases · Similar words

1 Introduction

Two texts (paragraphs, sentences, phrases) are called as a paraphrase if they are written in different ways, but their meanings are same. For instance, a paraphrasing sentence pair is posted in two various Vietnamese news websites:

© Springer Nature Switzerland AG 2018
T. K. Dang et al. (Eds.): FDSE 2018, LNCS 11251, pp. 172–182, 2018.
https://doi.org/10.1007/978-3-030-03192-3_13

s_1: "*Việt Nam được dự đoán sẽ có vị thế riêng của mình trong cuộc cách mạng Khoa học Công nghệ lần số 4 nhờ lực lượng nhân sự Công nghệ Thông tin trẻ và tài năng.*" (Translation: "*Vietnam is predicted to have its own place in the 4th Science and Technology Revolution based on young and talented Information Technology people.*")

s_2: "*Với lực lượng Công nghệ Thông tin dồi dào và có xu hướng trẻ hóa các chuyên gia dự đoán Việt Nam sẽ có ưu thế trong cuộc cách mạng Khoa học Công nghệ lần số 4.*" (Translation: "*With plentiful information technology people and a tending rejuvenate, experts predict that Vietnam will have the advantage of the 4th Science and Technology Revolution.*")

Paraphrasing identification has a major role in various significant natural languages processing applications such as text summarization, sentence simplification, header generation, and plagiarism detection. Such identification is used to filter and index numerous text data in a search engine or corpora creation. It is thus significant to develop an enhanced method for determining whether two texts are a paraphrase or not. Especially, various Vietnamese websites commonly post numerous news on the same topic. This redundant information makes web users actually confused.

However, there is only one research on identifying Vietnamese paraphrasing text suggested by Bach et al. [1]. Their method uses six common similarity distance (Jaro-Winkler, Levenshtein, Manhattan, Euclidean, Cosine, and N-gram) and three standard coefficients including Matching, Dice, and Jaccard. These metrics are considered as the nine features to create the classifier which is used to identify paraphrase texts. The drawback of this method is that it does not exploit the semantics of the text.

The state-of-the-art method in Vietnamese [1] is evaluated on vnPara corpus which includes 3000 paraphrase-labeled Vietnamese sentence pairs. 1500 pairs are annotated as paraphrase whereas the others are marked as non-paraphrase. Such pairs are limited in few Vietnamese websites. The corpus is extracted from the few Vietnamese news websites such as Vnexprees[1], Thanhnien[2]. Moreover, their topics are not mentioned in the corpus.

Most other researches have focused on English. These methods use sophisticated techniques such as deep learning [3,7,14], machine translation [8,10], parsing tree [5,13]. Our previous method detects the paraphrase by matching identical phrases and close words [12]. The similarity of close words is estimated by Wordnet, is a large English lexical database, with abundance and consistency semantic relationships between these items. However, the similar databases in Vietnamese are still restricted.

In this paper, we propose a method to distinguish a paraphrase with non-paraphrase text. Our contributions are listed as follows:

– We propose a *SimVN* metric to quantify the similarity of two Vietnamese words based on a restricted Vietnamese Wordnet database [4].

---

[1] http://vnexpress.net/.

[2] http://thanhnien.vn/.

- We develop a method to identify Vietnamese paraphrasing text by matching duplicate phrases and similar words using the $SimVN$ metric.
- We use the proposed method to create a high diversity Vietnamese paraphrase corpus from top fifteen news websites in eight primary topics.

We evaluated our proposed method on the vnPara corpus. The result archives better accuracy (97.78%) when comparing with the state-of-the-art method (accuracy = 89.10%). The method is used to identify the paraphrase sentence pairs from about 65000 news from top fifteen Vietnamese websites ranked by Alexa[3] in the most eight popular topics. 3134 identified pairs are annotated by a native speaker. 2748 pairs of the sentences (87.68%) are labeled as same-meaning whereas the others are tagged as different. This corpus is proved higher diversity than the vnPara one.

The remaining content is divided into four sections. Section 2 summarized some main related work. The proposed method is described in Sect. 3. Some experiments to evaluate our method are shown in Sect. 4. Finally, Sect. 5 points out some main key points and mentions future work.

2   Related Work

Paraphrase identification task has interested by numerous researchers. However, there is only one method suggested by Bach et al. for Vietnamese [1] according to the authors. They used nine similarity metrics including Levenshtein distance, Jaro-Winkler distance, Manhattan distance, Euclidean distance, cosine similarity, n-gram distance (with $n = 3$), matching coefficient, Dice coefficient, and Jaccard coefficient. These metrics are evaluated on vnPara corpus with support vector machine classifiers. The method archive the accuracy = 89.10%. The limitation of this method is that it is based on the statistical analysis. To overcome this weakness, the system needs to combine semantic resources such as ontology to increase the ability to handle ambiguity.

The first Vietnamese paraphrase corpus vnPara [1] which contains pairs of sentences is extracted from Vietnamese popular online webpages such as vnexpress.net, thanhnien.com.vn. Each page is preprocessed through several steps of natural language processing, including sentences separate (VnSentDetector1), words separate (vnTokenizer2) and detect part of the speech (VnTagger3). Finally, 3000 sentence pairs are annotated by a native speaker. Of these, 1500 pairs are labeled by paraphrasing whereas the remaining 1500 pairs are marked with non-paraphrase. This corpus is restricted by the few sources and topics.

Most methods for detecting paraphrase task have focused on the English text. The suggested methods are commonly evaluated on Microsoft paraphrasing (MSRP) corpus [6]. A standard corpus includes 5801 sentence pairs including 3900 paraphrasing pairs which are annotated by three native speakers.

The baseline of this task use vector-based similarity such as cosin [12], point-wise multiplication [4]. Other methods extracted features from parsing

---

[3]   http://www.alexa.com/topsites/countries/VN.

trees [5,13]. On the other hand, paraphrase is similar with machine translation in which the original and target text are the same languages. Therefore, many other methods use machine translation metric for quantifying the similarity of two sentences. For example, Finch et al. [8] extract features from four common machine translation metric including PER, BLUE, NIST, WER. Moreover, Madnani et al. extend the Finch's method for simply using a combination of eight metrics including TER, TERp, BADGER, SEPIA, BLEU, NIST, WER but it establishes the high performance on the MSRP corpus[4] (accuracy = 77.4%).

In recent years, deep learning is strongly developed caused by enhanced hardware infrastructure such as large memory and fast processing with GPU. Some methods are suggested to identify paraphrasing with significant improvements such as using autoencoders [14] recursive neuron network [3] or convolution neuron network [7]. However, these methods need a large labeled data. Therefore, they are not suitable for low resource languages such as Vietnamese. Our previous method of paraphrasing detection has high performance on English text (accuracy = 77.6). The method calculates the similarity of two sentences using matching identical phrases and similar words. The similarity of the two similar words is measured by using Wordnet similarity. Nevertheless, these ontologies are still restricted on Vietnamese. Therefore, we extend our previous work by proposing the $SimVN$ metric for estimate the similarity of two Vietnamese words. The detail of this metric is described below.

3   Proposed Method

Our proposed method to determine whether two sentences are paraphrase shown in Fig. 1:

- **Step 1 (*Separating words*):** Each word in sentences s_1 and s_2 are separated by maximum maching using a Vietnamese Wordnet.
- **Step 2 (*Matching duplicate phrases*):** Duplicate phrases including the separated words in Step 1 of the two sentences s_1 and s_2 are matched using the longest matching algorithm.
- **Step 3 (*Removing stop words*):** The remaining words after matching duplicate phrases are compared with candidate stop-words corpus to delete out of the sentences. These words are inserted into the sentences to improve the naturalness but not changing their meaning.
- **Step 4 (*Matching similar words*):** We propose a similarity metric $SimVN$ for two words based on a Vietnamese Wordnet. This metric is used to match remaining words in s_1 with the similar words in s_2.
- **Step 5 (*Calculating similarity matching metric*):** The matched duplicate phrases in Step 2 and similar words in Step 4 are used to quantify the similarity matching $SimMat$ metric. The $SimMat$ is presented to the similarity of the two input sentences s_1 and s_2.

---

[4] https://aclweb.org/aclwiki/index.php?title=Paraphrase_Identification_(State_of_th -e_art).

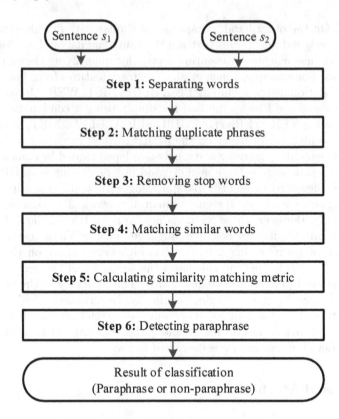

Fig. 1. The proposed schema for identifying paraphrase.

– **Step 6 (*Detecting paraphrase*):** The $SimMat$ metric estimated in Step 5 is used to decide the two input sentences are paraphrased or not.

Detail of each step is described as below with a pair sentence:

s_1: "*Nhiều hố sâu do quá trình khai thác cát gây ra.*" (Translation: "*Many deep holes are caused by sand mining.*")

s_2: "*Nhiều hố sâu bị đào nham nhở do hoạt động lấy cát.*" (Translation: "*Many deep holes were roughly dug due to sand mining.*")

3.1   Separating Words (Step 1)

Many Vietnamese tokenizer tools are proposed for separated words in the text. The most popular one is vnTokenizer[5] with accuracy reaching 97%. However, numerous Vietnamese words are separated by the tokenizer not containing in Vietnamese Wordnet [2] which includes 67,344 words. Therefore, these words cannot be identified their semantic relationships.

---

[5]   http://mim.hus.vnu.edu.vn/phuonglh/softwares/vnTokenizer.

To overcome this drawback, we applied the maximum matching algorithm using the Vietnamese Wordnet to separate words in the text. For examples used in here, words in s_1 and s_2 are split:

s_1: "*Nhiều hố_ sâu do quá_ trình khai_thác cát gây_ ra.*"
s_2: "*Nhiều hố_ sâu bị đào nham_ nhở do hoạt_ động lấy cát.*"

3.2 Matching Duplicate Phrases (Step 2)

The paraphrase sentences are often contained many duplicate phrases. Therefore, we inherit our previous algorithm to identify duplicate phrases in the two sentences [12]. The algorithm detects the copy phrases by heuristic matching longest candidate phrases in each iteration. The demonstration of matching for two sentences s_1 and s_2 is shown in Fig. 2. In result, three duplicate phrases involving "*Nhiều hố_ sâu,*" "*do,*" and "*cát*" are matched.

Fig. 2. Matching duplicate phrases.

3.3 Removing Stop Words (Step 3)

Stop words are added into the text to make it more natural. However, they rarely contribute to the text meaning. Therefore, we proposed a method to remove these words by comparing each word in a text with a list of candidate stop words of a corpus. The corpus[6] is suggested by VNG Corporation used in here. It contains 1951 famous Vietnamese candidate words. The result of removing the remaining words after matching duplicated phrases in Step 1 is shown in Fig. 3. Two crossed words "*bị*" and "*lấy*" are eliminated out of the s_2.

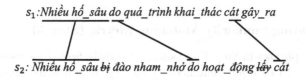

Fig. 3. Removing stop words.

---

[6] https://github.com/Tarrasch/vietnamese-stopwords.

3.4   Matching Similar Words (Step 4)

The similarity of two words are quantified in various approaches such as co-occurrence metric [12] or Wordnet-based metric [11] described in below:

The co-occurrence estimates the number of documents contains both of the two words. It needs a large corpus such as Google for estimation. However, the limitation of number queries in Google search makes this metric cannot apply for Vietnamese. Other corpora are not big enough for estimating this metric such as Vietnamese Wikipedia. The other drawback of co-occurrence is that this metric quantifies the semantics of two words.

Other semantic metrics such as Wordnet is used to measure the similarity using cohesion relationships between words. However, Wordnet does not support Vietnamese. A common Vietnamese Wordnet [2] is created by translating English words into Vietnamese ones. They contain 67,344 words with about twenty kinds of relationship such as synonym, antonym. We found out three of them are appropriate to quantify the similarity of two Vietnamese words. They include synonym, hyponym, and hypernym. We propose a $SimVN$ metric to measure the similarity of two Vietnamese words w_1 and w_2 shown in Eq. 1. This metric is based on Path metric of English Wordnet similarity metric.

$$SimVN(w_1, w_2) = \frac{1}{path(w_1, w_2) + 1} \tag{1}$$

where $path(w_1, w_2)$ is equal 0 if the two words is identical or synonyms. Otherwise, the path is estimated by 1 in the case of they are hyponym or hypernym. If we cannot find the path between two words using the synonyms, hyponym, and hypernym. The $SimVN$ metric is equal 0. For example, $SimVN(người_{human}, con\_người_{people}) = 1$ due to they are synonym words, $SimVN(người_{human}, học\_sinh_{student}) = \frac{1}{1+1} = 0.5$ because "$người_{human}$" is a hypernym of "$học\_sinh_{student}$."

The $SimVN$ metric is used to matching words in s_1 with similar words in s_2 after removing stop words in Step 3. The pair words are matched with a maximum summarization of their $SimVN$ metrics. The Hungarian algorithm suggested by Khun-Munkres [9] is used in here (Fig. 4). For example, two matched pairs are shown in Fig. 5 with maximum summarization $(1 + 0.3 = 1.3)$.

3.5   Calculating Similarity Matching Metric (Step 5)

The duplicate phrases in Step 1 and similar words in Step 4 are used to quantify the similarity of two input sentences s_1 and s_2 by similarity matching metric $SimMat$:

$$SimMat(s_1, s_2) = $$
$$= \frac{N_p + \sum_{i=1}^{N-1} len(p_i)^{0.2} + \sum_{j=1}^{M-1} SimVN_{w_j}^{0.2}}{N_p + N_w + \sum_{i=1}^{N-1} len(p_i)^{0.2} + \sum_{j=1}^{M-1} 1^{0.2}}$$

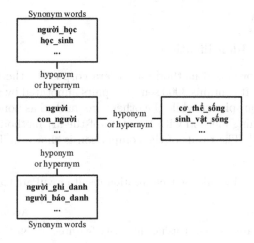

Fig. 4. Path length in Vietnamese Wordnet.

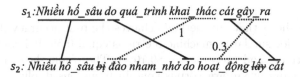

Fig. 5. Matching similar words with dotted lines.

where N_P is number of words in duplicate phrases in Step 2, N_w is number of words in matched words in Step 3, N and M is the number of duplicate phrases and similar words, correspondingly. p_i is the i-th matched phrases and w_j is the j-th matched words. This metric is inherited from our previous method on detecting English paraphrasing [12]. For examples used in here, the $SimMat$ metric of the two input sentences s_1 and s_2 is:

$$SimMat(s_1, s_2) =$$
$$= \frac{4 + (1^{0.2} + 2^{0.2} + 1^{0.2} + 1^{0.2}) + (1^{0.2} + 0.3^{0.2})}{4 + 2 + (1^{0.2} + 2^{0.2} + 1^{0.2} + 1^{0.2}) + (1^{0.2} + 1^{0.2})} = 0.819$$

3.6 Detecting Paraphrase (Step 6)

The similarity matching metric $SimMat$ calculated in the previous step is used to determine whether the two sentences are a paraphrase or not. If the $SimMat$ is greater than or equal a threshold $\alpha = 0.55$, they are decided as a paraphrase. Otherwise, the two sentences are not related about their meaning. The threshold α is estimated by logistic regression algorithm on the vnPara corpus.

4   Evaluation

4.1   Paraphrase Identification

We evaluated our proposed method on vnPara corpus [1], the first Vietnamese paraphrase corpus. It contains 3000 sentence pairs annotated by a native speaker. 1500 of them are paraphrase while the others are marked as non-paraphrase. We compare our matching method with the state-of-the-art method using similarity features method [1]. The result of the comparison is shown in Table 1.

Table 1. Paraphrase identification result on vnPara corpus

Method	Accuracy
The state-of-the-art using similarity features [1]	89.10%
Our method	**97.78%**

The result shows that our method achieves the best accuracy (97.78%) with threshold of α for similarity machine metric is equal 0.55. It is better than the state-of-the-art method (accuracy = 89.10%). The result shows that the most paraphrasing pairs in vnPara corpus contain numerous duplicate phrases and similar words. This fact comes from the most common news Vietnamese websites often post contents which mention the same topic. These contents often change positions of duplicate phrases or replace with similar words.

4.2   Paraphrase Corpus Creation

We use our method to create a new paraphrase corpus. Firstly, we collected about 65000 pages from top 15 Vietnamese news web pages (vnexpress[7], 24h[8], dantri[9], vietnamnet[10], etc.) with eight main topics including social, international, politics, economics, sports, culture, health, and technology. These web pages are ranked by Alexa[11]. 3134 sentence pairs having their similarity greater than the threshold α=0.55 are chosen for annotations by a native speaker. In result, 2748 of them are labeled by paraphrase. Other pairs are annotated by non-paraphrase. 2748 of 3134 pairs are marked with paraphrase (87.68%). It is lower than the accuracy of our method evaluation on vnPara corpus (97.78%). The result shows that our corpus has more diversity than the vnPara one.

---

[7]   http://vnexpress.net/.
[8]   http://www.24h.com.vn/.
[9]   http://dantri.com.vn/.
[10]   http://vietnamnet.vn/.
[11]   http://www.alexa.com/topsites/countries/VN.

5   Conclusion

Various Vietnamese news websites often post the same topic on the Internet. It makes the web users consuming the amount of time for reading the same information. These websites commonly use duplicate phrases and similar words in their contents. Therefore, we propose a method to identify paraphrase Vietnamese text by matching the identical phrases and related words. We also suggest a $SimVN$ metric for quantifying the similarity of two Vietnamese words using three relationships including synonym, hypernym, and hyponymy. Our method uses this metric achieve better accuracy (97.78%) when it is compared to the state-of-the-art method on the same vnPara corpus (accuracy = 89.10%).

We also used the proposed algorithm to build a new corpus consists of 3134 pairs of sentences labeled, diversified, collected from 15 Vietnamese top online news in eight topics such as Social, Economic, Technology. The pairs of sentences in the corpus are labeled by a native speaker including 2748 pairs of similar (88%) and 386 pairs of non-similar sentences (22%).

Future work includes pointing out other relationship of the Vietnamese Wordnet such as antonyms. We will apply the algorithm for building a web application which can classify Vietnamese web pages with the same topic or can identify the plagiarism in Vietnamese.

Acknowledgments. This work was supported by JSPS KAKENHI Grant Numbers JP16H06302 and 18H04120.

References

1. Bach, N.X., Oanh, T.T., Hai, N.T., Phuong, T.M.: Paraphrase identification in Vietnamese documents. In: Proceedings of the 7th Knowledge and Systems Engineering (KSE), pp. 174–179. IEEE (2015)
2. Blacoe, W., Lapata, M.: A comparison of vector-based representations for semantic composition. In: Proceedings of the Empirical Methods in Natural Language Processing (EMNLP), pp. 546–556. Association for Computational Linguistics (2012)
3. Cheng, J., Kartsaklis, D.: Syntax-aware multi-sense word embeddings for deep compositional models of meaning. In: Proceedings of the Empirical Methods on Natural Language Processing (EMNLP), pp. 1531–1542 (2015)
4. Corporation, N.: Vietnamese wordnet (2017). http://viet.wordnet.vn/wnms/. Accessed 10 June 2017
5. Das, D., Smith, N.A.: Paraphrase identification as probabilistic quasi-synchronous recognition. In: Proceedings of the Conference of the 47th Annual Meeting of the Association for Computational Linguistics (ACL), pp. 468–476. Association for Computational Linguistics (2009)
6. Dolan, B., Quirk, C., Brockett, C.: Unsupervised construction of large paraphrase corpora: exploiting massively parallel news sources. In: Proceedings of the 20th international conference on Computational Linguistics (COLING), pp. 350–356. Association for Computational Linguistics (2004)
7. Filice, S., Da San Martino, G., Moschitti, A.: Structural representations for learning relations between pairs of texts. In: Proceedings of the Annual Meeting of the Association for Computational Linguistics (ACL), pp. 1003–1013 (2015)

8. Finch, A., Hwang, Y.S., Sumita, E.: Using machine translation evaluation techniques to determine sentence-level semantic equivalence. In: Proceedings of the 3rd International Workshop on Paraphrasing (IWP), pp. 17–24 (2005)

9. Kuhn, H.W.: The Hungarian method for the assignment problem. Naval Res. Logist. Q. **2**(1–2), 83–97 (1955)

10. Madnani, N., Tetreault, J., Chodorow, M.: Re-examining machine translation metrics for paraphrase identification. In: Proceedings of the Conference of the North American Chapter of the Association for Computational Linguistics: Human Language Technologies (NAACL-HLT), pp. 182–190. Association for Computational Linguistics (2012)

11. Mihalcea, R., Corley, C., Strapparava, C., et al.: Corpus-based and knowledge-based measures of text semantic similarity. In: Proceedings of the AAAI Conference on Artificial Intelligence, vol. 6, pp. 775–780 (2006)

12. Nguyen-Son, H.Q., Miyao, Y., Echizen, I.: Paraphrase detection based on identical phrase and similar word matching. In: Proceedings of the 29th Pacific Asia Conference on Language, Information and Computation (PACLIC) (2015)

13. Qiu, L., Kan, M.Y., Chua, T.S.: Paraphrase recognition via dissimilarity significance classification. In: Proceedings of the Conference on Empirical Methods in Natural Language Processing (EMNLP), pp. 18–26. Association for Computational Linguistics (2006)

14. Socher, R., Huang, E.H., Pennin, J., Manning, C.D., Ng, A.Y.: Dynamic pooling and unfolding recursive autoencoders for paraphrase detection. In: Proceedings of the Advances in Neural Information Processing Systems (NIPS), pp. 801–809 (2011)

Advanced Studies in Machine Learning

Automatic Hyper-parameters Tuning
for Local Support Vector Machines

Thanh-Nghi Do[1,2(✉)] and Minh-Thu Tran-Nguyen[1]

[1] College of Information Technology, Can Tho University, Cantho 92100, Vietnam
dtnghi@cit.ctu.edu.vn
[2] UMI UMMISCO 209 (IRD/UPMC), Cantho, Vietnam

Abstract. In this paper, we propose the auto$k$SVM algorithm being possible to automatically tune hyper-parameters of k local SVMs for classifying large datasets. The auto$k$SVM is able to determine the number of clusters k to partition the large training data, followed which it learns the non-linear SVM model in each cluster to classify the data locally in the parallel way on multi-core computers. The auto$k$SVM combines the grid search, the *.632* bootstrap estimator, the hill climbing heuristic to optimize hyper-parameters in the local non-linear SVM training. The numerical test results on 4 datasets from UCI repository and 3 benchmarks of handwritten letters recognition showed that our proposal is efficient compared to the standard LibSVM and the original kSVM. An example of its effectiveness is given with an accuracy of 96.74% obtained in the classification of Forest covertype dataset having 581,012 datapoints in 54 dimensional input space and 7 classes in 334.45 s using a PC Intel(R) Core i7-4790 CPU, 3.6 GHz, 4 cores.

Keywords: Support vector machines · Large datasets
Local support vector machines · Automatic hyper-parameters tuning

1 Introduction

Support vector machines (SVM) proposed by Vapnik [1] is a state-of-the-art classification technique applied to many pattern recognition problems. Successful applications of SVMs include facial recognition, text categorization and bioinformatics [2]. In spite of the prominent properties, there are two main drawbacks while training a SVM model. The first problem is that the computational cost of a SVM approach is at least square of the number of training datapoints [3] making SVM impractical for massive datasets. Furthermore, the hyper-parameters optimization for SVM learning is tuned by hand to obtain a good model, this becomes intractable use for non-experts.

Our investigation is to extend the recent kSVM algorithm proposed by [4,5] for speeding-up non-linear classification of large datasets, to develop the auto$k$SVM algorithm being able to automatically search hyper-parameters. Instead of building a global SVM model, as done by the classical algorithm is very

© Springer Nature Switzerland AG 2018
T. K. Dang et al. (Eds.): FDSE 2018, LNCS 11251, pp. 185–199, 2018.
https://doi.org/10.1007/978-3-030-03192-3_14

difficult to deal with large datasets, the autokSVM is able to determine the number of clusters k (thanks to the performance analysis [6–8]) to partition the training data, followed which it learns a non-linear SVM in each cluster to classify the data locally in the parallel way on multi-core computers. The autokSVM combines the grid search (as proposed by [9–13]), the .632 bootstrap estimator [14], the hill climbing heuristic to optimize hyper-parameters in the local SVM training. The numerical test results on 4 datasets from UCI repository [15] and 3 benchmarks of handwritten letters recognition [16], MNIST [17,18] showed that our proposed autokSVM achieves good off-the-shelf classification performance compared to the standard LibSVM [19] and the original kSVM [4,5] in terms of training time and accuracy. Without any requirement of hyper-parameters tuning, the autokSVM classifies Forest covertype dataset having 581,012 datapoints in 54 dimensional input space and 7 classes with an accuracy of 96.74% in 334.45 s using a PC Intel(R) Core i7-4790 CPU, 3.6 GHz, 4 cores.

The paper is organized as follows. Section 2 briefly introduces the kSVM algorithm and our proposed autokSVM for automatic hyper-parameter tuning in the non-linear classification of large datasets. Section 3 shows the experimental results. Section 4 discusses about related works. We then conclude in Sect. 5.

2   Automatic Hyperparameter Tuning for k Local Support Vector Machines (autokSVM)

2.1   Support Vector Machines

Let us consider a linear binary classification of a given dataset with m datapoints x_i ($i = 1, \ldots, m$) in the n-dimensional input space R^n, having corresponding labels $y_i = \pm1$. The SVM algorithms [1] try to find the separating plane furthest from both class +1 and class −1. The standard SVMs pursue this goal with the quadratic programming (1).

$$min_\alpha(1/2) \sum_{i=1}^{m} \sum_{j=1}^{m} y_i y_j \alpha_i \alpha_j K\langle x_i, x_j \rangle - \sum_{i=1}^{m} \alpha_i$$

$$s.t. \begin{cases} \sum_{i=1}^{m} y_i \alpha_i = 0 \\ 0 \leq \alpha_i \leq C \quad \forall i = 1, 2, \ldots, m \end{cases} \tag{1}$$

where α_i are Lagrange multipliers, C is a positive constant used to tune the margin and the error and a linear kernel function $K\langle x_i, x_j \rangle = \langle x_i . x_j \rangle$.

The support vectors (for which $\alpha_i > 0$) are given by the solution of the quadratic program (1), and then, the separating surface and the scalar b are determined by the support vectors. The classification of a new data point x based on the SVM model is as follows:

$$predict(x, SVMmodel) = sign(\sum_{i=1}^{\#SV} y_i \alpha_i K\langle x, x_i \rangle - b) \tag{2}$$

SVM algorithms can be used for the non-linear classification tasks [20]. No algorithmic changes are required from the usual kernel function $K\langle x_i, x_j \rangle$ as a linear inner product, $K\langle x_i, x_j \rangle = \langle x_i \cdot x_j \rangle$ other than the modification of the kernel function evaluation. According to [21], the RBF (Radial Basis Function) is most general function, as follows:

$$K\langle x_i, x_j \rangle = e^{-\gamma \|x_i - x_j\|^2} \tag{3}$$

SVMs are accurate models for facial recognition, text categorization and bioinformatics [2]. Nevertheless, the study in [3] illustrated that the computational cost requirements of the SVM solutions in (1) are at least $O(m^2)$ (where m is the number of training datapoints) making SVM impractical for massive datasets. Furthermore, the hyper-parameters optimization for SVM learning is tuned by hand to obtain a good model, this becomes intractable use for non-experts.

2.2   Parallel Algorithm of k Local Support Vector Machines (kSVM)

Instead of learning a global SVM model, as done by the classical algorithm which is very difficult to deal with large datasets, the kSVM algorithm proposed by [4,5] performs the training task with two main steps as described in Algorithm 1. The first one is to use $k$means algorithm [22] to partition the full dataset into k clusters, and then it is easily to learn a non-linear SVM in each cluster to classify the data locally in parallel way on multi-core computers (based on the shared memory multiprocessing programming model OpenMP [23]). Figure 1 shows the comparison between a global SVM model (left part) and 3 local SVM models (right part), using a RBF kernel function with $\gamma = 10$ and a positive constant $C = 10^6$. The hyper-parameter γ of the kernel function and the positive constant C are denoted by $\theta = \{\gamma, C\}$.

The class of a new datapoint x is predicted by the local SVM model $lsvm_{NN}$ obtained on the cluster D_{NN} whose center c_{NN} is nearest x.

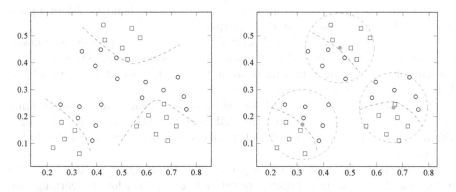

Fig. 1. Global SVM model (left part) versus local SVM models (right part)

Algorithm 1. kSVM classification algorithm for large datasets

---
    input :
                training dataset D with m datapoints
                number of local models k
                hyper-parameter θ including the parameter γ of the kernel function and
                cost constant C for tuning margin and errors of SVMs
    output:
                k local support vector machines models

1 **begin**
2   /*$k$means performs the data clustering on dataset D;*/
3   creating k clusters denoted by $D_1, D_2, \ldots, D_k$ and
4   their corresponding centers $c_1, c_2, \ldots, c_k$
5   #pragma omp parallel for schedule(dynamic)
6   **for** $i \leftarrow 1$ **to** k **do**
7   /*learning local support vector machine model from D_i;*/
8         $lSVM_i = SVM(D_i, \theta = \{\gamma, C\})$
9   **end**
10   return kSVM-model $= \{(c_1, lSVM_1), (c_2, lSVM_2), \ldots, (c_k, lSVM_k)\}$
11 **end**

As illustrated in [4,5], the complexity of parallel training k local SVM models on a P-core processor is $O(\frac{k}{P}(\frac{m}{k})^2) = O(\frac{m^2}{kP})$. The kSVM is kP times faster than building a global SVM model (the complexity is at least $O(m^2)$). The studies in [4–8] show that the parameter k gives a trade-off between the generalization capacity (the classification correctness) and the computational cost (the training time).

2.3 Automatic Hyper-parameters Tuning for k Local Support Vector Machines (auto$k$SVM)

Although the kSVM can reduce the computational cost of a SVM training task but the hyper-parameters optimization is still tuning by hand to obtain a good model. This becomes hard use for non-experts.

A training task of the kSVM described in Algorithm 1 requires three parameters including the number of local models k, the hyper-parameters θ including the parameter γ of the kernel function and the cost constant C (the trade-off between the margin size and errors of SVMs).

As illustrated in [21], the RBF kernel function is general and efficient function. Therefore, we propose to use the RBF kernel function in the learning algorithm kSVM. Figure 2 shows different flexible separation forms according to the hyper-parameter γ.

And then, we propose to develop the auto$k$SVM algorithm being able to automatically search these hyper-parameters k, $\theta = \{\gamma, C\}$ for the learning task of kSVM.

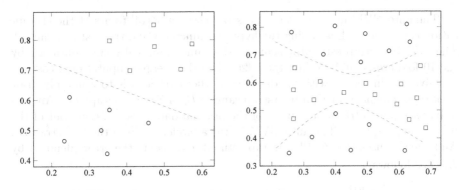

Fig. 2. RBF kernel with $\gamma = 10^{-5}$ (left part) versus $\gamma = 1$ (right part)

Parameter k.

According to the performance analysis in terms of the algorithmic complexity and the generalization capacity studied in [4–8], the role of the parameter k in kSVM is related to a trade-off between the generalization capacity and the computational complexity. If k is large then the kSVM reduces significant training time (the speed up factor of local SVMs over the global one is k). And then, the size of a cluster is small; the locality is extreme with a very low generalization capacity. When $k \to m$, the kSVM is closed to the 1 nearest neighbor classifier. If k is small then the kSVM reduces insignificant training time. However, the size of a cluster is large; it improves the generalization capacity. When $k \to 1$, the kSVM is approximated the global one.

It leads to set k so that the cluster size is large enough (e.g. 200 proposed by [6,7]). The empirical results in [4,5,24,25] show that the cluster size is about 500. Therefore, the auto$k$SVM guarantees to obtain the cluster size $\sim$500 by setting $k = \frac{m}{500}$.

Hyper-parameters $\theta = \{\gamma, C\}$ for Local SVMs Using the RBF Kernel Function.

Our proposal is to use different kernel parameter γ and cost constant C for k local SVMs, it means that the auto$k$SVM has to tune to find best kernel parameter γ_i and cost constant C_i for each cluster D_i.

The auto$k$SVM searches the hyper-parameter γ of RBF kernel (RBF kernel of two datapoints x_i and x_j, $K[i,j] = exp(-\gamma \|x_i - x_j\|^2)$) and the cost constant C (the trade-off between the margin size and the errors) to obtain the best correctness. The cost constant C is chosen in $G = \{1, 10^1, 10^2, 10^3, 10^4, 10^5, 10^6\}$ and the hyper-parameter γ of RBF kernel is tried among $P = \{10^{-5}, 5.10^{-5}, 10^{-4}, 5.10^{-4}, 10^{-3}, 5.10^{-3}, 10^{-2}, 5.10^{-2}, 10^{-1}, 5.10^{-1}, 1, 5, 10^1, 5.10^1\}$. Furthermore, the local model is less complex than the global one (ref. Fig. 1). And then, the auto$k$SVM starts with the smallest values of the hyper-parameter γ and the cost constant C in the grid and stops when the classification correctness can not be improved in two next trials (the hill climbing heuristic).

The auto$k$SVM uses the *.632* bootstrap estimator [14] to assert the classification correctness while tuning the hyper-parameter γ and the cost constant C. The *.632* bootstrap estimator achieves the same empirical results obtained by the 10-fold cross-validation protocol [26] with the cheap computational cost.

Given a cluster D_i with m_i datapoints, the main idea is to randomly draw with replacement from D_i a bootstrap sample B_i of size m_i datapoints. An out-of-bag sample OOB_i about 36.8% of the original data cluster D_i are out of the bootstrap sample B_i. The auto$k$SVM learns a model $\widehat{M_i}$ from the bootstrap B_i. And then, it uses the *.632* bootstrap estimator to assert the error (denoted by $\widehat{Err}$) as follows:

$$\widehat{Err}^{.632} = .632 \times \widehat{Err}(OOB_i, \widehat{M_i}) + .368 \times \widehat{Err}(B_i, \widehat{M_i}) \tag{4}$$

The auto$k$SVM in Algorithm 2 presents automatic hyper-parameters tuning for k local support vector machines.

Figure 3 is an example of the automatic classification done by the auto$k$SVM.

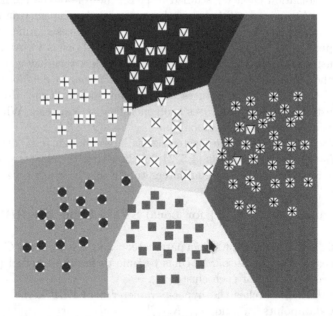

Fig. 3. Example of the automatic classification given by the auto$k$SVM

3   Numerical Test Results

We are interested in the performance evaluation of our proposed auto$k$SVM algorithm for automatic hyper-parameters tuning in classification problems. We have implemented auto$k$SVM in C/C++, OpenMP [23], using the highly efficient standard library SVM, LibSVM [19].

Algorithm 2. auto$k$SVM learning algorithm for the classification

---

input :
        training dataset D with m datapoints
output:
        k local support vector machines models

1 **begin**
2   /*$k$means performs the data clustering on dataset D;*/
3     $k = \lceil \frac{m}{500} \rceil$
4   creating k clusters denoted by $D_1, D_2, \ldots, D_k$ and
5   their corresponding centers $c_1, c_2, \ldots, c_k$
6   #pragma omp parallel for schedule(dynamic)
7   **for** $i \leftarrow 1$ **to** k **do**
8   /*learning local support vector machine model from D_i;*/
9   randomly drawing with replacement from D_i a bootstrap sample B_i of size m_i datapoints (corresponding an out-of-bag sample OOB_i of D_i are out of B_i)
10   best hyper-parameters $\theta* = \oslash$
11   smallest estimated error $\widehat{Err}* = 1.0$
12   number of trials $j = 0$
13   **for** $\theta_t = \{\gamma_t, C_t\}$ in $T = G \bowtie P$ **do**
14            $lSVM_t = SVM(B_i, \theta_t = \{\gamma_t, C_t\})$
15            $\widehat{Err}_t^{.632} = .632 \times \widehat{Err}(OOB_i, \widehat{lSVM}_t) + .368 \times \widehat{Err}(B_i, \widehat{lSVM}_t)$
16   **if** $(\widehat{Err}_t^{.632} < \widehat{Err}*)$ **then**
17                $\theta* = \{\gamma_t, C_t\}$
18                $\widehat{Err}* = \widehat{Err}_t^{.632}$
19   **else**
20                $j + +$;
21   **if** $(j > 1)$ **then**
22   break;
23   **end**
24   **end**
25   **end**
26   /*retraining local support vector machine model from D_i using $\theta*$;*/
27        $lSVM_i = SVM(D_i, \theta* = \{\gamma*, C*\})$
28   **end**
29   return kSVM-model $= \{(c_1, lSVM_1), (c_2, lSVM_2), \ldots, (c_k, lSVM_k)\}$
30 **end**

Our evaluation of the classification performance is reported in terms of correctness and training time. The comparative study includes classification results obtained by LibSVM, kSVM and auto$k$SVM.

All experiments are run on machine Linux Fedora 20, Intel(R) Core i7-4790 CPU, 3.6 GHz, 4 cores and 32 GB main memory.

Datasets.

Experiments are conducted with the 4 datasets from UCI repository [15] and 3 benchmarks of handwritten letters recognition, including USPS [16], MNIST [17], a new benchmark for handwritten character recognition [18].

Table 1 presents the description of datasets. The last column of Table 1 presents evaluation protocols. Datasets are already divided in training set (Trn) and testing set (Tst). We used the training data to build the SVM models. Then, we classified the testing set using the resulting models.

Table 1. Description of datasets

ID	Dataset	Individuals	Attributes	Classes	Evaluation protocol
1	Opt. Rec. of handwritten digits	5620	64	10	3832 Trn–1797 Tst
2	Letter	20000	16	26	13334 Trn–6666 Tst
3	Isolet	7797	617	26	6238 Trn–1559 Tst
4	USPS handwritten digit	9298	256	10	7291 Trn–2007 Tst
5	A New bench. hand. char. rec	40133	3136	36	36000 Trn–4133 Tst
6	MNIST	70000	784	10	60000 Trn–10000 Tst
7	Forest cover types	581012	54	7	400000 Trn–181012 Tst

Tuning Hyper-parameters for LibSVM, kSVM.

We propose to use RBF kernel type in SVM models because it is general and efficient [21]. We also tried to tune the hyper-parameter γ of the RBF kernel and the cost constant C (the trade-off between the margin size and the errors) to obtain a good accuracy. Furthermore, the kSVM uses the parameter k local models (number of clusters). The cross-validation protocol [26, 27] is used to tune these hyper-parameters. And then, the optimal parameters in Table 2 give the highest accuracy for datasets.

Table 2. Hyper-parameters for LibSVM and kSVM

ID	Datasets	γ	C	k
1	Opt. Rec. of handwritten digits	0.0001	100000	10
2	Letter	0.0001	100000	30
3	Isolet	0.0001	100000	10
4	USPS handwritten digit	0.0001	100000	10
5	A new bench. hand. char. rec	0.001	100000	50
6	MNIST	0.05	100000	100
7	Forest cover types	0.0001	100000	500

Classification Results.

Table 3, Figs. 4, 5, 6 and 7 show the classification results of LibSVM, kSVM, auto$k$SVM on the 7 datasets.

Although our proposed auto$k$SVM is an off-the-shelf classification tool (automatic hyper-parameters tuning) but it achieves very competitive performances compared to LibSVM and kSVM. In terms of the averaging classification correctness, the superiority of LibSVM on kSVM and auto$k$SVM corresponds to 1.00% and 1.30%, respectively.

Table 3. Classification results in terms of accuracy (%) and training time (s)

ID	Datasets	Classification accuracy(%)			Training time(s)		
		LibSVM	kSVM	auto$k$SVM	LibSVM	kSVM	auto$k$SVM
1	Opt. Rec. of handwritten digits	98.33	97.05	96.88	0.58	0.21	0.29
2	Letter	97.40	96.14	95.54	2.87	0.5	1.21
3	Isolet	96.47	95.44	95.45	8.37	2.94	8.32
4	USPS handwritten digit	96.86	95.86	95.37	5.88	3.82	9.08
5	A new bench. hand. char. rec	95.14	92.98	92.72	107.07	35.7	105.51
6	MNIST	98.37	98.11	97.84	1531.06	45.50	86.87
7	Forest cover types	NA	97.06	96.74	NA (>1987200)	223.7	334.45
	Average	97.10	96.09	95.79	NA (>284122.26)	44.62	78.20

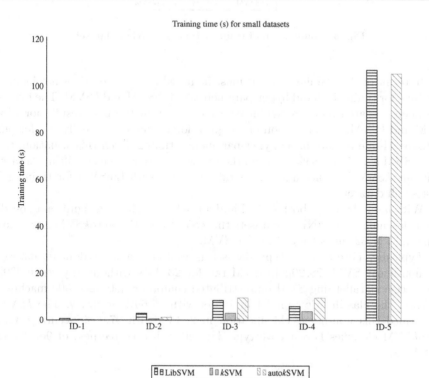

Training time (s) for small datasets

LibSVM kSVM   auto$k$SVM

Fig. 4. Comparison of training time on 5 small datasets

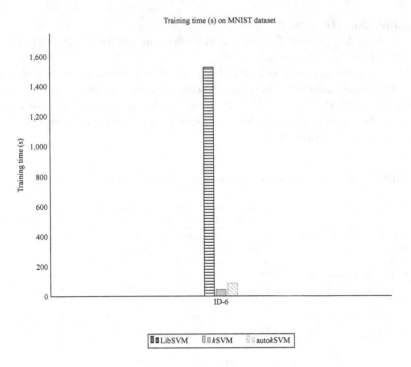

Fig. 5. Comparison of training time on MNIST dataset

In terms of the training time, it must be noted that Table 3 does not include the time for tuning by hand hyper-parameters of LibSVM and kSVM. The hyper-parameters search requires significant time to obtain the best results for Lib-SVM and kSVM. A comparison of computational time is not really fair for our auto$k$SVM due to automatic hyper-parameters tuning. Even this situation, the auto$k$SVM is 1.75 times slower than the kSVM to perform the classification task. The auto$k$SVM has the same computational time with LibSVM for training 5 first small datasets.

With large datasets, both kSVM and auto$k$SVM achieve a significant speed-up in learning. For MNIST dataset, the kSVM and the auto$k$SVM are 33.64 times and 17.62 times faster than LibSVM.

Typically, Forest cover type dataset is well-known as a difficult dataset for non-linear SVM [28,29]; LibSVM ran for 23 days without any result [29]. The recent parallelizing SVM on distributed computers [30] uses 500 machines to train the classification model in 1655 s with 97.69% accuracy. The kSVM performed this non-linear classification in 223.7 s with 97.06% accuracy. Our auto$k$SVM classifies Forest covertype dataset with a correctness of 96.74% in 334.45 s.

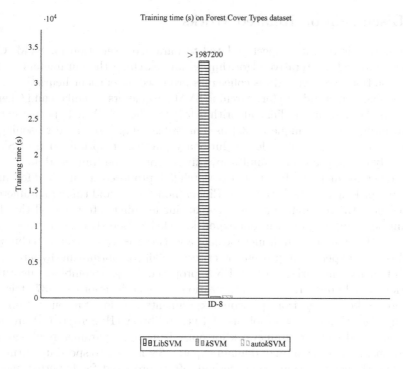

Fig. 6. Comparison of training time on forest cover types dataset

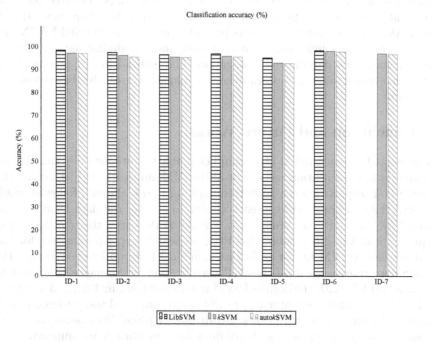

Fig. 7. Comparison of classification correctness

4   Discussion on Related Works

Our proposal is in some aspects related to parameter selection for SVM. Carl Staelin proposed an iterative algorithm [9] for searching the parameters on the grid resolution. Keerthi and his colleagues proposed an efficient heuristic [11], a gradient-based method [10] for searching SVM parameters. Adankon and Cheriet proposed the descent gradient algorithm [31] for the LS-SVM hyperparameter optimization. The technique in [32] uses meta-learning, case-based reasoning to propose good staring points for evolutionary parameter optimization of SVM. The study in [33] presented a multi-objective optimization framework for tuning SVM hyperparameters. More recent research [34] proposed the method for automatically picking out the kernel type (linear/non-linear) and tuning parameters.

Recently, the research of automatic machine learning is to make off-the-shelf machine learning methods without expert knowledge. Bergstra and his colleagues proposed the random search method [35] and two greedy sequential techniques based on the expected improvement criterion [36] for optimizing hyperparameters of neural networks. Auto-WEKA proposed by [12] combines algorithm selection and hyperparameter optimization. A generic approach [37] tries to incorporate knowledge from previous experiments for simultaneously tuning a learning algorithm on new problems. A Python library (Hyperopt [13]) provides algorithms and software framework for optimizing hyperparameters of learning algorithms. A strategy human domain experts [38] is used to speed up optimization. The Bayesian optimization method [39] is proposed for hyperparameters selection of many learning algorithms. A comparative studies of three Bayesian optimization techniques (Spearmint, Tree Parzen Estimator, Sequential Model-based Algorithm Configuration) was presented in [40]. AUTO-SKLEARN [41] automatically takes into account past performance on similar datasets during the Bayesian hyperparameter optimization. The method in [42] presented the Bayesian hyperparameters optimization technique for ensemble-based learning algorithms.

5   Conclusion and Future Works

We presented the auto$k$SVM algorithm being able to automatically tune hyperparameters in classification tasks. The auto$k$SVM automatically determines the parameter k to partition the large training data into k clusters, followed which it learns the non-linear SVM in each cluster to classify the data locally in the parallel way on multi-core computers. The auto$k$SVM uses the $.632$ bootstrap estimator and the hill climbing heuristic to search hyper-parameters for the local non-linear SVM in the grid. The numerical test results showed that the auto$k$SVM achieves good off-the-shelf classification performance compared to the standard LibSVM, the orginal kSVM in terms of training time and accuracy.

In the near future, we intend to provide more empirical test on large benchmarks and comparisons with other algorithms. A promising avenue for future research aims at developing a distributed implementation for improving the training time.

References

1. Vapnik, V.: The Nature of Statistical Learning Theory. ISS, 2nd edn. Springer, New York (2000). https://doi.org/10.1007/978-1-4757-3264-1
2. Guyon, I.: Web page on SVM applications. http://www.clopinet.com/isabelle/Projects/SVM/app-list.html
3. Platt, J.: Fast training of support vector machines using sequential minimal optimization. In: Schölkopf, B., Burges, C., Smola, A. (eds.) Advances in Kernel Methods–Support Vector Learning, pp. 185–208 (1999)
4. Do, T.-N.: Non-linear classification of massive datasets with a parallel algorithm of local support vector machines. In: Le Thi, H.A., Nguyen, N.T., Do, T.V. (eds.) Advanced Computational Methods for Knowledge Engineering. AISC, vol. 358, pp. 231–241. Springer, Cham (2015). https://doi.org/10.1007/978-3-319-17996-4_21
5. Do, Thanh-Nghi, Poulet, François: Parallel learning of local SVM algorithms for classifying large datasets. In: Hameurlain, Abdelkader, Küng, Josef, Wagner, Roland, Dang, Tran Khanh, Thoai, Nam (eds.) Transactions on Large-Scale Data- and Knowledge-Centered Systems XXXI. LNCS, vol. 10140, pp. 67–93. Springer, Heidelberg (2017). https://doi.org/10.1007/978-3-662-54173-9_4
6. Bottou, L., Vapnik, V.: Local learning algorithms. Neural Comput. 4(6), 888–900 (1992)
7. Vapnik, V., Bottou, L.: Local algorithms for pattern recognition and dependencies estimation. Neural Comput. 5(6), 893–909 (1993)
8. Vapnik, V.: Principles of risk minimization for learning theory. In: Advances in Neural Information Processing Systems 4, NIPS Conference, Denver, Colorado, USA, 2–5 December 1991, pp. 831–838 (1991)
9. Staelin, C.: Parameter selection for support vector machines. Technical report, Hp Laboratories (2002)
10. Keerthi, S.S., Sindhwani, V., Chapelle, O.: An efficient method for gradient-based adaptation of hyperparameters in SVM models. In: Proceedings of the 19th International Conference on Neural Information Processing Systems, NIPS 2006, pp. 673–680. MIT Press, Cambridge (2006)
11. Keerthi, S.S., Lin, C.J.: Asymptotic behaviors of support vector machines with Gaussian kernel. Neural Comput. 15(7), 1667–1689 (2003)
12. Thornton, C., Hutter, F., Hoos, H.H., Leyton-Brown, K.: Auto-WEKA: combined selection and hyperparameter optimization of classification algorithms. In: Proceedings of the 19th ACM SIGKDD International Conference on Knowledge Discovery and Data Mining, KDD 2013, pp. 847–855. ACM (2013)
13. Bergstra, J., Komer, B., Eliasmith, C., Yamins, D., Cox, D.D.: Hyperopt: a python library for model selection and hyperparameter optimization. Comput. Sci. Discov. 8(1), 014008 (2015)
14. Efron, B., Tibshirani, R.J.: An Introduction to the Bootstrap. Softcover reprint of the original 1st edn, 1993 edition. Chapman and Hall/CRC, Boca Raton (1994)
15. Lichman, M.: UCI machine learning repository (2013)
16. LeCun, Y., Boser, B., Denker, J., Henderson, D., Howard, R., Hubbard, W., Jackel, L.: Backpropagation applied to handwritten zip code recognition. Neural Comput. 1(4), 541–551 (1989)
17. LeCun, Y., Bottou, L., Bengio, Y., Haffner, P.: Gradient-based learning applied to document recognition. Proc. IEEE 86(11), 2278–2324 (1998)
18. van der Maaten, L.: A new benchmark dataset for handwritten character recognition (2009). http://homepage.tudelft.nl/19j49/Publications_files/characters.zip

19. Chang, C.C., Lin, C.J.: LIBSVM : a library for support vector machines. ACM Trans. Intell. Syst. Technol. **2**(27), 1–27 (2011)
20. Cristianini, N., Shawe-Taylor, J.: An Introduction to Support Vector Machines: and other Kernel-Based Learning Methods. Cambridge University Press, New York (2000)
21. Lin, C.: A practical guide to support vector classification (2003)
22. MacQueen, J.: Some methods for classification and analysis of multivariate observations. In: Proceedings of 5th Berkeley Symposium on Mathematical Statistics and Probability, vol. 1, pp. 281–297. University of California Press, Berkeley, January 1967
23. OpenMP Architecture Review Board: OpenMP application program interface version 3.0 (2008)
24. Do, T.-N., Poulet, F.: Random local SVMs for classifying large datasets. In: Dang, T.K., Wagner, R., Küng, J., Thoai, N., Takizawa, M., Neuhold, E. (eds.) FDSE 2015. LNCS, vol. 9446, pp. 3–15. Springer, Cham (2015). https://doi.org/10.1007/978-3-319-26135-5_1
25. Do, T.N., Poulet, F.: Classifying very high-dimensional and large-scale multi-class image datasets with Latent-LSVM. In: IEEE International Conference on Cloud and Big Data Computing (2016)
26. Hastie, T., Tibshirani, R., Friedman, J.: The Elements of Statistical Learning: Data Mining, Inference, and Prediction. SSS, 2nd edn. Springer, New York (2009). https://doi.org/10.1007/978-0-387-84858-7
27. Pádraig, C.: Evaluation in machine learning. Tutorial (2009)
28. Yu, H., Yang, J., Han, J.: Classifying large data sets using SVMs with hierarchical clusters. In: Proceedings of the ACM SIGKDD International Conference on Knowledge Discovery and Data Mining, pp. 306–315. ACM (2003)
29. Do, T.N., Poulet, F.: Towards high dimensional data mining with boosting of PSVM and visualization tools. In: Proceedings of the 6th International Conference on Enterprise Information Systems, pp. 36–41 (2004)
30. Zhu, K., et al.: Parallelizing support vector machines on distributed computers. In: Platt, J.C., Koller, D., Singer, Y., Roweis, S.T. (eds.) Advances in Neural Information Processing Systems 20, pp. 257–264. Curran Associates, Inc. (2008)
31. Adankon, M.M., Cheriet, M.: Model selection for the LS-SVM. Application to handwriting recognition. Pattern Recogn. **42**(12), 3264–3270 (2009)
32. Reif, M., Shafait, F., Dengel, A.: Meta-learning for evolutionary parameter optimization of classifiers. Mach. Learn. **87**(3), 357–380 (2012)
33. Chatelain, C., Adam, S., Lecourtier, Y., Heutte, L., Paquet, T.: Non-cost-sensitive SVM training using multiple model selection. J. Circ. Syst. Comput. **19**(1), 231–242 (2010)
34. Huang, H., Lin, C.: Linear and kernel classification: when to use which? In: Proceedings of the 2016 SIAM International Conference on Data Mining, pp. 216–224. Society for Industrial and Applied Mathematics, June 2016
35. Bergstra, J., Bengio, Y.: Random search for hyper-parameter optimization. J. Mach. Learn. Res. **13**(1), 281–305 (2012)
36. Bergstra, J., Bardenet, R., Bengio, Y., Kégl, B.: Algorithms for hyper-parameter optimization. In: Proceedings of the 24th International Conference on Neural Information Processing Systems, NIPS 2011, USA, pp. 2546–2554. Curran Associates Inc. (2011)
37. Bardenet, R., Brendel, M., Kégl, B., Sebag, M.: Collaborative hyperparameter tuning. In: Proceedings of The 30th International Conference on Machine Learning, pp. 199–207 (2013)

38. Feurer, M., Springenberg, J.T., Hutter, F.: Initializing Bayesian hyperparameter optimization via meta-learning. In: Proceedings of the Twenty-Ninth AAAI Conference on Artificial Intelligence, AAAI 2015, Austin, Texas, pp. 1128–1135. AAAI Press (2015)
39. Snoek, J., Larochelle, H., Adams, R.P.: Practical Bayesian optimization of machine learning algorithms. In: Proceedings of the 25th International Conference on Neural Information Processing Systems, NIPS 2012, USA, pp. 2951–2959. Curran Associates Inc. (2012)
40. Eggensperger, K., et al.: Towards an empirical foundation for assessing Bayesian optimization of hyperparameters. In: NIPS Workshop on Bayesian Optimization in Theory and Practice (2013)
41. Feurer, M., Klein, A., Eggensperger, K., Springenberg, J.T., Blum, M., Hutter, F.: Efficient and robust automated machine learning. In: Advances in Neural Information Processing Systems 28: Annual Conference on Neural Information Processing Systems 2015, 7–12 December 2015, Montreal, Quebec, Canada, pp. 2962–2970 (2015)
42. Lévesque, J.C., Gagné, C., Sabourin, R.: Bayesian hyperparameter optimization for ensemble learning. In: Proceedings of the Thirty-Second Conference on Uncertainty in Artificial Intelligence, UAI 2016, Arlington, Virginia, United States, pp. 437–446. AUAI Press (2016)

Detection of the Primary User's Behavior for the Intervention of the Secondary User Using Machine Learning

Deisy Dayana Zambrano Soto[1], Octavio José Salcedo Parra[1,2(✉)], and Danilo Alfonso López Sarmiento[1]

[1] Faculty of Engineering, Intelligent Internet Research Group, Universidad Distrital "Francisco José de Caldas", Bogotá D.C., Colombia
ddzambranos@correo.udistrital.edu.co,
{osalcedo,dalopezs}@udistrital.edu.co
[2] Department of Systems and Industrial Engineering, Faculty of Engineering, Universidad Nacional de Colombia, Bogotá D.C., Colombia
ojsalcedop@unal.edu.co

Abstract. The predictive analysis for the spectral decision with automatic Learning is a task that is currently challenging. Some automatic Learning techniques are shown in order to predict the presence or absence of a primary user (PU) in Cognitive Radio. Four machine learning methods are examined including the K-nearest neighbors (KNN), the support vector machines (SVM), logistic regression (LR) and decision tree (DT) classifiers. These predictive models are built based on data and their performance is compared with the purpose of selecting the best classifier that can predict spectral occupancy.

Keywords: Cognitive Radio · KNN · LR · Machine learning · Primary users (PU) SVM

1 Introduction

Cognitive radio authorizes users with a license (PUs) to Exchange the exceeding spectrum and temporarily transfer the use of spectrum to non-licensed users (SUs). In consequence, the PU need a resource management scheme that allows them to optimally assign a fixed amount of spectrum offered between different types of services and adapt to the changes in the network conditions.

To detect its environment in a Cognitive Radio node, it examines the parameters in the air and makes decisions for the allocation and dynamic management of resources in time-frequency-space to improve the use of the radio spectrum. For an efficient real time process, cognitive radio is combined with artificial intelligence and automatic learning techniques to allocate adaptive and intelligently [1]. By using the spectrum detection techniques, the inactive channels are located and allocated to the secondary users.

Spectrum detection techniques are classified into three types such as: cooperative system, non-cooperative system and interference-based system [3] as seen in Fig. 1.

© Springer Nature Switzerland AG 2018
T. K. Dang et al. (Eds.): FDSE 2018, LNCS 11251, pp. 200–213, 2018.
https://doi.org/10.1007/978-3-030-03192-3_15

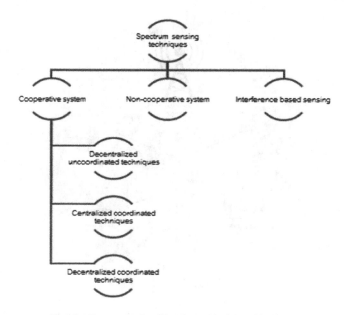

Fig. 1. Spectrum classification techniques (Author).

The advantages of cooperative techniques include: reduction of the sensibility requirements, minimization of costs and power, etc. The main disadvantage of this technique lies in the periodical demand in terms of the detection of the spectrum.

The centralized detection has a cognitive radio controller (Fig. 2). When the main user is found, the cognitive radio informs the controller (Fig. 2). Then, the controller transmits all the cognitive radio users over the main user. The centralized coordinated techniques have two types: partially cooperative techniques and fully cooperative techniques. In the partially cooperative techniques, all nodes cooperate between them on the channel detection while the fully cooperative technique the nodes cooperate between them both in the channel detection and in the transference of information between nodes [3].

The purpose of the cognitive controller is to coordinate the actions of the CR using Machine Learning algorithms. However, only in recent years, has there been a growing interest in the application of learning algorithms in Cognitive Radio [4]. The main challenge is the complexity and convergence of these techniques under a limited amount of time. With the purpose of reducing the complexity and achieving an efficient real time resource allocation, the CRs use automatic learning and artificial intelligence to make decisions based on models built using the cognitive radio's learning capacity. However, this is not complete or precise due to limited training data [2].

For machine learning in CR, the main steps involve the observation of the environment and analyzing its responses in terms of feedback, learning, keeping decisions and observations which update the model. This leads to highly accurate decision-making over management of resources and adjusting the transmission errors as can be seen in Fig. 3.

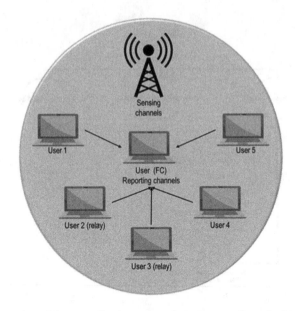

Fig. 2. Representation of the centralized spectrum detection coordinated with Cognitive Radio users (Author).

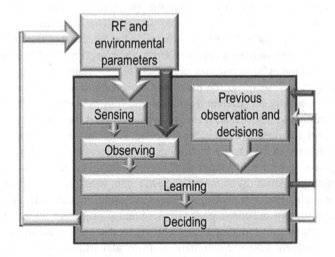

Fig. 3. Principales pasos de Machine Learning en RC [1].

The contributions and results of the document are hereby summarized: A new learning strategy is introduced to analyze spectrum behavior by applying ANFIS and time series to obtain prediction results based on comparisons of the obtained data.

The rest of the document is organized as follows: Sect. 2 offers the architecture of the CR model; the recollection process of data is described in Sect. 3, Sect. 4 presents the ANFIS model, Sect. 5 shows the results and Sect. 6 discusses them. Conclusions are shown in Sect. 7.

2 Background

In the "Least-square support vector machine-based learning and decision making in cognitive radios" [5] a model for decision-making is exposed in a Cognitive Radio scenario based on learning and Support Vector Machines (SVM). For the methodology, the authors focus on pre-processing the data which implies their normalization, reduction, grouping and selection of the learning method which determines the setting parameters. A peer-to-peer wireless communication is built with orthogonal frequency division multiplexing (OFDM) technology. C is established as $C' = d(K_l, K_R, E', R')$ where $d(*)$ indicates the decision-making function, E' is the current condition of the environment and R' is the user requirement. The decision model or function uses the learned knowledge (Kl) and the previous knowledge (Kr) to obtain adequate setting parameters. By recollecting communication instances from MATLAB simulations and taking into account the calculated parameters, it is concluded that their selection and configuration is highly important in smart decision-making in CR systems so that there is no insufficient or excessive adjustment of data. This leads to a comparison-based multiclass classification method and a parameter search through Genetic Algorithms (GA).

In the article exposed in 2013 by Bkassiny et al., a variety of learning problems in CR is shown as well as the importance of the application in artificial intelligence in solving them; The two main categories of learning in CR are established such as decision-making (where rules and decision policies are made for CR) and classification (where different observation models are identified). The three basic conditions for intelligence in a Cognitive Radio system are: perception (obtained through the sensorial observations to classify and organize them into adequate categories) and finally reasoning from what has been learned. The main classification methods in the CRs are exposed such as Support Vector Machines used in pattern recognition and the classification of objects characterized by the absence of local minimum and the mapping of input vectors in a linear-based separated space of high dimension. The authors establish a protocol where every secondary user can independently bet on each main channel and the offer that saves the most energy is chosen. There is also a distributed algorithm where each secondary user updates his strategy based on local information in order to converge to the equilibrium point; this is seen within a decentralized Cognitive Radio architecture [4].

In the article called "Machine Learning Techniques with Probability Vector for Cooperative Spectrum Sensing in Cognitive Radio Networks" [6] a low dimension probability vector (n-dimensional energy vector in a CR network with one primary user and n secondary users) is proposed as a characteristics vector for classification based on automatic learning, improving the spectrum detection and establishing the duration of the training and the classification delay. By testing on some scenarios (varying the number of SU and simulating with some learning algorithms), the spectrum detection's accuracy is highlighted with short training duration and a lower classification time in comparison to the energy vector and the two-dimensional probability vector.

Sharma and Bohara in the article "Exploiting Machine Learning Algorithms for Cognitive Radio" discuss the application of Machine Learning algorithms (such as the Genetic Algorithm, the Artificial Neural Networks (ANN) and the Hidden Markov Models (HMM) in the resolution of specific problems in diverse CR scenarios; so the spectrum management can focus on different algorithms where prediction depends on the size of the data as well as the complexity of the space [7].

In "MAC Protocol Selection Based on Machine Learning in Cognitive Radio Networks", Qiao et al. gather external and internal network parameters by building a dataset that will be trained with a classification algorithm. It will be used to decide which MAC protocol will adapt to the network's current situation in dynamic networks; The candidates for classification protocols are the DFC (Distributed Function Coordination) as the competitive one and the TDMA (Time Division Multiple Access) as the non-competitive one. This assures that the established classifier can swiftly and robustly select an adequate MAC protocol. For each protocol, when the performance curve descends, the DFC classifies the samples as modifiable. When the curve has not reach the saturation region, the TDMA is modifiable.

The simulation results show that the average probability of correct classification (PoCC) of the proposed method exceeds the baseline. The classification model manages to choose the appropriate MAC protocol that fits the current circumstance of the network [8].

In an article published in 2017 [9], the authors present an architecture for the cognitive management of 5G networks with two approaches: 'SLA enforcement' (level of service) and 'Mobile Quality Predictor'. The first approach applies LSTM (Long Short Term Memory) which basically works as an automatic learning technique that contains cells with three doors to manage the state of the memory (forget, remember and update). The purpose is to anticipate and make sure that recovery actions are followed which can be seen as an early alarm system. The second approach uses automatic learning to precisely predict the bandwidth in real time and improve the quality of the service; this establishes that although the strategy is likely, it requires some modifications in the mobile nodes since the servers must support the transport protocol.

For "Profit optimization in multi-service cognitive mesh network using machine learning", the authors analyze the capacity of cognitive wireless mesh networking (CWMN) to maximize the PU's income, maintain the QoS and attend to the majority of SU. The CWMN uses the spectrum rented from the PU to overlap its traffic. The inferred CWMN was modeled, analyzed and simulated. The authors propose a scheme so that primary users control the spectrum's traffic for the spectrum's emergent state. The objective was to adjust the size and the price of the spectrum in order to maximize the net income of the PU while still keeping the QoS of the PU. Simulations revealed the capacity of the algorithm to establish the requirements of the SU and achieve the potential profit through the application of cognitive radio. This confirms that the cognitive scenarios can have additional users without affecting the QoS. The capacity of this architecture was also revealed when maintaining the QoS for users by adapting the size and price of the offered spectrum under different conditions. The Primary Users share a spectrum based on demand so they need to borrow spectrum from their neighbors while complying with interference rules; the proposed scheme leads to higher earnings [10].

3   Methodology

The GSM 850 MGZ band is used in this study. The measures correspond to one day in the week of March 2016. Since the channel distribution goes from 128 to 151 for GSM, there are 124 channels. In this research, two days are analyzed which includes patterns of use in one day of the week and the weekend.

To keep track of the previously required data, specialized equipment was used. Records of the power of the absolute frequency of a radio-electric channel in dB mW obtained in 290 ms intervals throughout 15 h per day. For the aspects regarding this document, 35725 data will be averaged every 5 min, i.e. 1034 records will be averaged to determine the following probabilities [14].

3.1   Activity Percentage

Activity Probability

$$P(Occupied) = \frac{Number\ of\ times\ occupied}{Total\ PU\ activities\ in\ the\ 0,300\ s\ interval}$$

Inactivity Probability

$$P(Availability) = \frac{Number\ of\ times\ available}{Total\ PU\ activities\ in\ the\ 0,300\ s\ interval}$$

3.2   Permanence Probability

It is the probability of permanence within a time interval. It determines the percentage of occupied tendency and comes from averaging the activity and inactivity data where 40 represents the total number of users found.

$$P(Activity\ Tendency) = \frac{\#\ Duration\ tendencies}{40}$$

$$P(Inactivity\ Tendency) = \frac{\#\ Duration\ tendencies}{40}$$

3.3   Probability of the "Amplitude Node" (Presence or Absence of the PUs)

It is the probability of the "Amplitude" node to represent the presence or absence of the PUs which is represented as the probability of having a high value of P(High) and the probability of having a low value P(Low) based on a threshold established as 89. The percentage of high and low amplitude are defined as:

$$P(High) = \frac{Average\ of\ high\ frequencies}{Total\ activities\ of\ the\ PU\ in\ the\ 0,300\,s\ interval}$$

$$P(Low) = \frac{Average\ of\ low\ frequencies}{Total\ activities\ of\ the\ PU\ in\ the\ 0,300\,s\ interval}$$

The probabilities of the nodes "State", "Amplitude" and "Occupation tendency" are generated in this way which complement the Machine Learning techniques established as follows.

K-nearest Neighbors

The K-nearest neighbors classifies an object based on the vote of the majority of its closest neighbors. In other words, the class of a new instance is predicted based on some distance metrics. The distance metric used in the nearest neighbors for numerical attributes can be a simple Euclidian distance [15]:

$$d(x, y) = \sqrt{\sum_{i=1}^{k} (x_i - y_i)^2}$$

The Euclidian distance $d(x, y)$ is used to measure the distance to find the closest examples in the pattern's space [16].

Support Vector Machine

The vector machine models are defined as vectorial spaces of finite dimension where each dimension represents a "feature" of a particular object. It has been proven to be an efficient approach in highly-dimension spatial problems [15]:

$$maxQ(a) = \sum_{i=1}^{n} a_i - \frac{1}{2} \sum_{i=1}^{n} \sum_{j=1}^{n} a_i a_j d_i d_j x_i x_j$$

where $0 \leq a_i \leq C$ for $i = 1, 2, \ldots, n$.

SVM uses the decision function $f(x)$ defined as a kernel function to calculate the output as:

$$f(x) = sign\left[\sum_{i=1}^{i} a_i d_i K(x, x_i) + b \right]$$

where $K(x, x_i)$ is the kernel function.

Decision Tree

A decision tree is a diagram flow structure includes a root node, branches and leaf nodes. The attributes of the dataset are defined with the internal nodes. The branches are the result of every test against every node. The data is divided in classes based on the value of the attribute found in the training sample described in Algorithm 1.

Algorithm 1: Decision Tree

---

1. create a node N;

2. **if** tuples in D are all of the same class, C **then**

    return N as a leaf node labeled with the class C;

3. **if** $attribute\_list$ is empty **then**

    return N as a leaf node labeled with the majority class in D;

4. apply **Attribute_selection_method**$(D, attribute\_list)$ to find the "best" $splitting\_criterion$;

5. label node N with $splitting\_criterion$;

6. **if** $splitting\_attribute$ is   discrete-valued   and multiwaysplits allowed **then**

    $attribute\_list \leftarrow$
    $attribute\_list - splitting\_attribute$;

7. **for each** outcome j of $splitting\_criterion$

    let D_j be the set of data tuples in D satisfying outcome j;

    if D_j is empty **then**

        Attach a leaf labeled with the majority class in D to node N;

8. **else**   attach   the   node   returned   by **Generate_decision_tree** $(Dj, attribute\_list)$ to node N;

9. return N;

The purpose of the decision tree is to from a set of rules to analyze a dataset sample with the purpose of making decisions on the classification of unknown data. The real operation flow of the decision tree algorithm is shown in Fig. 4.

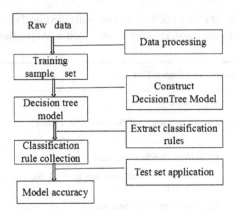

Fig. 4. Flow of the decision tree algorithm for Machine Learning [Taken from [11].

Logistic Regression

Logistic regression (LR) is a model where the result variable in the logic regression is binary and this is reflected both in the selection of a parametric model and the suppositions [12]. LR calculates the distribution between the example X and the boolean class label Y by P(X). Logistic regression classifies the Boolean class label as:

$$P(Y = 1|X) = \frac{1}{1 + \exp\left(w_0 + \sum_{i=1}^{n} w_i X_i\right)}$$

$$P(Y = 0|X) = \frac{\exp\left(w_0 + \sum_{i=1}^{n} w_i X_i\right)}{1 + \exp\left(w_0 + \sum_{i=1}^{n} w_i X_i\right)}$$

4   Proposed Method

This article compares four types of mechanical learning techniques to predict the absence or presence of primary users (PUs). The proposed construction processes of the predictive models are shown in Fig. 5.

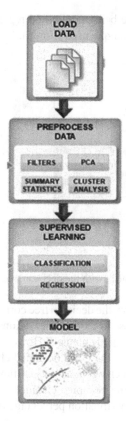

Fig. 5. Data training process [17].

This training flow includes two steps:

1. Data training which contains the data insertion where pre-processing is carried out through feature extraction. Then, the classification methods will be chosen to learn from our data. This must be iterative and regression-based since it requires to go back to the data pre-processing phase. Different automatic learning algorithms are tested with different parameters.
2. Using the data model where the pre-processing steps will be reused and the model will allow the prediction over the new data.

5   Results

The performance of the tests based on verisimilitude can be hard to estimate since large amount of data have to be simulated or recorded [13]. This research generates a confusion matrix which is used in Machine Learning to summarize the performance of classification algorithm. The calculation of a confusion matrix might be helpful during feature selection since it offers a better idea of what the classification method is doing right and what type of errors are being committed regarding the classifiers [18]. This matrix can be seen for every technique used (Fig. 6):

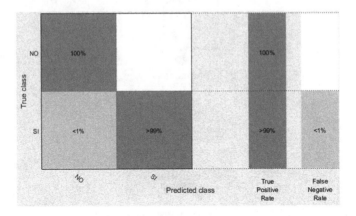

Fig. 6. Confusion matrix with the decision tree [Matlab by author].

The previous figure shows how the decision tree correctly classifies the channels that are available which represent close to 18593 observations while it makes 1 mistake over the 1873 observations in the occupied channels. However, the algorithm classifies does have a 100% classification rate for the GSM occupied channels. There is a clear difference with the SVM algorithm which wrongfully classifies 132 observations out of 18593 for the available channels. For the occupied channels, 1655 observations out of 1874 are correctly classified with a classification percentage of 98.3% as shown in Fig. 7.

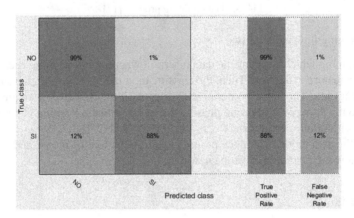

Fig. 7. Confusion matrix using SVM [Made in MatLab by author].

In terms of the KNN and logistic regression techniques, percentages of 99.9% and 97.8% are established respectively which implies that the best classification technique for spectral occupancy are the decision trees followed by KNN, SVM and logistic regression where KNN classifies 11 channels as occupied within the 18593 observations.

This means that at some time instance where the channel is unoccupied, the algorithm establishes that it is not case. Out of 1873 occupied channels, it shows that 13 are unoccupied which is not true and can be seen in Fig. 8.

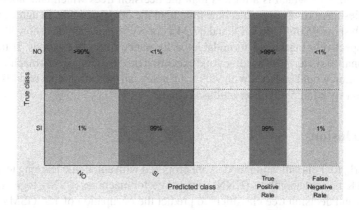

Fig. 8. Confusion matrix using KNN [Made in Matlab by author].

The technique that is least advised with a 97.8% accuracy is the Logistic Regression that detects 147 observations as occupied out of 18593 and 305 as available out of 1873 which implies a high risk in an embedded application. If the algorithm makes these types of mistakes, it will lead to interventions of the SU when the channel is really occupied by the PU. The confusion matrix for this technique is shown (Fig. 9).

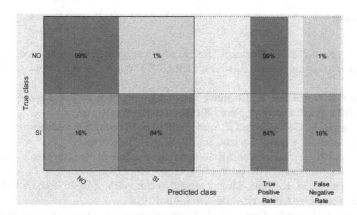

Fig. 9. Confusion matrix using Logistic Regression [Made in Matlab by author].

6   Discussion of Results

In consequence and according to [5] and [4], the Support Vector Machine algorithm is a good option to classify the occupation in Cognitive Radio channels with a 98.3% accuracy. As stated by [7], it is important to establish that spectrum management can be approached with different algorithms by establishing the architectural needs in CR. The results show that the best classification is developed by the decision tree algorithm. However, in terms of the execution time of the algorithm, SVM, KNN and LR spend an average of 42.18 s which is a lot less than the decision trees which take about 184 s. This implies three times more machine time than the other methods. In agreeance with [19] which show 95.96% for KNN and 87.34% for SVM, our methods show 99.9% and 98.3% respectively marking a formidable performance for both methods. If there is a rigorous analysis in terms of accuracy and execution time (resources on a machine level), the best option would be KNN with 99.9% accuracy, an execution time of 44.073 s and a prediction speed of 5900 observations/second.

7 Conclusions

In this work, four classification models are proposed with automatic learning techniques which are K-nearest neighbors (KNN), support vector machines (SVM), logistic regression (LR) and decision tree classifiers to predict the occupation of 124 GSM channels by establishing 7 predictors and 2 types of responses: whether the channel is occupied or not so that it can be used by the secondary user. Based on experimental results, the decision trees offer 100% accuracy on channel occupancy on behalf of the primary user which strictly accomplishes the initial goal of this study which is the classification of the GSM channels. The prediction speeds reach 46000 observations/second with a training time of 185.59 s. This research contributes to the development of embedded applications hat can maximize the efficient use of the Cognitive Radio spectrum.

References

1. Abbas, N., Nasser, Y., El Ahmad, K.: Recent advances on artificial intelligence and learning techniques in cognitive radio networks. EURASIP J. Wirel. Commun. Netw. **2015**(1), 174 (2015)
2. Elhachmi, J., Guennoun, Z.: Cognitive radio spectrum allocation using genetic algorithm. EURASIP J. Wirel. Commun. Netw. **2016**(1), 133 (2016)
3. Senthilkumar, S., GeethaPriya, C.: A review of channel estimation and security techniques for CRNS. Autom. Control Comput. Sci. **50**(3), 187–210 (2016)
4. Bkassiny, M., Li, Y., Jayaweera, S.K.: A survey on machine-learning techniques in cognitive radios. IEEE Commun. Surv. Tutor. **15**(3), 1136–1159 (2013)
5. Wu, C., Yu, Q., Yi, K.: Least-squares support vector machine-based learning and decision making in cognitive radios, pp. 2855–2863, April 2012
6. Lu, Y., Zhu, P., Wang, D., Fattouche, M.: Machine learning techniques with probability vector for cooperative spectrum sensing in cognitive radio networks. In: WCNC, pp. 1–6 (2016)
7. Sharma, V.: Exploiting machine learning algorithms for cognitive radio, pp. 1554–1558 (2014)
8. Qiao, M., Zhao, H., Wang, S., Wei, J.: MAC Protocol selection based on machine learning in cognitive radio networks, pp. 453–458 (2016)
9. Grida, I., Yahia, B., Bendriss, J.: Cognitive 5G net works: comprehensive operator use cases with machine learning for management operations, pp. 252–259 (2017)

10. Alsarhan, A., Agarwal, A.: Profit optimization in multi-service cognitive mesh network using machine learning, pp. 1–14 (2011)
11. Liu, X., Li, B., Shen, D., Cao, J., Mao, B.: Analysis of grain storage loss based on decision tree algorithm. Procedia Comput. Sci. **122**, 130–137 (2017)
12. Hosmer Jr., D.W., Lemeshow, S., Sturdivant, R.X.: Applied Logistic Regression. Wiley Series in Probability and Statistics, pp. 15–19 (2013)
13. Lingenfelter, D.J., Fessler, J.A., Scott, C.D., He, Z.: Predicting ROC curves for source detection under model mismatch, pp. 1092–1095. IEEExplorer (2010)
14. Ordoñez, J.: Caracterización de usuarios primarias para la implementación de un modelo predictor para la toma de decisiones en redes inalámbricas de radio cognitiva. In: 2016 Trabajo de Grado, Universidad Distrital Francisco José de Caldas (2016)
15. Pouriyeh, S.: A comprehensive investigation and comparison of machine learning techniques in the domain of Herat disease. In: 2017 IEEE International Conference Signal Processing, Informatics, Communication and Energy Systems (SPICES), February 2015
16. Charleonnan, A.: Predictive analytics for chronic kidney disease using machine learning techniques. In: 2014 International Conference on Advanced Computing and Communication Technologies (ICACACT 2014), pp. 1–5 (2014)
17. Mathworks: MatLab (2017), vol. 21, no. 8, pp. 680–693 (2010). https://es.mathworks.com/products/matlab.html
18. Teixera, H.: Contextual game design: from interface development to human activity recognition. Facultad de Ingeniería Universidad de Porto, Tesis maestría en Ingeniería Electrónica y de computadores (2017)
19. Bernal, C., Distrital, U., José, F., Hernández, C., Distrital, U., José, F.: Intelligent decision-making model for spectrum in cognitive wireless networks, vol. 10, no. 15, pp. 721–738 (2017)

Text-dependent Speaker Recognition System Based on Speaking Frequency Characteristics

Khoa N. Van[1], Tri P. Minh[1], Thang N. Son[1(✉)], Minh H. Ly[1(✉)],
Tin T. Dang[1], and Anh Dinh[2]

[1] Faculty of Electrical and Electronics Engineering,
Ho Chi Minh City University of Technology, Vietnam National University,
Ho Chi Minh City, Vietnam
{1413647,1411312}@hcmut.edu.vn
[2] Department of Electrical and Computer Engineering, University of Saskatchewan,
Saskatoon, Canada

Abstract. Voice recognition is one of the various applications of Digital Signal Processing and has many important real-world impacts. The topic has been investigated for quite a long time and is usually divided into two major divisions which are speaker recognition and speech recognition. Speaker recognition identifies the person who is speaking based on characteristics of the vocal utterance. On the other hand, speech recognition focuses on determining the content of the spoken message. In this project, we designed and implemented a speaker recognition system that identifies different users based on their previously stored voice samples. The samples were gathered and its features were extracted using the Mel-frequency Cepstrum Coefficient feature extraction method. These coefficients, which characterize its corresponding voice, would be stored in a database for the purpose of later comparison with future audio inputs to identify an unknown speaker. The module is currently designed to be used as a standalone device. In the future, the module is equipped with the Internet of Things (IoT) for various security systems based on human biometrics.

Keywords: Voice recognition · Mel-frequency cepstrum coefficient
Gaussian mixture model · Discrete fourier transform

1 Introduction

Speech is recognized based on evidence that exists at various levels of the speech knowledge, ranging from acoustic phonetics to syntax and semantics [1]. While the field of speech recognition mainly focuses on converting speech into a sequence of words using computer [2] to extract the underlying linguistic message in an utterance, the field of speaker recognition is concerned with the identification of a person from the characteristics of his/her voice [3]. The speaker

© Springer Nature Switzerland AG 2018
T. K. Dang et al. (Eds.): FDSE 2018, LNCS 11251, pp. 214–227, 2018.
https://doi.org/10.1007/978-3-030-03192-3_16

recognition task falls into two categories which are text-dependent recognition and text-independent recognition. Each speaker recognition system is subdivided into two steps which are identification step and verification step [4]. In identification step, the goal is to record the speaker's voice and then extract a number of features from this voice to form a speaker's voice model. On the other hand, in the verification step, a speech sample is passed through all the previously created speaker's voice models for the purpose of comparison. If the lexicon inputs in two phases are the same, the system is called text-dependent speaker recognition as opposed to text-independent speaker recognition in which no conditional constraint is put on the input lexicon [5]. Successes in the tasks depend on extracting and modeling the speaker-dependent characteristics of speech signal which can effectively distinguish one speaker from another.

In this paper, a new speaker model combines the Mel-frequency Cepstrum Coefficient (MFCC) feature extraction methodology and Gaussian mixture speaker model will be introduced for text-dependent speaker recognition. The MFCC features are the most commonly used features in speaker recognition [6]. On the other hand, because of the accuracy in extracting and representing some general speaker-dependent spectral shapes and the capability to model arbitrary densities, the Gaussian mixture models (GMM) have been used largely for modeling speaker identity. For example, in [7], Leonardo Gongora has developed a system for extracting speech features, based on the MFCC and in [8], Reynol has used GMM to build a robust system for text-dependent speaker identification and achieved highly accurate result. Moreover, several different methods for voice recognition such as Unimodal Gaussian model [9], vector quantization (VQ) codebook [10] have been compared with GMM, and GMM demonstrates its outperformance for voice recognition (VR) tasks over other methods [8].

In this paper, we present an embedded system that uses the Elechouse V3 module integrated with the two algorithms mentioned above for the purpose of speaker recognition. The paper is organized into five sections. After the introduction, Sect. 2 depicts the theoretical aspect of the method and mechanism to implement an effective algorithm to identify a speaker. Section 3 proposes the structure and design of the module. Section 4 shows the experimental results, the corresponding accuracy as well as the comparison with other systems. Finally, Sect. 5 presents some observations and conclusions of the work and then proposes further improvements.

2   Speaker Recognition System

Voice is arguably the most basic form of human communication [11]. Human voice or speech is an information-rich signal that conveys a wide range of information such as the content of the speech, the feelings of the speaker, the tone of the speech, etc. [12]. The goal of the Speaker Recognition (SR) is to extract, describe and identify the speaker based on the voice characteristics. There are many systems which can facilitate the process of speaker identification. Generally, these systems contain two phases which are identification phase and verification phase. In the identification phase, each speaker's voice is collected and used to form that speaker's corresponding model. The set of all speakers' voice models is called the speaker database (Fig. 1).

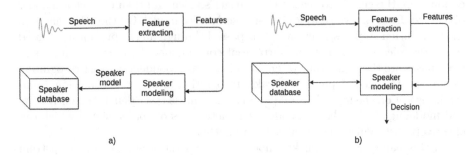

Fig. 1. Speaker recognition system: (a) The identification phase; (b) The verification phase.

Finally, in verification phase, the unknown speaker's voice data is put into the system and compared with all the models in the speaker database.

The two phases have two steps in common. The first step is to collect the voice, which can be collected through the microphone and digitalized to become discrete signals or digital signals. The second step is extraction, aiming to reduce the size of the data but still ensures sufficient information to identify the speaker. In the last step of the identification phase, the speaker's data is modeled using the GMM method and then stored in the database. Finally, in the last step of the verification phase, the extracted data is compared with all the models in the database to output the prediction about the speaker based on the likelihood. We present the above ideas of MFCC and GMM algorithms with more insights in the following sections.

2.1   Mel-Frequency Cepstrum Coefficient

Voice Signal Encoding. The simplest way of voice signal encoding is to encode the voice signal by approximating sound waves with a sequence of bytes representing the corresponding oscillation amplitude at equally spaced time intervals which are sufficiently small to maintain the sound information. This time interval unit is called the sampling rate. Figure 2(a) describes the different sampling rates that can affect the amount of information being captured which is used later in the verification phase. The value at each sampling is expressed in a specified, discrete value range called bit depth. The larger the bit depth range, the higher the identity between the sampled information and the real information (Fig. 2(b)).

Voice Feature Extraction. Speech signal covers a wide variety of information about the speaker which includes 'high-level' information such as language,

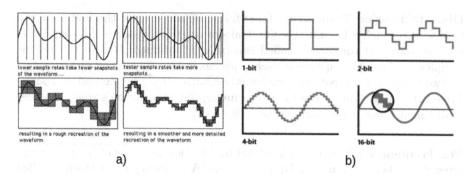

Fig. 2. Voice signal encoding: (a) The different sample rates [13] (b) The different bit depths.

context, spoken language, mood, etc. High level features contain more specific speaker-dependent information, but their extraction is very complicated [13]. Instead, low-level information such as pitch, intensity, frequency, band, audio spectrum, etc. can be easily extracted and are found to be very effective for the implementation of automatic SR systems [13]. The MFCC algorithm proposes an efficient way to extract sufficient information to distinguish one person from another.

Mel-Frequency Cepstrum Coefficients (MFCC) Feature Extraction.
MFCC is based on the evidence that information carried by the low-frequency components is more phonetically important than the high-frequency sounds [14]. Figure 3 describes the step-by-step algorithm to extract the MFCC features.

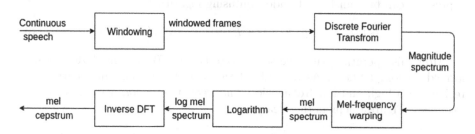

Fig. 3. MFCC feature extraction algorithm.

Windowing. In the first step, the original voice signal is divided into consecutive frames where these frames are sampled at equally spaced time intervals. Each frame's features will then be extracted using the MFCC model. In reality, the voice varies slowly, therefore if the analysis is implemented on sufficiently short time (20–30 ms) then the voice's characteristic will be stable. Features extracted at these time intervals characterize the speaker's voice. This process is called the short-term analysis.

Discrete Fourier Transform. Each frame obtained after the first step will be passed through a Discrete Fourier Transform (DFT). After performing DFT, another voice's characteristic called cepstrum is obtained. Due to the unique complexity of one person's voice, each voice has its own fingerprint qualities, i.e., no two people's voice patterns are exactly similar [16]. The method of Fourier Transform analysis maps the frequency fingerprint of a person's voice, in order to distinguish one voice from another [16].

Mel-Frequency Filtering. The signal information in terms of frequency and intensity is obtained after the DFT transform. A frequency scale which is called the mel-frequency is used to measure the perception of human ear. One of the most widely used formulas to convert from Hz to mel is from Lindsay and Loman [17]:

$$m = 2410 \log_{10}(1.6 \times 10^{-3} f + 1) \tag{1}$$

where f is the value of the frequency in Hz and m is the value of the frequency in mel.

Logarithm and Inverse Discrete Fourier Transform. Voice signal can be represented by two components which are fast-changing E and slow-changing H components [18]. It is possible to express the correlation of these two fast and slow information as follows:

$$|S(x)| = |E(x) * H(x)| \tag{2}$$

Where $E(x)$ is the high-frequency (fast-changing) component, $H(x)$ is the low-frequency (slow-changing) component and $S(x)$ is the original signal. The above expression can be translated to addition using logarithm:

$$\log_{10}(|S(x)|) = \log_{10}(|E(x)|) + \log_{10}(|H(x)|) \tag{3}$$

After this operation, the inverse Discrete Fourier Transform (IDFT) is performed on $\log_{10}(|S(x)|)$. As a result of this transform, one can separate two regions with high and low frequencies. The frequency region needed is the low frequency. Figure 4 depicts the idea.

2.2   Gaussian Mixture Model (GMM) and Speaker Recognition

Gaussian Distribution Model and Gaussian Mixture Model. The Gaussian mixture model is defined as the weighted sum of M components:

$$p(\boldsymbol{x}|\lambda) = \sum_{i=1}^{M} b_i p_i(\boldsymbol{x}) \tag{4}$$

Where b_i is the mixture weight, $p_i(\boldsymbol{x})$ is the probability density of the ith component with vector $\boldsymbol{x}$ whose length is N and M is the total number of components.

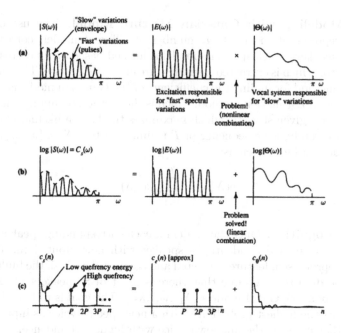

Fig. 4. Inverse Discrete Fourier Transform and separating the low and high frequencies components [15].

The sum of b_i equals 1. In other words, $\sum_{i=1}^{M} b_i = 1$. Each component density $p_i, i = 1, \ldots, M$ is a multivariate Gaussian normal distribution as follows:

$$f_x(x_1, x_2, \ldots, x_N) = p_i(x) =$$

$$\frac{1}{\sqrt{|\Sigma_i|(2\pi)^N}} \exp\left(\frac{-(x - \mu_i)^T(\Sigma_i^{-1}(x - \mu_i))}{2}\right) \tag{5}$$

Where x is the random vector and μ_i is the expected vector (mean vector) both with length N. Σ_i is the covariance matrix of size N x N. The right hand side product:

$$(x - \mu_i)^T(\Sigma_i^{-1}(x - \mu_i)) \tag{6}$$

would produce a 1×1 matrix, i.e., a real number value.
The Gaussian Mixture Model is characterized by three parameters [8]:

$$\lambda = \{b_i, \mu_i, \Sigma_i\} \tag{7}$$

where

$$i = 1, 2, \ldots, M$$

For speaker recognition, each speaker is represented by a GMM and is referred to by his/her model λ.

Speaker Modeling with Gaussian Mixture Model. The use of GMM allows the representation of a large number of different models corresponding to different speakers. Each speaker's model is formed based on its corresponding MFCC's vector which is extracted in the feature extraction phase. The most commonly used method for finding the coefficients of the Gaussian model is the Maximum Likelihood Estimation method, which is the method of finding one or more parameters for a given statistic which maximizes the known likelihood distribution [19]. Concretely, for a sequence of T training vectors $X = \{x_1, x_2, ..., x_T\}$, the likelihood can be written as:

$$p(X|\lambda) = \prod_{t=1}^{T} p(x_t|\lambda) \tag{8}$$

Speaker Recognition. Once the model for each corresponding speaker is built, the system can be used to identify a speaker with new input data. The data should be preprocessed, feature extracted and compared with all the built models stored in the database. Assume that there is a set of S speakers $\{1, 2, ..., S\}$ with S corresponding GMM models $\{\lambda_1, \lambda_2, ..\lambda_S\}$. The goal is to find the model which outputs the highest probability with a new specific voice as input. This is the model that matches the unknown voice with highest confidence.

$$\hat{S} = \operatorname*{argmax}_{1 \leq k \leq S} \Pr(\lambda_k|X) \tag{9}$$

3   Embedded System Design

3.1   Speaker Recognition Module and Microcontroller

Speaker Recognition (VR) Module Elechouse V3. In this VR embedded system, the Elechouse V3 module, currently one of the most compact voice control modules, is used. The V3 board can support up to 80 voice samples each with a duration of 1500 milliseconds. The module would compare a speaker's voice with a set of recorded voices. The module can obtain results with 99% recognition accuracy under ideal conditions [20]. The choice of microphone and the noise conditions can substantially affect the performance of the module.

Arduino Microcontrollers. The Arduino Nano and Arduino Uno R3 boards are used for the proposed embedded system. These boards are equipped with a set of digital and analog inputs and outputs where serial communications interfaces, including Universal Serial Bus (USB), are supported [21,22].

3.2   Speaker Recognition Embedded System Design

From the perspective of hardware design, the system design is divided into four units including the transmitting operation unit, the processing unit, the sampling unit, and the output unit. The system design schematic is as illustrated in Fig. 7 (Figs. 5 and 6).

Fig. 5. Voice recognition module Elechouse V3 [20].

a)   b)

Fig. 6. (a): Arduino Nano microcontroller and (b): Arduino Uno R3 microcontroller [23].

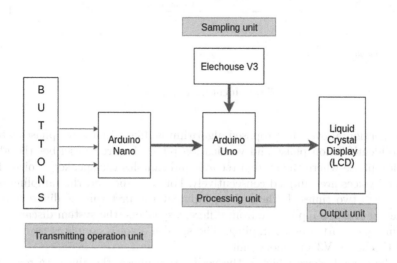

Fig. 7. Speaker recognition embedded system design.

Transmitting Operation Unit. There are three main operations which are speaker's voice sampling operation, speaker's sample loading operation and sample deleting operation. Each operation can be implemented by pushing the corresponding button on the Arduino Nano board.

Sampling Unit. Elechouse V3 is tasked to record every speaker's voice sample using a microphone. It is required that for each speaker, the sampling process repeats two times. In order to have good results, the environment noise should be reduced when the Elechouse module is recording the speaker's voice.

Processing Unit. The Arduino Uno R3 is used as the main controller, which receives commands from the Arduino Nano and data sent from the Elechouse V3. The controller would handle the received data and then display the output on the LCD (liquid crystal display) screen.

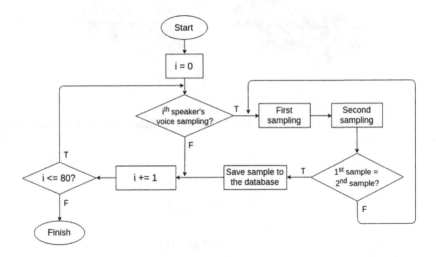

Fig. 8. Identification phase.

As mention above, the proposed algorithm is divided into two phases which are the identification phase and verification phase. Figure 8 describes the identification phase where the system records and samples the speaker's voice. The speakers' voices are sampled consecutively. For each speaker, the sampling process repeats two times. If the lexicon input at the first time is different from the second or these inputs are from different speakers, the system discards this sampling and start a new sampling. The speaker's voice sample is then stored in the Elechouse V3's memory unit.

In the second phase which is the verification phase, the aim is to recognize the speaker's voice. Firstly, all voice samples from the memory are loaded into the microcontroller. After that, the speaker whose voice needs to be recognized is recorded and sampled. This sample is then compared with all the voice samples stored in the database. If there exists a high similarity between two voice samples, the system would output the corresponding speaker's ID on the screen, otherwise it would return a null value. Figure 9 depicts this verification phase. The dataset size was limited to at most 80 individuals' voice sample because the Elechouse Module V3 can support up to 80 voice samples [20].

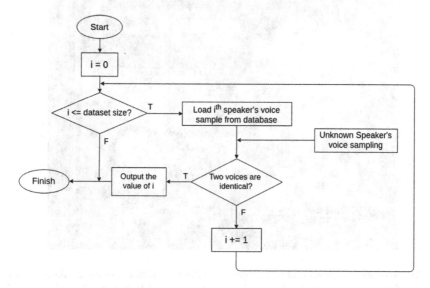

Fig. 9. Verification phase.

4   Results

4.1   Integrated System

The system was designed and implemented as shown in Fig. 10. In addition to 2 Arduino microcontrollers and the Elechouse V3 module, one LCD screen, one microphone and one battery power supply are required.

4.2   Experimental Result

The system was tested on 10 speakers (5 men and 5 women with the ages between 15 and 60), all under noisy and close-to-ideal (CTI) environment circumstance. Each person is tested 50 times. Then we evaluated the received results by comparing it with the other current systems, results were taken from [8] as the reference.

Table 1 shows that the proposed system achieved a very good accuracy (approximately 90%) in close-to-ideal environmental circumstances (i.e. without noise). This highly accurate result can be substantially affected by noise,

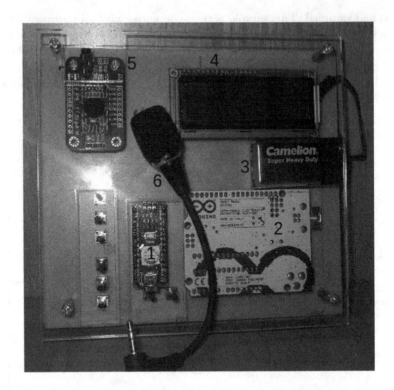

Fig. 10. The implemented embedded system with 1 - Arduino Nano; 2 - Arduino Uno R3; 3 - The power supply; 4 - The LCD screen; 5 - The Elechouse V3 module and 6 - The microphone.

Table 1. Embedded system test results

Speaker	Noisy environment		CLI environment	
	*Successes**	*Accuracy*	*Successes**	*Accuracy*
1	35	70%	45	90%
2	36	72%	44	88%
3	35	70%	45	90%
4	38	76%	45	90%
5	34	68%	45	90%
6	36	72%	44	88%
7	35	70%	46	92%
8	37	74%	44	88%
9	36	72%	45	90%
10	37	74%	45	90%

*Successes over 50 trials

which could decrease the accuracy to about 70%. However, in both noisy and close-to-ideal environment circumstance, the proposed system gave a small standard deviation value (σ) in the overall result compared to other methods, as shown in Table 2.

Table 2. Comparison to other SR methods, results taken from [8]

Methods	Accuracy (%)
GMM-nv	94.5 ± 1.8
VQ-100	92.9 ± 2.0
Our system under CTI environment circumstance[†]	89.6 ± 0.86
GMM-gv	89.5 ± 2.4
RBF	87.2 ± 2.6
TGMM	80.1 ± 3.1
Our system under noisy environment circumstance[†]	71.8 ± 1.85
GC	67.1 ± 3.7

[†]99% confidence interval

5   Conclusion

This paper has introduced and evaluated the design and operation of an embedded system using MFCC feature extraction method along with GMM model for robust text-dependent speaker recognition. The primary focus of this work is for real-world applications, such as for home security. The experimental results evaluated several aspects of using GMM models along with MFCC method for speaker recognition. Some observations and conclusions are:

- The proposed system achieved high accuracy compared to other contemporary methods. However, the overall result depends on the environment circumstance. In other words, this accuracy rate is affected by the noise condition
- To improve the accuracy of the system, state-of-the-art machine learning algorithms can be implemented to extract the voice's features along with GMM model instead of using MFCC feature extraction method.
- For the purpose of future improvement, the authors suggest programming the module to be able to take voice sample of the speaker from audio file through various media devices such as smartphone, tablet or laptop, using IoT technology. This would help the sampling step to be implemented even when the speaker is not present and help discard the necessity of using a microphone.

References

1. O'Shaughnessy, D., Deng, L., Li, H.: Speech information processing: theory and applications. Proc. IEEE **101**(5), 1034–1037 (2013)
2. Huang, X., Deng, L.: An overview of modern speech recognition. In: Handbook of Natural Language Processing, pp. 339–366. CRC Press, New York (2010)
3. Poddar, A., Sahidullah, M., Saha, G.: Speaker verification with short utterances: a review of challenges, trends and opportunities. IET Biom. **7**(2), 91–101 (2018)
4. Sara Rydin page. http://www.speech.kth.se/~rolf/gslt_papers/SaraRydin.pdf. Accessed 3 June 2018
5. Larcher, A., Lee, K.A., Ma, B.: Text-dependent speaker verification: classifiers, databases and RSR2015. Speech Commun. **60**, 56–77 (2017)
6. Yankayis, M.: Feature Extraction Mel Frequency Cepstral Coefficient (MFCC). http://citeseerx.ist.psu.edu/viewdoc/download?doi=10.1.1.701.6802&rep=rep1& type=pdf. Accessed 3 June 2018
7. Gongora, L., Ramos, O., Amaya, D.: Embedded mel frequency cepstral coefficient feature extraction system for speech processing. Int. Rev. Comput. Softw. **11**(3) (2016)
8. Reynolds, D.A., Rose, R.C.: Robust text-dependent speaker identification using Gaussian mixture speaker models. IEEE Trans. Speech Audio Process. **3**(1), 72–83 (1995)
9. Soong, F., Rosenberg, A., Rabiner, L., et al.: A vector quantization approach to speaker recognition. In: Proceedings of the IEEE International Conference on Acoustics, Speech and Signal Processing, Florida, vol. 1, pp. 387–390 (1985). https://doi.org/10.1109/ICASSP.1985.1168412
10. Gish, H., Karnofsky, K., Krasner, M., et al.: Investigation of text-independent speaker identification over telephone channels. In: Proceedings of the IEEE International Conference on Acoustics, Speech and Signal Processing, Florida, vol. 1, pp. 379–382 (1985). https://doi.org/10.1109/ICASSP.1985.1168410
11. Truax, B.: Voices in the soundscape: from cellphones to soundscape composition. In: Electrified Voices: Medial, Socio-Historical and Cultural Aspects of Voice Transfer, pp. 61–79. V&R Unipress, Gottingen (2013)
12. Johar, S.: Emotion, Affect and Personality in Speech: The Bias of Language and Paralanguage. Springer, New York (2016). https://doi.org/10.1007/978-3-319-28047-9
13. Animemusicvideos's Understanding Audio Homepage. https://www.animemusicvideos.org/guides/avtech/audio1.html. Accessed 3 June 2018
14. Nayana, P.K., Mathew, D., Thomas, A.: Comparison of text independent speaker identification systems using GMM and i-vector methods. Procedia Comput. Sci. **115**, 47–54 (2017)
15. Deller, J.R., Hansen, J.H.L., Proakis, J.G.: Discrete-Time Processing of Speech Signals. IEEE Press, New Jersey (2000)
16. Kohanski, M., Lipski, A.M., Tannir, J., Yeung, T.: Development of a voice recognition program. https://www.seas.upenn.edu/~belab/LabProjects/2001/be310s01t2.doc. Accessed 3 June 2018
17. Lindsay, P.H., Norman, D.A.: Human Information Processing: An Introduction to Psychology, 2nd edn. Academic Press, New York (1977)
18. Feng, L.: Speaker Recognition, Master Thesis at Technical University of Denmark, Kongens Lyngby (2014). http://www2.imm.dtu.dk/pubdb/views/edoc_download.php/3319/pdf/imm3319.pdf. Accessed 3 June 2018

19. Wikipedia's Maximum likelihood estimation Homepage. https://en.wikipedia.org/wiki/Maximum_likelihood_estimation. Accessed 3 June 2018
20. Elechouse Voice Recognition Module V3 datasheet. https://www.elechouse.com/elechouse/images/product/VR3/VR3_manual.pdf. Accessed 3 June 2018
21. Arduino Nano V2.3 datasheet. https://www.arduino.cc/en/uploads/Main/ArduinoNanoManual23.pdf. Accessed 3 June 2018
22. Arduino Uno datasheet. https://www.farnell.com/datasheets/1682209.pdf. Accessed 3 June 2018
23. Arduino store. https://store.arduino.cc/. Accessed 3 June 2018

Static PE Malware Detection Using Gradient Boosting Decision Trees Algorithm

Huu-Danh Pham[1], Tuan Dinh Le[2], and Thanh Nguyen Vu[1(✉)]

[1] University of Information Technology, Vietnam National University Ho Chi Minh City,
Ho Chi Minh City, Vietnam
14520134@gm.uit.edu.vn, nguyenvt@uit.edu.vn
[2] Long An University of Economics and Industry, Tan An, Long An Province, Vietnam
le.tuan@daihoclongan.edu.vn

Abstract. Static malware detection is an essential layer in a security suite, which attempts to classify samples as malicious or benign before execution. However, most of the related works incur the scalability issues, for examples, methods using neural networks usually take a lot of training time [13], or use imbalanced datasets [17, 20], which makes validation metrics misleading in reality. In this study, we apply a static malware detection method by Portable Executable analysis and Gradient Boosting Decision Tree algorithm. We manage to reduce the training time by appropriately reducing the feature dimension. The experiment results show that our proposed method can achieve up to *99.394%* detection rate at *1%* false alarm rate, and score results in less than *0.1%* false alarm rate at a detection rate *97.572%*, based on more than 600,000 training and 200,000 testing samples from Endgame Malware BEnchmark for Research (EMBER) dataset [1].

Keywords: Malware detection · Machine learning · PE file format
Gradient boosting decision trees · EMBER dataset

1 Introduction

Malware is typically used as a catch-all term to refer to any software designed to cause damage to a single computer, server, or computer network. A single incident of malware can cause millions of dollars in damage, i.e., zero-day ransomware WannaCry has caused world-wide catastrophe, from knocking U.K. National Health Service hospitals offline to shutting down a Honda Motor Company in Japan [3]. Furthermore, malware is becoming more sophisticated and more varied every day. Accordingly, the detection of malicious software is an essential problem in cybersecurity, especially as more of society becomes dependent on computing systems.

Malware detection methods can be classified in either static malware detection or dynamic malware detection [7]. Static malware detection classifies samples as malicious or benign without executing them, in contrast to dynamic malware detection which detects malware based on its runtime behavior. In theory, dynamic malware detection provides the direct view of malware action, is less vulnerable to obfuscation, and makes it harder to reuse existing malware [11]. But in reality, it is hard to collect a dataset of

© Springer Nature Switzerland AG 2018
T. K. Dang et al. (Eds.): FDSE 2018, LNCS 11251, pp. 228–236, 2018.
https://doi.org/10.1007/978-3-030-03192-3_17

malware behavior because the malwares can identify the sandbox environment and prevent itself from performing the malicious behavior. In contrast, although static malware detection is known to be undecidable in general [5], it has enormous datasets which can be created by aggregating the binaries files. It also is a critical layer in a security suite because when successful, it allows identifying malicious files before execution.

In this study, we present and optimize a static malware detection method using hand-crafted features derived from parsing the PE files and Gradient Boosting Decision Trees (GBDT), a widely-used powerful machine learning algorithm. Rather than using raw binary files, our proposed method uses the statistical summaries to decrease the privacy concerns of various benign files and makes it easy to request the balanced dataset. GBDT also shows its efficiency and accuracy by taking less time for training and achieving impressive evaluation results.

2   Related Works

Malware detection has grown over the past several years, due to the more rising threat posed by malware to large businesses and governmental agencies. Various machine learning-based static Portable Executable (PE) malware detection methods have been proposed since at least 1995 [9, 10, 13, 17, 18]. In 2001, Schultz et al. represented PE files by features that included imported functions, strings, and byte sequences [18]. In 2006, Kolter et al. used byte-level N-grams and techniques from natural language processing, including TFIDF weighting of strings to detect and classify malicious files [10]. In 2015, Saxe and Berlin leveraged novel by using a histogram of byte entropy values for input features and a multi-layer neural network for classification [17]. In 2017, Edward Raff et al. showed that fully connected and recurrent networks could be applied in the malware detection problem [14]. They also extend those results by training end-to-end deep learning networks on entire, several million-byte long executables, and encounters a wide breadth of potential byte content [13].

In Vietnam, Nguyen Van Nhuong et al. proposed a semantic set method to detect metamorphic malware in CISIM 2015 [19]. Vu Thanh Nguyen et al. proposed a combined method of Negative Selection Algorithm and Artificial Immune Network for virus detection in FDSE 2014 [12], and a metamorphic malware detection system by Portable Executable Analysis with the Longest Common Sequence in FDSE 2017 [20].

3   Proposed Method

3.1   The Issues of Using Imbalanced Dataset

Most of the related work use the imbalanced dataset [17, 20]. For examples, Saxe and Berlin used the dataset of 431,926 binaries which consists 350,016 malicious files [17], Vu Thanh Nguyen et al. used the dataset of 9690 files which only has 300 benign files [20]. In fact, the number of malicious files is often much more massive than the number of benign files because almost benign binaries are often protected by the copyright laws which do not allow for sharing. This makes malware identification problem become different from other machine learning classification problems, which commonly have fewer samples in important classes. Further, the size of the dataset is usually not large enough because the malware analysis and data labeling are consuming processes that required well-trained security engineers. There are also many risks in publishing a large dataset that includes malicious binaries.

Using imbalanced datasets can make validation metrics misleading. For examples, with 96.9% of data is malicious files, a model that labels all samples as malware achieves 96.9% accuracy, 96.9% precision (P), 100% recall (R) and 0.9843 F-score ($F = 2PR/(P + R)$ [4]). It also gives way to false positives, which cause negative user experiences. According to a survey of IT administrators in 2017, 42% of companies assume that their users lost productivity as an issue of false-positive results, which creates a choke point for IT administrators in the business life cycle [6].

3.2   Feature Extraction

By using the simple feature extraction methods inspired by the EMBER dataset owners rather than raw binary files, collecting data is not affected by privacy policies and it is much easier to get a balanced dataset. By conducting many experiments, we decrease the feature dimension by 30% (1711 instead of 2351) to reduce the training time but still manage to achieve a better evaluation result. In detail, we extracted each Portable Executable file into eight feature groups which can be classified into two types: format-agnostic features (byte-entropy histogram, byte histogram and string information) and parsed PE features. File-format agnostic feature groups decrease privacy concerns while parsed PE feature groups encapsulates the information related to executable code.

Byte-Entropy Histogram. The work [17] shows that, in practice, representing byte values in the entropy "context" in which they occur separates byte values from the context effectively. For example, x86 instruction data from byte values occurring in compressed data. To compute the byte-entropy histogram, we slide a 2048-length window over all the input bytes with a step size of 1024 bytes. Use a simple trick to calculate the entropy H faster, i.e., reducing the information by half, and pairing it with each byte within the window. Then, we compute a two-dimensional histogram with 16×16 bins that quantize entropy and the byte value. Finally, row vectors in the matrix are concatenated and normalized the final 256-value vector.

Byte Histogram. The byte histogram is a 256-value vector which represents the distribution of each byte value within the file.

String Information. The final format-agnostic group of features is string information. These features are derived from the printable sequences of characters in the range 0x20 to 0x7f, that have at least five characters long. We use the number of strings, the average length of these strings, the numbers of lines that may sequentially indicate a path (begin with C:\), an URL (start with http:// or https://), a registry key (the occurrences of HKEY_) and a bundled executable (the short string MZ). Also, we use a histogram of the printable characters within these strings.

Parsed PE Features. Five other feature groups are delivered from parsing the Portable Executable file format. The first group is the general information group, which includes the file size and necessary information collected from the PE header (the virtual size of the file, the number of imported and exported functions, the number of symbols, whether the data has a debug section, thread local storage, resources, relocations, or a signature). The second group is the information from Common Object File Format header (the timestamp in the header, the target machine and a list of image characteristics) and from the optional header (target subsystem, DLL characteristics, the file magic as a string, major and minor image versions, linker versions, system versions and subsystem versions, and the code size, header size and commit size). The next two feature groups are imported and exported functions. we apply 512-bin hashing trick to capture individual imported functions, by representing it in format library:function, for example, kernel32.dll:CreateFileMappingA. Similarly, a list of the exported functions is hashed into a list of 128-value vectors. The last group contains section information, i.e., the name, size, entropy, virtual size, and a list of strings representing section characteristics.

3.3   Gradient Boosting Decision Trees

For classification, we propose the method which uses the traditional Gradient Boosting Decision Trees (GBDT) algorithm with 400 iterations and 64 leaves in one tree. In the training step, we configure that there must be at least 200 samples in one child and set learning rate at 5%. The reasons for choosing GBDT are below.

Scalability. There are many features that is necessary to classify malicious and benign files. Even after feature selection, the massive number of features causes scalability issues for many machine learning algorithms. For example, non-linear SVM kernels require $O(N^2)$ multiplication during each iteration, and k-Nearest Neighbors (k-NN) requires significant computation and storage of all label samples during prediction. Accordingly, scalable alternatives are used in malware detection, and the two most popular algorithms are neural networks and ensemble decision trees.

Training Time. One of the most challenges that anti-malware systems usually have is the enormous amounts of data which needs to be evaluated for possible malicious intent [16]. Those results will be used to train the model to automatically evaluate samples in future. Hence, the training time becomes the essential criteria. Using neural networks usually take a long time for training. For examples, MalConv, the end-to-end deep learning model [13], took 25 h for each epoch in training with the binaries from EMBER dataset [1]. In contract, tree ensemble algorithms handle very well a large number of training samples in moderate training times. One of popular tree ensemble algorithms, which has shown its effectiveness in several challenges, is gradient boosting decision tree. GBDT is widely-used and achieves state-of-the-art performances in many tasks, such as ranking [2] and click prediction [15]. Besides, Ke et al. recently release LightGBM, which is a highly efficient gradient boosting framework that speeds up the training process up to over 20 times while achieving almost the same accuracy [8].

4   Experiment

4.1   Dataset

In our experiment, we use 600,000 labeled training samples and 200,000 testing samples from Endgame Malware BEnchmark for Research (EMBER) dataset [1] (Fig. 1).

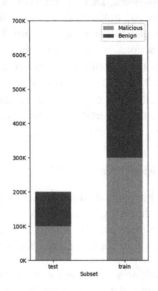

Fig. 1. Distribution of samples in EMBER dataset.

4.2   Evaluation Criteria

False Alarm Rate. False positives, or false alarms, happen when a detector mistakes a malicious label for a benign file. We intend to make the false positive rate as low as possible, which is untypical for machine learning application. It is important because even one false alarm in a thousand benign files can create severe consequences for users. This problem is complicated by the fact that there are lots of clean files in the world, they keep appearing, and it is more challenging to collect these files. We evaluate the accuracy of our method at two specific false alarm rate values: at less than *0.1%*, and at less than *1%*.

$$False\ alarm\ rate = \frac{\sum False\ positive}{\sum Condition\ negative}$$

Detection Rate. The detection rate, (eqv. with recall or true positive rate), measures the ratio of malicious programs detected out of the malware files used for testing. With higher recall, fewer actual cases of malware go undetected.

$$Detection\ rate = \frac{\sum True\ positive}{\sum Condition\ positive}$$

Area Under the ROC Curve. The Area Under the ROC curve, AUROC or AUC for short, provides an aggregate measure of performance across all possible classification thresholds. AUC is scale-invariant and measures how well predictions are ranked, rather than their absolute values. Besides, AUC is classification-threshold-invariant, so that it can measure the quality of the predictions irrespective of what threshold is chosen. A model whose predictions are *100%* wrong has an AUC of *0.0*, and the one whose predictions are *100%* correct has an AUC of *1.0*.

4.3   Experimental Results

The proposed GBDT-based malware detection method is implemented with LightGBM framework [8], and the input feature vectors have dimension of 1711. All our experiments were run on an instance which has 24 vCPUs and 32 GB memory. Using parallel programming, it took about 10 min to vectorize the raw features and about 5 min to train the model. The ROC curve of the final model is shown in Fig. 2. Tables 1 and 2 show the training time and evaluation results of our proposed model in comparison with MalConv model and the dataset owners' baseline model.

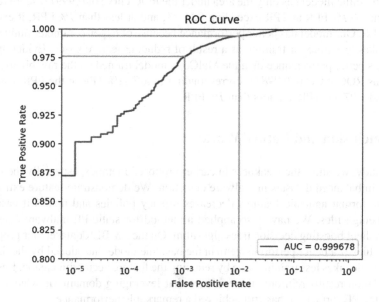

Fig. 2. The ROC curve of proposed model.

Table 1. The training time of our proposed model in comparison with MalConv model and the dataset owners' baseline model

Model	Input	Specifications	Training time
MalConv	Raw binaries	2 NVDIA TITAN X (Pascal) GPUs	10 days (25 h/epoch)
EMBER	2351-value vectors	8 vCPUs (2015 MacBook Pro i7)	20 h
Our model	**1711-value** vectors	24 vCPUs (Google Compute Engine)	**5 min**

Table 2. The evaluation results of our proposed model in comparison with MalConv model and the dataset owners' baseline model

Model	False alarm rate (FPR)	Detection rate (TPR)	Area under the ROC curve (AUC)
MalConv	0.1%	92.200%	0.998210
	1.0%	97.300%	
EMBER	0.1%	92.990%	0.999110
	1.0%	98.200%	
Our model	0.1%	**97.572%**	**0.999678**
	1.0%	**99.394%**	

The area under ROC curve exceeds *0.999678*, which means that almost all the predictions are correct. With a threshold of *0.828987*, the model score results in less than *0.1%* false alarm rate at a detection rate *97.5720%*. At less than *1%* false positive rate, the model exceeds *99.3940%* detection rate with a threshold of *0.307897*.

The baseline model has only the area under the ROC curve of *0.99911*, score results in less than *0.1%* FPR at TPR exceeding *92.99%*, and at less than *1%* FPR, it exceeds *98.2%* TPR. Our model has better performance because of hyper-parameter tuning, and it also takes less time for training as a result of reducing feature space. Evidently, the model has better performance than the MalConv model trained on the raw binaries [1], which has ROC AUC is *0.99821*, corresponding to a *92.2%* TPR at the FPR less than *0.1%*, and a *97.3%* TPR at a less than *1%* FPR.

5   Conclusion and Future Works

In this study, we show the weakness in earlier proposed methods, especially the issues in using imbalanced datasets in malware detection. We demonstrate feature extraction using file-format agnostic features decreases privacy policies and makes it easier to collect benign files. We have also applied a method for static PE malware detection using gradient boosting decision trees algorithm. On the EMBER dataset, our proposed model achieves a better performance than the baseline model introduced by the dataset owners while takes less training time by reducing the feature vector dimension. Besides, the validation results reinforce the evidence that leveraging domain knowledge from parsing the PE format file has still achieved a remarkable performance.

The study conducted in this project was a proof-of-concept, and we can identify some future developments related to the practical implementation.

Reduce the Feature Space. It is possible to reduce the dimension of feature vectors. Input vectors with smaller size boost the model and take less training time.

Use Other Datasets. Although the EMBER dataset is broad, covering most of the malware species, it does not include all possible kinds. Collecting a dataset is a task that requires a lot of time and efforts, especially in malware detection domain. With using format-agnostic features, we can receive more samples from security organizations in future.

Acknowledgment. This research is funded by Vietnam National University, Ho Chi Minh City (VNUHCM) under grant number C2018–26–06.

References

1. Anderson, H.S., Roth, P.: Ember: an open dataset for training static PE malware machine learning models. arXiv preprint arXiv:1804.04637 (2018)
2. Burges, C.J.: From ranknet to lambdarank to lambdamart: an overview. Technical report, June 2010
3. Chen, Q., Bridges, R.A.: Automated behavioral analysis of malware a case study of WannaCry ransomware. CoRR (2017)
4. Chinchor, N.: MUC-4 evaluation metrics. In: Proceedings of the Fourth Message Understanding Conference (MUC-4), p. 22. Morgan Kaufman Publishers (1992)
5. Cohen, F.: Computer viruses: theory and experiments. Comput. Secur. **6**(1), 22–35 (1987)
6. Crowe, J.: Security false positives cost companies $1.37 million a year on average (2017)
7. Egele, M., Scholte, T., Kirda, E., Kruegel, C.: A survey on automated dynamic malware-analysis techniques and tools. ACM Comput. Surv. (CSUR) **44**(2), 6 (2012)
8. Ke, G., et al.: LightGBM: a highly efficient gradient boosting decision tree. In: Advances in Neural Information Processing Systems, pp. 3146–3154 (2017)
9. Kephart, J.O., et al.: Biologically inspired defenses against computer viruses. In: IJCAI (1), pp. 985–996 (1995)
10. Kolter, J.Z., Maloof, M.A.: Learning to detect and classify malicious executables in the wild. J. Mach. Learn. Res. **7**, 2721–2744 (2006)
11. Moser, A., Kruegel, C., Kirda, E.: Limits of static analysis for malware detection. In: Twenty-Third Annual Computer Security Applications Conference. ACSAC 2007, pp. 421–430. IEEE (2007)
12. Nguyen, V.T., Nguyen, T.T., Mai, K.T., Le, T.D.: A combination of negative selection algorithm and artificial immune network for virus detection. In: Dang, T.K., Wagner, R., Neuhold, E., Takizawa, M., Küng, J., Thoai, N. (eds.) FDSE 2014. LNCS, vol. 8860, pp. 97–106. Springer, Cham (2014). https://doi.org/10.1007/978-3-319-12778-1_8
13. Raff, E., et al.: Malware detection by eating a whole EXE (2017). arXiv preprint arXiv: 1710.09435
14. Raff, E., Sylvester, J., Nicholas, C.: Learning the PE header, malware detection with minimal domain knowledge. In: Proceedings of the 10th ACM Workshop on Artificial Intelligence and Security, pp. 121–132. ACM (2017)
15. Richardson, M., Dominowska, E., Ragno, R.: Predicting clicks: estimating the click through rate for new ads. In: Proceedings of the 16th International Conference on World Wide Web, pp. 521–530. ACM (2007)

16. Ronen, R., Radu, M., Feuerstein, C., Yom-Tov, E., Ahmadi, M.: Microsoft malware classification challenge (2018). arXiv preprint arXiv:1802.10135
17. Saxe, J., Berlin, K.: Deep neural network-based malware detection using two-dimensional binary program features. In: 2015 10th International Conference on Malicious and Unwanted Software (MALWARE), pp. 11–20 (2015)
18. Schultz, M.G., Eskin, E., Zadok, E., Stolfo, S.J.: Data mining methods for detection of new malicious executables. In: Proceedings of the 2001 IEEE Symposium on Security and Privacy, pp. 38–49 (2001)
19. Van Nhuong, N., Nhi, V.T.Y., Cam, N.T., Phu, M.X., Tan, C.D.: Semantic set analysis for malware detection. In: Saeed, K., Snášel, V. (eds.) CISIM 2014. LNCS, vol. 8838, pp. 688–700. Springer, Heidelberg (2014). https://doi.org/10.1007/978-3-662-45237-0_62
20. Vu, T.N., Nguyen, T.T., Phan Trung, H., Do Duy, T., Van, K.H., Le, T.D.: Metamorphic malware detection by PE analysis with the longest common sequence. In: Dang, T.K., Wagner, R., Küng, J., Thoai, N., Takizawa, M., Neuhold, E.J. (eds.) FDSE 2017. LNCS, vol. 10646, pp. 262–272. Springer, Cham (2017). https://doi.org/10.1007/978-3-319-70004-5_18

Comparative Study on Different Approaches in Optimizing Threshold for Music Auto-Tagging

Khanh Nguyen Cao Minh[✉], Thinh Dang An, Vu Tran Quang, and Van Hoai Tran

Ho Chi Minh City University of Technology, VNU-HCM, Ho Chi Minh City, Vietnam
nguyencaominhkhanh@gmail.com

Abstract. In multi-label classification applied to music auto-tagging, the classification threshold is simply set to a constant value (called static threshold), which is usually unsuitable for the classification on imbalanced datasets. There are many approaches to solve this problem. Some find an appropriate threshold for the whole dataset, while the others find one for each tag or for each individual musical instance. In this paper, we present a method for finding an appropriate classification threshold for each individual track using multiple techniques. The ranking model used to experiment with the thresholding model is built based on fully convolutional neural network structure. The performance of the classifier including the thresholding strategy is evaluated against the classifier using static threshold on various evaluation metrics. The results show that the proposed method helps to improve the classification quality of classifier to testing instances.

Keywords: Multi-label classification · Music tagging
Thresholding strategy

1 Introduction

Tagging is a process of assigning a label (or tag) or multiple labels to a certain object. In digital era, tagging is primarily utilized to tag digital contents. This process is mainly executed by human, based on their inherent knowledge. Music tagging is similar to general tagging, instead it assigns more specific tags in music to audio clips, tracks, or albums. These tags can be genres, instrumentation, rhythmic structure, and so on. Also, the quality of music tagging depends mostly on music knowledge of listeners, or by the intention of composers.

A rising question is "Can music tagging be performed automatically?". An experienced listener can listen to audios and annotate accurately tags to those very quickly. So, it is possible to build a classifier to predict music tags by using audio signal. In this process, the classifier automatically assigns a tag to an

© Springer Nature Switzerland AG 2018
T. K. Dang et al. (Eds.): FDSE 2018, LNCS 11251, pp. 237–250, 2018.
https://doi.org/10.1007/978-3-030-03192-3_18

instance, based on its own characteristics. Music auto-tagging brings us a lot of benefits. For example, it can provide some useful information to listeners, such as genres, instrumentation, rhythmic structure. Moreover, based on characteristics of a listener's favorite tracks, music auto-tagging can help commercial products like Spotify, iTunes to recommend relevant tracks to him, which improves user experience.

To build a classifier, it is necessary to extract features from audio. However, feature extraction process is hard to be conducted because hand-crafted features must be aggregated, so determining which features are good and suitable for tagging is difficult. To overcome this problem, using deep learning is a possible approach. Deep learning has recently become widely used in audio analysis research. The advantage of deep learning is to learn the hierarchical representations of the data. Besides, deep learning scale much better with more data than traditional algorithms. Convolutional neural networks (CNN) [7] represents features in different levels of hierarchy, so the model itself can learn from low-level feature to high-level feature. Especially, many hybrid models are built based on CNN such as convolutional recurrent neural network, fully convolutional network (FCN). FCN is built on CNN but replaces the last layer by a convolutional layer instead of a fully connected one. Therefore, FCN maximizes the advantage of convolutional networks. It not only reduces the number of parameters by sharing weights but also aggregates features. Nevertheless, the biggest problem of deep learning is about the availability of labeled datasets. Recently, many large-scale music datasets are released, such as Million Song Dataset (MSD) [1], MagnaTagATune (MTAT) [6], Free Music Archive (FMA) [4]. These datasets are used in many studies because of their quality. Another challenge is the noise in the datasets.

In single-label classification, the label that has the best score can be chosen to be the output, while in multi-label classification, many labels can be chosen to form the result. Area under the curve (AUC) value is often used as a reliable metric to reflect how well a multi-label classifier performs. Recently, many researches only focus on AUC to evaluate the performance of classifiers because of the accuracy paradox. However, if AUC is used as the only evaluation metric, some limitation still remains. Specifically, high value in AUC does not assure high value in accuracy. Currently, most of classification models use a same static threshold for all testing instances and all labels. AUC is not affected by this threshold, but accuracy is. Some tracks are hard to be classified to specific tags (labels) because the scores of all tags are not higher than the static threshold, though there is at least one score higher than the others. Therefore, the classifier predicts that those tracks do not belong to any tag. This leads to a significant decline in value of accuracy.

The first approach to think of is based on scores in validation set to configure a static threshold for all track instances. For example, it can be done by performing a brute-force search for the best of threshold t under a given criterion. t will then be applied to the scores produced for verification set to evaluate the overall performance of the classifier. However, this approach has two problems. First, the score distribution on validation set does not represent the counter part in testing data. For example, although the score distribution on validation set (Fig. 1a) and that on verification set (Fig. 1b) show many similarities, there are still differences between them in detail. Second, all scores of a specific track can be lower than the static threshold, which results in no label belonging to this track, although one of its scores is much higher than the others. So, assigning a constant threshold to all instances is not appropriate.

There are researches that propose algorithms to generate a reasonable threshold for each instance instead of using a static one. RCUT [8] generates a threshold to guarantee that a fixed proportion of tags is predicted as positive, which is not common with music tagging problems where a number of tags assigned to each instance are not constant. Beside RCUT, MCUT [8] is a wonderful method to produce the threshold, it splits positive predictions and negative ones in a position where two scores have the largest distance. In general, it plays an important role to choose an appropriate threshold based on the characteristics of each instance. There are various metrics to assess this threshold. Among them, F_β [9] is the most suitable, with β is used to adjust weight between precision and recall.

The basic idea of above researches is to separate the classifier into two individual parts, the ranking model and the thresholding model. For each musical instance, the ranking model generates scores of all tags whereas the thresholding model generates an appropriate threshold value. The overall result of the classifier recommends tags whose scores are higher or equal to the generated threshold. However, the approaches mentioned above do not perform well in practice. There are two main reasons for this. First, they are hard-coded (for RCUT it is the fixed proportion of tags that are predicted as positive). Second, they do not utilize the features of each instance. To deal with these limitations, we propose a method to implement a thresholding model which can learn from labeled instances and then generates a good threshold for testing ones based on their characteristics.

The rest of the paper is organized as follows. Section 2 mentions researches working on determining an appropriate threshold value, their pros and cons. Section 3 describes our proposed method in details while Sect. 4 presents datasets used in this study, the structure of experimented models, the optimization and the evaluation metrics as well as experiment results. Finally, Sect. 5 summarizes the achievement of the research.

2   Related Work

Currently, all of the work uses ranker's score, which is output of the ranking model, as input to the threshold generator. Some algorithms require training or optimizing on training set before deployment. The training set may be the validation set of ranker, or can be created by cross validation when training ranker on its training set. That training set has the same label set as ranker's dataset but its instance set is ranker's scores.

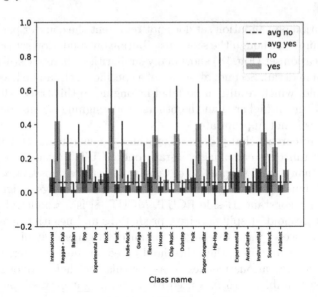

(a) On validation set

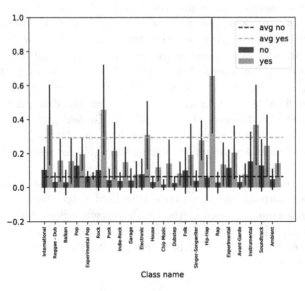

(b) On verification set

Fig. 1. Mean and standard deviation of score by genre. Yellow color shows the set of instances having that genre, while blue shows set of instances that do not. (Color figure online)

The study [8] provides basic method for threshold selection. SCut finds threshold to minimize specific loss function, TCut finds threshold to guarantee the ratio of positive prediction on training set. TCut and SCut are label-level thresholding algorithms while RCut, MCut are instance-level thresholding ones. RCut returns the threshold that lets classifier predict k labels for each individual instance. For MCut, label scores are sorted first, resulting threshold is the average value of two consecutive scores, whose gap is maximum.

Some works research on label-based thresholding [5,10]. For each class, look for the threshold that optimize specific label-level classification metric. [5] provides method for mapping from threshold optimized in training set to verification set, but this method will not work on different instances. In general, all of the optimization methods above tend to be overfit on training set.

Machine learning is the statistical method that work well on testing data by optimizing on training set. For each instance, the proposed method generates instance-level threshold that optimize specific instance-level classification metric, this method involves using k-nearest neighbors (KNN) [2].

In this paper, besides machine learning technique (in particular, KNN), we propose another instance-based approach to determine threshold that uses deep learning and transfer learning. Moreover, our model is optimized using various metrics Subsect. 4.2. Finally, we experiment our method on two popular large-scale music dataset MSD and MTAT, then compare it with machine learning one and classifier using static threshold.

3   Different Approaches in Determining Instance-Based Threshold

As mentioned above, the ranking model f takes its input as a waveform x of a track. Its output is a score vector $s \in [0,1]^q$ of that track. However, the main purpose is to classify tags, which means the expected output is a binary vector $p \in \{0,1\}^q$. Each element of p indicates the predicted classification of a certain tag. If $p_i = 0$, that track does not belong to tag i, and vice versa for $p_i = 1$. So, it is necessary to have a thresholding model l whose output is a threshold value t. If the score s_i of a certain tag i is higher than t, $p_i = 1$. And if not, $p_i = 0$.

To implement l, we propose an instance-based thresholding model. This means, instead of estimating a static threshold for all instances or all labels, the instance-based method will generate threshold value for each particular instance x. The value is used to decide the classification result p. Our proposed method to implement the instance-based thresholding model is described as follows.

3.1   Splitting the Dataset

Note that, the proposed classifier consists of two individual parts - the ranking model and the thresholding model. They both need to be trained, so it is necessary to split the dataset appropriately for our usage. Given the dataset is $D = \{(x,y)\} = D^{Tr} \cup D^{Va} \cup D^{Ve}$, with $x \in \mathbb{R}^d$ is waveform of track instance and $y \in [0,1]^q$ is the corresponding score vector of all tags. Because y is the true label of x, so each of its element is standardized to 0 or 1, which guarantees that the track belongs or does not belong to a certain tag. Since the ranker is trained on the training set $D^{Tr} = \{(x^{Tr}, y^{Tr})\}$, it is not used to train the thresholding model. Neither is the verification set $D^{Ve} = \{(x^{Ve}, y^{Ve})\}$ because it is used to evaluate the performance of both model. Validation set $D^{Va} = \{(x^{Va}, y^{Va})\}$ is the last resort. D^{Va} is used to generate $D_t = D_t^{Tr} \cup D_t^{Va}$, which are used to train and validate the thresholding model, respectively. Generating method of D_t will be described later.

3.2 Building and Training the Ranking Model

The ranking model f is built and trained on the training set D^{Tr}, detailed block diagram of the model is shown in Fig. 2.

The ranking model is a convolutional neural network which is designed based on [3] (Fig. 2). Its input is the waveform which is then transformed to a spectrogram. The spectrogram has a fixed size $96 \times 1366 \times 1$ corresponding to frequency, time and channel dimension, respectively. It goes through batch normalization layer along the frequency axis. Then, the output of this layer is put through five convolutional blocks. Each block consists of a 2-D convolutional layer to extract features, a batch normalization layer along the channel axis to stabilize the value domain, followed by a ReLU activation layer, a dropout layer with ratio 0.2 to prevent overfit and a max pooling layer at last to aggregate features and reduce feature map size. The output of these blocks are vectorized and connected to a sigmoid dense layer whose number of neurons is equal to the number of tags. The loss function is binary cross entropy.

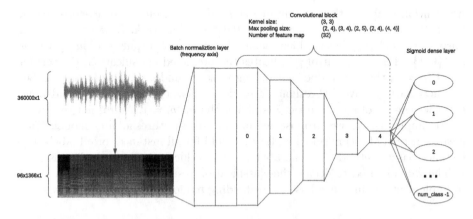

Fig. 2. Block diagram of the ranking model [3]: Audio of a track is first preprocessed to create its corresponding mel-spectrogram. This image is the input of the ranking model whose first layer is the batch normalization layer along the frequency axis, followed by five convolutional blocks. The last layer is a sigmoid dense layer.

3.3   Generating the Dataset for the Thresholding Model

The dataset D_t is used to train and validate the thresholding model. Label set of D_t is formed by using Algorithm 1 called g_m. The purpose of this algorithm is to generate an appropriate threshold that optimizes a certain instance-based metric. For each instance x^{Va} in D^{Va}, the trained ranker f^{Tr} from step 2 will predict its corresponding score vector s^{Va}. Given s^{sorted} is the descending-sorted version of s^{Va}, sequentially use each score e in s^{sorted} as the classification threshold. Consequently, the chosen label y_t for the dataset of the thresholding model l_m that optimizes metric m is the mean of two values. The first is the score $s^{sorted}_{i_m}$ that makes the predicted classification p^{Va} achieve highest $m(p^{Va}, y^{Va})$. The second is $s^{sorted}_{i_m+1}$. In general, the above algorithm g_m needs two parameters, instance x and its true label y, to generate the threshold $y_t = gm(x, y)$. When deployed, l_m is not provided with the true label $\hat{y}$ of testing instance $\hat{x}$, so it needs to learn how to predict the best threshold $\hat{y}_t$ that optimizes metric m for $\hat{x}$.

3.4   Designing the Thresholding Model

Building the thresholding model, training and validating it on D_t^{Tr} and D_t^{Va}, respectively. The thresholding generating problem is a regression one, so it is necessary to design a regressor r. The paper proposes five approaches to solve it. All of these are described as follows.

1. KNN: The regressor uses k-nearest neighbors regression with $k = \sqrt{|X_t^{Tr}|}$, $|X_t^{Tr}|$ is the number of instances in D_t^{Tr}. For each instance in verification data, r will calculate its similarity to the training instances by certain distance metric (e.g. Euclidean distance). The mean of threshold values of k closest instances is set as the threshold of verification instance, with k is the parameter of KNN. Its dataset consists of pairs (x_t, y_t), with x_t is the score vector $s^{Va} = f(x^{Va})$ of instance $x^{Va} \in D^{Va}$ and $y_t = g_m(x^{Va}, y^{Va})$. So, the dataset for the regressor r is $\{(s^{Va}, y_t)\}$.
2. KNNS: Similar to KNN, but includes sorting the input score vector in descendant. Because Algorithm 1 sorts scores to generate labels for the dataset, it is expected that adding the sorting step will help the model determine the threshold better.
3. MLP: Deep learning regressor is built based on multi-layer perceptron structure. The used structure consists of two fully-connected layers, each has 50 neurons. The loss function is the mean square error (MSE) between the predicted threshold t and the true label y_t. r is trained and validated on the same dataset as KNN and KNNS. For each instance in test set, r will predict the corresponding threshold value.
4. MLPS: Similar to MLP with the sorting step added.
5. TL: This approach uses transfer learning technique. While the instance set of above approaches is the score vector s^{Va} of $x^{Va} \in D^{Va}$, the instance set of TL is the learned features of the track extracted from the ranker that was

trained on MSD dataset (f^{MSD}) [3]. So, for this approach only, the dataset for the regressor is $\{(f^{MSD}(x^{Va}), y_t)\}$. The classification algorithm used for TL is k-nearest neighbors regression.

6. T0.5: The thresholding model that always generates a threshold value 0.5 for all music instances is also experimented as a baseline to compare with the five mentioned approaches.

Figure 3 visualizes how the data is used in five proposed approaches.

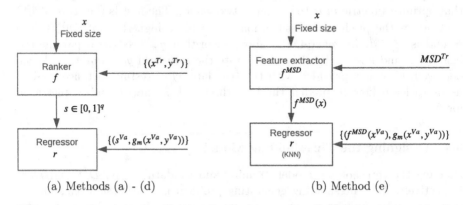

(a) Methods (a) - (d)   (b) Method (e)

Fig. 3. Data usage in two training method classes of thresholding models l. Figure 3a describes methods 1–4. Figure 3a describes method 5. Horizontal arrows show data used in training while vertical arrows show data in verification. In all five approaches, the ranker f is trained on D^{Tr}, and MSD^{Tr} specifically in approach TL. For the regressor r, approaches 1–4 use the dataset containing pairs of score vector as instance and threshold value as label. Approach TL uses the dataset containing pairs of extracted feature from f^{MSD} as instance and threshold value as label.

4   Experiment

4.1   Dataset

To train and evaluate performance of the proposed method, we use three datasets Million Song Dataset (MSD) [1], MagnaTagATune (MTAT) [6] and Free Music Archive (FMA) [4]. These datasets appear to be well suited for training Deep learning ranking model because of their large collection of tracks and tags.

While it is very easy to download both data and metadata of MTAT and FMA dataset, MSD forces researchers to download audio clips from online services. But thanks to Mr. Keunwoo Choi[1], we managed to get almost all data of MSD. We really appreciate his enthusiastic help.

For MSD, we split the dataset into training/validation/test set based on convention proposed in[2]. Specifically, only track instances that have tags in

---

[1] https://keunwoochoi.wordpress.com/.

[2] https://github.com/keunwoochoi/MSD_split_for_tagging.

Input: An instance x
      A label vector y that associates with x
      A ranking model f that was trained on training set D^{Tr} and can
      generate score $s = f(x)$ as a vector for instance x
      An instance-based metric m, assumption of m is that greater is better
Output: A threshold y_t that optimizes metric m
$s \leftarrow f(x)$;
$s^{sorted} \leftarrow sort\_descending(s)$;
Initialize M as empty list to save metric result $m(y, p)$;
for $i \leftarrow 1$ **to** $size(s^{sorted})$ **do**
    | $e = s_i^{sorted}$;
    | `// Compute bit vector p indicate the prediction after applying`
    | `threshold e to score vector s`
    | $p \leftarrow unit(s, e)$;
    | $M.append(m(y, p))$;
end
`// Find index of maximum metric result`
$i_m \leftarrow index\_max(M)$;
$y_t = average(s_{i_m}^{sorted}, s_{i_m+1}^{sorted})$;
return y_t;

Algorithm 1. Generate labels for the dataset used to train and evaluate the thresholding model.

50 most popular tags are kept. After this process, training set consists of 201680 instances, validation set 12605 instances and verification set 25904 instances.

For MTAT, this dataset has a problem of synonymous tags, so it needs to be preprocessed based on[3]. Then, MTAT is split into training/validation/verification set with ratio 12/1/3. Similar to MSD, only track instances that have tags in 50 most popular tags are kept.

For FMA, it has three versions - large, medium and small. Based on its small version which consists of 8 genres, 8000 track instances, 1000 per genre, we generate a new version called FMAS2 that includes some children of these 8 genres. The reason behind this is to support the multi-label classification problem. Consequently, FMAS2 consists of 22 genres and 8000 track instances. Then, it is splited into training/validation/verification set with ratio 8:1:1.

4.2 Optimization and Evaluation Measures

The proposed method develops multiple thresholding models, each optimizes a certain instance-based metric. In particular, these metrics are *Accuracy*, *Precision*, *Recall* and F_β. They are defined as follows:

- $Accuracy_{inst} = \frac{TP}{TP+FP+FN}$,
- $Precision_{inst} = \frac{TP}{TP+FP}$,

---

[3] https://github.com/keunwoochoi/magnatagatune-list.

- $Recall_{inst} = \frac{TP}{TP+FN}$,
- $F_{inst}^{\beta} = \frac{(1+\beta^2)Precision_{inst}Recall_{inst}}{\beta^2 Precision_{inst}+Recall_{inst}} = \frac{(1+\beta^2)TP}{(1+\beta^2)TP+\beta^2 FN+FP}$,

with TP, FP, TN and FN are True Positive, False Positive, True Negative and False Negative, respectively. β is a parameter used to adjust the weight between $Precision$ and $Recall$. In our experiment, we use $\beta = 0.5, 1, 2$.

For evaluation, above metrics are used but calculated by two ways, instance-based and label-based. Additionally, $SubsetAccuracy$ is also calculated. Moreover, they are averaged to evaluate the performance of the thresholding model on the whole testing set. Given the number of instances in testing set is D and the number of labels is q, their formulas need to redefined as below:

- $\overline{Accuracy_{inst}} = \frac{1}{D}\sum_{i=1}^{D} \frac{TP_i}{TP_i+FP_i+FN_i}$,
- $\overline{Precision_{inst}} = \frac{1}{D}\sum_{i=1}^{D} \frac{TP_i}{TP_i+FP_i}$,
- $\overline{Recall_{inst}} = \frac{1}{D}\sum_{i=1}^{D} \frac{TP_i}{TP_i+FN_i}$,
- $\overline{F_{inst}^{\beta}} = \frac{1}{D}\sum_{i=1}^{D} \frac{(1+\beta^2)TP_i}{(1+\beta^2)TP_i+\beta^2 FN_i+FP_i}$,
- $\overline{Accuracy_{label}} = \frac{1}{q}\sum_{j=1}^{q} \frac{TP_j+TN_j}{TP_j+FP_j+TN_j+FN_j}$,
- $\overline{Precision_{label}} = \frac{1}{q}\sum_{j=1}^{q} \frac{TP_j}{TP_j+FP_j}$,
- $\overline{Recall_{label}} = \frac{1}{q}\sum_{j=1}^{q} \frac{TP_j}{TP_j+FN_j}$,
- $\overline{F_{label}^{\beta}} = \frac{1}{q}\sum_{j=1}^{q} \frac{(1+\beta^2)TP_j}{(1+\beta^2)TP_j+\beta^2 FN_j+FP_j}$,
- $\overline{subset\_acc} = \frac{1}{D}\sum_{i=1}^{D}(p_i = y_i)$.

4.3 Experiment Result

To evaluate the performance of the thresholding model, we use the evaluation metrics mentioned in Subsect. 4.2. Table 1 shows the overall performance of five proposed approaches when trained and verified on MSD, MTAT, FMAS2 datasets. The experimented thresholding models $l_{F_{inst}^1}$ focus on optimizing F_{inst}^1 and the evaluated metric is $\overline{F_{label}^1}$. All five approaches give equivalent result and much better than that of T0.5. KNN, KNNS, MLP and MLPS show the best result, though KNNS and MLPS need more computational cost because of the sorting step. Moreover, the bigger the dataset is, the more slowly KNN and KNNS execute. The computational cost of transfer learning approach is much higher than that of k-nearest neighbor ones as the size of its optimal input feature vector is 64, whereas the score vector has fewer elements. In conclusion, multi-layer perceptron without sorting step approach is an appropriate choice to deploy. The rest of this section focuses on analyzing the MLP thresholding model $l_{F_{inst}^1}$.

Looking at Fig. 4, the thresholding model that optimizes $Recall_{inst}$ always generates threshold value 0, which is as expected. With this value, recall metrics are always equal to 1. So, this model is useless. We just experiment it to verify that our proposed method work well. However, the other five models generate meaningful threshold values. The standard deviation is about 0.15.

Table 1. $\overline{F_{label}^1}$ of classifiers using proposed thresholding strategies on datasets MSD, MTAT, FMAS2. All thresholding models optimize F_{inst}^1 metric. The gray cell indicates that approach TL is not experimented on MSD dataset.

Classifier	MSD dataset	MTAT dataset	FMAS2 dataset
T0.5	0.0426	0.2450	0.1385
KNN	0.1977	0.3424	0.2446
MLP	0.1912	0.3340	0.2337
KNNS	0.1948	0.3371	0.2341
MLPS	0.1914	0.3422	0.2480
TL		0.3401	0.2529

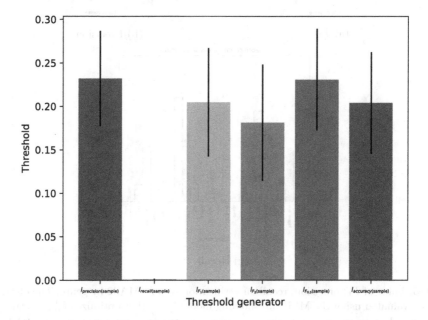

Fig. 4. Mean and standard deviation of the generated threshold from each thresholding model l_m. There are six models optimizing different instance-level measures m. They generate threshold value y_t for each instance.

The threshold values lie between 0.15 and 0.3. Referring back to Fig. 1, the score mean of instances that are predicted to have a tag is around 0.3, whereas the score mean of instances that are not is around 0.07. So, it is reasonable that the generated threshold is valued from 0.15 to 0.3. The thresholding models that prioritize recall generate lower threshold values for the same reason as $l_{Recall_{inst}}$. $l_{Precision_{inst}}$ generates highest threshold, followed by $l_{F_{inst}^{0.5}}$, $l_{F_{inst}^1}$, $l_{F_{inst}^2}$, $l_{Recall_{inst}}$.

It is obvious that $\overline{F_{label}^1}$ improves when using MLP model that optimizes F_{inst}^1 compared to T0.5 (Fig. 6). However, for each tag, not every F_{label}^1 improves

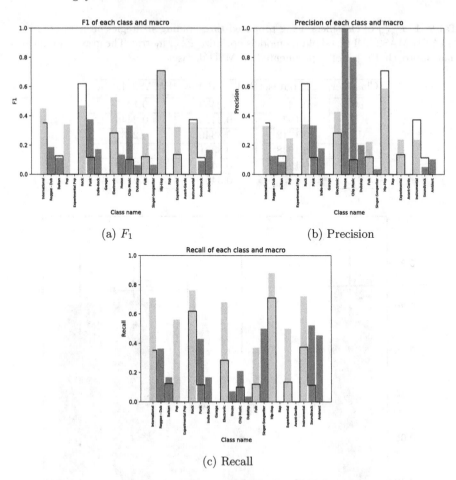

(a) F_1 (b) Precision

(c) Recall

Fig. 5. Value of F_1, precision, recall on verification set of FMAS2 by genre. The classifier is evaluated using the MLP thresholding model $l_{F_{inst}^1}$ that optimizes F_{inst}^1. Orange columns show parents of tags that are demonstrated by right blue columns. (Color figure online)

(Fig. 5a). Specifically, only F_{label}^1 of rock reduces significantly; balkan, hip-hop, instrumental and soundtrack show the same value as their corresponding baseline, the other 16 tags improve substantially compared to the baseline. F_{label}^1 is the harmonic mean of $Precision_{label}$ and $Recall_{label}$. Based on Fig. 5c, all recall values are increased, so the model has precision values reduced (Fig. 5b). Since $l_{F_{inst}^1}$ generates average threshold value less than 0.5 (Fig. 4), the classifier produces more positive predictions, which leads to the increase in recall. However, the number of false positives increasing in some tags results in the reduction in their corresponding precision.

Figure 6 visualizes results of many metrics on verification set of FMAS2 when using five models optimizing $Precision_{label}$, F_{inst}^1, F_{inst}^2, $F_{inst}^{0.5}$ and $Accuracy_{inst}$.

All these five perform significantly better than the baseline. Specifically, all evaluation results are increased except for $\overline{Precision_{label}}$, $\overline{Accuracy_{label}}$, $\overline{subset\_acc}$. The reason why $\overline{Accuracy_{label}}$ and $\overline{subset\_acc}$ decrease is because all optimization metrics do not care about true negative, while these two have true negative in their formulas. Additionally, optimizing instance-level metrics also help to increase result of label-level metrics. However, label-level metrics show less improvement than instance-level ones. Precision is an exception. For baseline T0.5, $\overline{Precision_{label}}$ is higher than $\overline{Precision_{inst}}$. While for the thresholding models, $\overline{Precision_{label}}$ is much lower than $\overline{Precision_{inst}}$, even lower than baseline's $\overline{Precision_{label}}$.

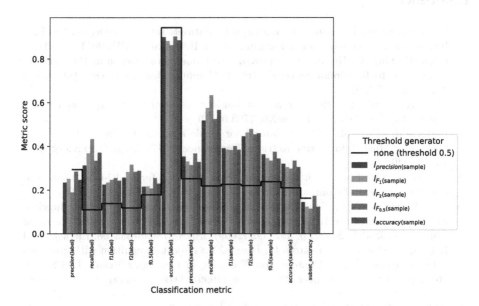

Fig. 6. Performance evaluation of the classifier with five different thresholding models. Horizontal axis consists of evaluation metrics whereas vertical axis shows performance value of each metric.

5   Conclusion

From Sect. 4, it is obvious that the classifier combined with a thresholding model achieves better performance than that using a static threshold. Through experiment, it is an appropriate choice to use multi-layer perceptron structure without sorting step to deploy a thresholding model in real projects. The improvement in the performance of the classifier gained from experiment is also significant. However, there is some aspect to consider that can help the improvement even better. A recommendation to further enhance the performance of the thresholding models is described as follows. Beside determining the target threshold based on only each individual instance, it is possible to put label-based thresholding method together with instance-based one. Consequently, for each instance, the thresholding model generates a vector, each value in that vector demonstrates the threshold for a label in label set. With this approach, instances are independent of each other, as well as labels.

Acknowledgements. The authors would like to thank Faculty of Computer Science and Engineering, HCMC University of Technology for providing computing facilities to this study. The experiments presented in this paper are tested on the High Performance Computing Lab (HPC Lab) of the faculty.

References

1. Bertin-Mahieux, T., et al.: The million song dataset. In: Proceedings of the 12th International Conference on Music Information Retrieval (ISMIR 2011) (2011)
2. Chen, B., Gu, W., Hu, J.: An improved multi-label classification method and its application to functional genomics. Int. J. Comput. Biol. Drug Des. (IJCBDD) **3**(2), 133–145 (2010)
3. Choi, K., et al.: Transfer learning for music classification and regression tasks. CoRR abs/1703.09179 (2017). arXiv:1703.09179
4. Defferrard, M., et al.: FMA: a dataset for music analysis. In: 18th International Society for Music Information Retrieval Conference (2017). https://arxiv.org/abs/1612.01840
5. Fan, R.-E., Lin, C.-J.: A Study on Threshold Selection for Multi-label Classification. Department of Computer Science, National Taiwan University (2005)
6. Law, E., et al.: Evaluation of algorithms using games: the case of music tagging. In: Proceedings of the 10th International Society for Music Information Retrieval Conference, ISMIR 2009, pp. 387–392 (2009). ISBN 9780981353708. English (US)
7. LeCun, Y., Haffner, P., Bottou, L., Bengio, Y.: Object recognition with gradient-based learning. Shape, Contour and Grouping in Computer Vision. LNCS, vol. 1681, pp. 319–345. Springer, Heidelberg (1999). https://doi.org/10.1007/3-540-46805-6_19
8. Al-Otaibi, R., Flach, P., Kull, M.: Multi-label classification: a comparative study on threshold selection methods (2014)
9. Zhang, M., Zhou, Z.: A review on multi-label learning algorithms. IEEE Trans. Knowl. Data Eng. **26**(8), 1819–1837 (2014). https://doi.org/10.1109/TKDE.2013.39. ISSN: 1041-4347
10. Zou, Q.: Finding the best classification threshold in imbalanced classification. Big Data Res. **5**, 2–8 (2016)

Using Machine Learning for News Verification

Gerardo Ernesto Rolong Agudelo[1], Octavio José Salcedo Parra[1,2(✉)],
and Javier Medina[3]

[1] Faculty of Engineering, Universidad Distrital "Francisco José de Caldas",
Bogotá D.C., Colombia
grolong@correo.udistrital.edu.co, osalcedo@udistrital.edu.co

[2] Department of Systems and Industrial Engineering, Faculty of Engineering,
Universidad Nacional de Colombia, Bogotá D.C., Colombia
ojsalcedop@unal.edu.co

[3] Faculty of Engineering, GEFEM Research Group, Universidad Distrital "Francisco José
de Caldas", Bogotá D.C., Colombia
rmedina@udistrital.edu.co

Abstract. The news fakes are issued with the intention of misleading, manipulating personal decisions, discredit or exalt an institution, entity or person or obtain economic gains or political revenue. They are related to propaganda and post-truth. Fake news, by presenting falsehoods as if they were real, are considered a threat to the credibility of serious media and professional journalists. The dissemination of false news in order to influence the behavior of a community has antecedents since antiquity, but given that its scope is directly related to the means of reproduction of information specific to each historical stage, its area and speed of propagation was scarce in the historical stages prior to the appearance of the mass media.

Keywords: Fake news · Machine learning · NLTK · Sklearn

1 Introduction

It is not possible to make a formal definition of a false news [1] since false news has become a place in today's society, from apparently innocuous publications on social networks [2] to web pages completely dedicated to the production of false information, but made in such a way that they can masterfully imitate some of the most recognized newspapers and news channels.

2 Related Work

Promising results were obtained with the previous research taking into account the existing limitations in terms of the reduced volume of the available information. This is only one of the aspects that affect this complex problem since this is a classification task that seeks optimal balance between the accuracy of the obtained classifications and the

© Springer Nature Switzerland AG 2018
T. K. Dang et al. (Eds.): FDSE 2018, LNCS 11251, pp. 251–257, 2018.
https://doi.org/10.1007/978-3-030-03192-3_19

computational cost [3]. It can be approached in two different ways since the dataset provides two types of information: it provides visual readings from the patient's vital signs that show the states of the body's various systems. Since these signals are so complex, it is necessary to train the system to recognize the current situation of the patient based on those images. In [4] two previously trained convolutional neural networks (CNN) were used and the machine managed a success rate of 83.2% when trying to classify images in 10 special categories: "ceremony", "concert", "demonstration", "football", "picnic", "race cars", "reunion", "swimming", "tennis", "traffic".

The other type of information that the dataset provides is a logbook that contains the records from the performed procedures to stabilize the patient. In previous work, where it is sought to predict future results by studying only the present that changes, a statistical analysis has been proposed in order to deliver success or failure rates [5]. This is only being explored up to now.

On the other hand in [6] present a review of several existing methods for the detection of false news, on the one hand there are works focused on the processing of news content and its form, those based on knowledge use external sources to verify the information exposed in the news. Those based on style seek to find within the news signs of language that demonstrates subjectivity or disappointment.

The study done in [7] makes an analysis of how bots have been used to spread false news on social networks like twitter and facebook.

In [8] they present a study of Deep Learning using natural language processing for the detection of false news; thus, different models are presented, and an assessment is made of which may be the best option to obtain adequate results.

3   Methodology

In the study carried out natural language processing (PLN) is used as a Python computational tool; This programming language uses different libraries and platforms, among them its PANDAS natural language processing library (Python Data Analysis Library) which is an open source library with BSD license that provides data structures and data analysis tools. Additionally, NLTK was used, which is a set of libraries and programs oriented to natural language processing and Scikit-learn which is a specialized machine learning library for classification, regression and clustering. The three libraries mentioned above have been designed to operate in conjunction with the other Numpy and Scipy libraries which were also included in the program.

To obtain news for the study, a public data set located in a github repository was used https://github.com/GeorgeMcIntire/fake_real_news_dataset compiled in equal parts for ten thousand five hundred and fifty-eight (10558) news items collected in total between the years 2015 and 2017 written in English with their title, full text and false or true label which were taken from different media, making scrapping processes in news web portals for half of real news and news from a published dataset in Kaggle conformed only by false news.

So once having the dataset, the methodology consisted of three fundamental stages; the pre-processing that involved transforming the dataset from a .csv file to a Python object belonging to Pandas; a data frame to be able to deal with it efficiently. Subsequently, for processing, the data was changed so that the first half of the data with false label and the second half with a true label were not simply what would cause impartiality when applying the machine learning methods. Once this is done, groups of data are taken to make training and test sets with which tokenisation algorithms are executed so that the result is processed by the Multinomial Naive Bayes algorithm of the Scikit-Learn package and finally an array was made in analysis. of confusion to make analysis of the results obtained.

4 Design

To begin with the processing of the data, it was necessary to use the read_csv() function of the Pandas library, passing the path of the file in which the.csv file is located, which converts to the Data Frame format. For the creation of the test and training sets, the train_test_split() function of the sklearn library was used, which takes as parameters the column with which the learning will be done, the type of classification that must be determined, the size with which will be the test set and a random to scramble the data.

Subsequently, sets "bags" of features are gathered, which are words or subsets of words with which you can extract the frequencies that have the word within the paragraphs belonging to the news texts with two different functions CountVectorizer(), but first it is necessary to do a new cleaning, since at the time of applying machine learning one looks for to see a relation between the veracity of the news and the words that more frequently appear in this one; as is logical there will be many occurrences of "stop words" is words like "that, in, on" these words that serve as connectors and to give structure to the sentences but semantically does not have a great meaning, so it becomes necessary to get rid of those "stop words" and then if you can proceed to build structures made up of all the words that are part of the news.

Having the sets of words conformed, the NaiveBayes() function is sent as arguments so that it makes the process of determining from the word bag and the training sets if the news should be classified as false or true and subsequently an open source function is used to graph a confusion matrix in which the main diagonal shows the quantity of correctly classified news and the ones that are not seen outside the diagonal.

5 Implementation

The first thing that was done so that the program functions correctly is to import the necessary libraries so that all the functions used are recognized by the interpreter.

```python
import pandas as pd
from sklearn.model_selection import train_test_split
from sklearn.metrics import accuracy_score
from sklearn.metrics import confusion_matrix
from sklearn.feature_extraction.text import CountVector-
izer
from sklearn.feature_extraction.text import TfidfVector-
izer
from sklearn.naive_bayes import MultinomialNB
import pickle
import nltk
import numpy as np
import matplotlib.pyplot as plt
import itertools
```

The file is then imported into a DataFrame from the Pandas library and formatted to make it easy to manipulate the data using the following commands.

```python
features =
pd.read_csv("fake_or_real_news.csv",usecols=['text', 'la-
bel'])
```

Thus, the parts that we will be using for this study of the news, the text and its actual classification have been stored in features; First, mixing them is done to avoid that the classification is affected by the order of the news. These two columns are separated and used to create the training and test sets.

```python
features.sample(frac=1)

trainig_set = features.text[:1900]
label_train = features.label[:1900]
test_set = features.text[1900:]
label_test = features.label[1900:]
```

The following commands make the word arrays to be generated do not contain stop words.

```python
count_vectorizer = CountVectorizer(stop_words='english')
tfidf_vectorizer = TfidfVectorizer(stop_words='english',
max_df=0.7)
```

Then other counVectorizer functions are used to do a tokenization and frequency count of the tokens and the result is put into matrices made up of the tokens of the test set and the evaluation set.

```
count_train = count_vectorizer.fit_transform(trainig_set)
count_test = count_vectorizer.transform(test_set)
```

To contrast, another way of counting the frequency of the tokens is used and again applied to the test and training sets.

```
tfidf_train = tfidf_vectorizer.fit_transform(trainig_set)
tfidf_test = tfidf_vectorizer.transform(test_set).
```

6   Discussion and Results Analysis

After running the Naive Bayes algorithm with the two forms of tokenization Count-Vectorizer and TfidfVectorizer the following percentages of certainty were obtained:

- Count Vectorizer: certainty: 0.881
- TfidfVectorizer: certainty 0.848

Likewise, the confusion matrices were plotted with the results of both classifications, producing the following results:

TfidfVectorizer (Fig. 1):

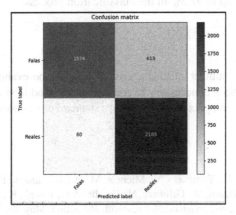

Fig. 1. Confusion matrix: TfidfVectorizer. Source: Authors

Count Vectorizer (Fig. 2):

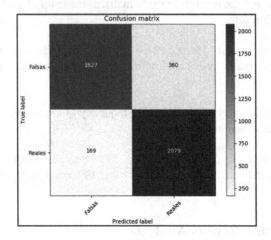

Fig. 2. Confusion matrix: Count Vectorizer. Source: Authors

This shows that it was more effective to use CountVectorizer as a classification method, since it successfully classified 89.3% of the news correctly classified as false 1827 and as true 2079.

The model used, however, proves not to have the same effectiveness as others, since in (Chiu, Gokcen, Wang, & Yan, nd) for example they use a model based on Support Vector Machines and achieve an average success rate of 95% and in (Chaudhry, Baker, & Thun-Hohenstein, nd) make an approximation using deep neural networks and achieve a certainty of up to 97.3% in the classification process.

7   Conclusions

The news has a large number of characteristics that can be evaluated and to reach a certainty greater than 95% is necessary to consider them to address an objective such as news classification is a complex task even using a standard procedure of text classification.

References

1. Mauri, M., Jonathan, G., Tommaso, V., Michele, M.: A field guide to fake news (2017)
2. Mele, N., Lazer, D., Baum, M., Grinberg, N., Friedland, L., Joseph, K., Hobbs, W., Mattsson, C.: Combating fake news: an agenda for research and action, May 2017
3. Gerazov, B., Conceicao, R.C.: Deep learning for tumour classification in homogeneous breast tissue in medical microwave imaging. In: IEEE EUROCON (2017)
4. Affonso, C., Rossi, A.L.D., Vieira, F.H.A., de Carvalho, A.C.P.D.L.F.: Deep learning for biological image classification. Expert Syst. Appl. **85**, 114122 (2017). https://doi.org/10.1016/j.eswa.2017.05.039
5. Yudin, D., Zeno, B.: Event recognition on images by fine-tuning of deep neural networks (2018). https://doi.org/10.1007/978-3-319-68321-8_49

6. Shu, K., Wang, S., Sliva, A., Tang, J., Liu, H.: Fake news detection on social media: a data mining perspective. ACM SIGKDD Explor. Newslett. **19** (2017)
7. Shao, C., Ciampaglia, G.L., Varol, O., Flammini, A., Menczer, F.: The spread of fake news by social bots (2017). arXiv:1707.07592
8. Bajaj, S.: The Pope Has a New Baby! Fake News Detection Using Deep Learning (n.d.). https://web.stanford.edu/class/cs224n/reports/2710385.pdf

Deep Learning and Applications

A Short Review on Deep Learning for Entity Recognition

Hien T. Nguyen[1]([⊠]) and Thuan Quoc Nguyen[2]

[1] Artificial Intelligence Laboratory, Faculty of Information Technology, Ton Duc Thang University, 19 Nguyen Huu Tho St., Tan Phong Ward, District 7, Ho Chi Minh City 700000, Vietnam
nguyenthanhhien@tdtu.edu.vn
[2] Ho Chi Minh City Open University, 97 Vo Van Tan St., Ward 6, District 3, Ho Chi Minh City 700000, Vietnam
thuan.nq@ou.edu.vn

Abstract. Deep learning is a kind of representation learning − a subfield of machine learning. While most machine learning methods work well thanks to feature engineering, deep learning automatically learns good feature representations of input data at multiple levels. In this paper, we present distributed representations and deep learning models that automatically learn features for coarse- and fine-grained entity recognition. The former recognizes entities with very few types, whereas the latter identifies entities and classifies them into a large number of types. Until now, most of research on entity recognition has focused on the former. However, the latter is more challenging and has attracted much research attention recently. This paper presents state-of-the-art methods for both coarse- and fine-grained entity recognition until late 2017.

Keywords: Deep learning · Named entity recognition
Entity extraction

1 Introduction

Entity recognition is a task of identifying precisely entity mentions and classifying them into the corresponding entity types according to their context. It is the most important task in the area of information extraction (IE) − a key component in human language technology. IE systems that extract entities of a small set of types (*e.g.* person, location, organization, and MISC) are referred to as performing coarse-grained entity recognition (CGER). Those systems that extract entities of a broad set of types (*e.g.* actor, director, coach, doctor, government, military, airport, city, food, *etc.*) are known as performing fine-grained entity recognition (FGER).

Previous approaches to CGER range from hand-crafted rules [54] to classical machine learning models such as Maximum Entropy Models (MEM) [49],

© Springer Nature Switzerland AG 2018
T. K. Dang et al. (Eds.): FDSE 2018, LNCS 11251, pp. 261–272, 2018.
https://doi.org/10.1007/978-3-030-03192-3_20

Conditional Random Fields (CRF) [46] or Support Vector Machines [47]. The CoNLL-2003 shared task of language-independent CGER [48] introduced a commonly public benchmark dataset and attracted many different classical learning systems. In 2007, Nadeau and Sekine [44] presented a good survey on entity recognition and showed that employed classical machine learning models are not only supervised such as [46–48] but also semi-supervised [45] and unsupervised [51]. Classical learning systems achieved good performance, nevertheless they relied on manual feature engineering processes that are costly, time-consuming, dependent on the domain and language.

Recent advances in machine learning proved that deep learning is successful in sequence modeling tasks such as Speech Synthesis [2,22], Sexual Orientation Detection [15], Machine Translation [27], or Question Answering [28]. Deep learning automatically learns features or representations of data with multiple levels of abstraction using computational models composing of multiple processing layers [32]. Its representational power is to stack multilayers of basic modules based on Multi-layer Perceptron (MLP), Convolutional Neural Networks (CNN) [52] or Recurrent Neural Networks (RNN), *i.e.*, Long-Short Term Memory (LSTM) [53] or Gated Recurrent Unit (GRU) [36]. In recent years, stacking RNN taking word embeddings and character-level representation of words as input, in combination with a classical machine learning model, *e.g.*, CRF or SVM, at the final layer, is the state-of-the-art approach to coarse-grained entity recognition [19,21,23,31].

Until now, most of research on entity recognition has focused on CGER, but coarse-grained types of entities are not enough for some Natural Language Processing (NLP) tasks such as Relation Extraction, Sentiment Analysis or Question Answering. Therefore, IE systems that extract entities of fine-grained entity types have been proposed [5,6,11,50]. Those fine-grained systems identify entity mentions and classify them into hundreds or more of entity types, *e.g.*, FINET [29] (16K types), FIGER [40] (112 types), HYENA [41] (505 types).

Unfortunately, manual annotation of the training data for FGER is extremely time-consuming, expensive and error-prone. One way to tackle this problem is to employ distant supervision paradigm to automatically generate the training data based on a knowledge base. Assigning type labels to entity mentions is inherently based on their local context; however, distant supervision assigns the same set of type labels, no matter what the local context, to all mentions of an entity, which lead to add label noise to the generated training data. Figure 1 shows an example of entity typing based on the local context, in which three different mentions *Munich* are mapped to different type labels. If three sentences S_1, S_2, and S_3 in Fig. 1 appear in a dataset, distant supervision will map the mentions *Munich* in these sentences to the same label, which causes noise in the results.

In recent years, among proposed methods for FGER, learning embeddings of mentions, context, and type labels to reduce the label noise in the training data has been the state-of-the-art technique [1,25,26]. Besides, unsupervised learning is another choice to avoid manually building a training set for FGER [5].

Furthermore, attention mechanism has been also shown significant contribution to achieve state-of-the art results [6,11].

The rest of this paper is organized as follows. Section 2 presents distributed representation. Coarse- and fine-grained entity recognition are presented in Sects. 3 and 4, respectively. Finally, we draw conclusion in Sect. 5.

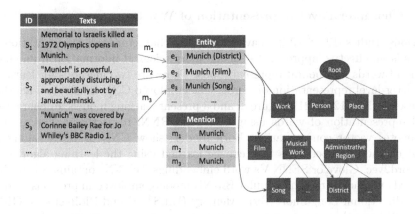

Fig. 1. An example of entity typing based on the local context

2   Distributed Representation

In the area of NLP, distributed representation has recently shown tremendous success in many tasks such as Machine Translation [27], Question Answering [28], and Dependency Parsing [35]. According to Hinton *et al.* [55], a distributed representation means many-to-many relationship between two types of representations, such as entities and processing units, where an entity is represented by many processing units and each processing unit participates in the representation of many entities. Some examples of such distributed representations are word vectors and entity vectors. While word vector captures the meaning of a word by encoding its collocates in its entries, an entity is represented as an entity vector whose entries encode its relevant information, *i.e.*, its attributes and relations. This section presents distributed representation at multiple levels for coarse- and fine-grained entity recognition.

2.1   Word Embedding

Learning distributed representations of words, called word embeddings, is an effective approach to word representation. Word embedding is a feature learning technique that applies some transformations to map a word into a lower dimensional space. In such a lower dimensional space, each vector associated with a word, namely a word vector, is "dense, low-dimensional, and real-valued" [43]. Recently, pre-trained word embeddings have become a standard component in neural networks architecture for entity recognition [19]. Three popular pre-trained models are SENNA's word embeddings [42], word2vec [39], and Glove [38], of which SENNA's word embeddings and Glove were evaluated better than word2vec. Besides, there are some approaches training word embedding from scratch [5,8] instead of using pre-trained word vectors.

2.2   Character-Level Representation of Words

Previous studies [12,18,19,23] have shown that character-level representation of words is an effective approach to encode morphological information from characters of words into neural representations. There are two methods for learning character-level representation of words. One employed CNN [19,23] and another did use BiRNN [12,18,21]. Figures 2 and 3 present learning models for character-level representation of words using CNN and RNN, respectively. The symbol $\oplus$ denotes vector concatenation. Both models show CNN or RNN taking as input character embeddings that can be learned using the learning algorithms for word2vec, Glove or SENNA's word embeddings. BiRNN contains two RNNs: forward and backward. The popular BiRNN models employed in previous studies are Bidirectional Long-Short Term Memory (BiLSTM) and Bidirectional GRU. Figure 3 shows that character representation of a word is a concatenation of two final outputs of forward and backward RNNs.

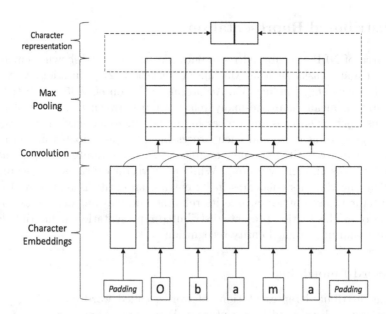

Fig. 2. A CNN model for character-level representation of a word

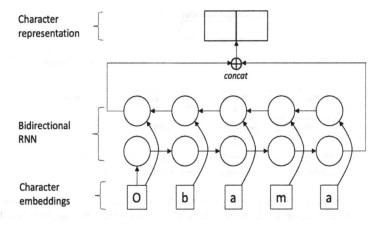

Fig. 3. BiRNN for character-level representation of a word

2.3 Mention and Context Representation

Figure 4 presents a general model for learning feature representation of an entity mention in which FNN stands for Feed-forward Neural Network. Each entity mention has words on the left and on the right. We call left context and right context, respectively. In Fig. 4, the embeddings layer performs word embeddings (Word Embed.) or character-level representation of words (Character Rep.) or both of them as shown in the first layer of Fig. 5. Note that the embeddings layer for entity mention may be different from those for the left context and the right context. In particular, in [1] the authors employed word embeddings for learning context representation and character-level representation of words for learning representation of entity mentions, whereas in [11] word embeddings are employed for learning representations of both context and mentions. In [3] and [30] the authors proposed a method that learns context representations using MLP and mention representations using RNN, respectively. Regarding to context representation, this method is different from that of [1] and [11] in that it only takes into account entity mentions, instead of all words, in the left and the right context. The attention layer is optional and it was only used in [11]. It means that the methods proposed in [1] and [3] directly concatenate the outputs of RNN layers.

2.4 Knowledge Representation of Entities

Existing open domain knowledge bases such as DBpedia or YAGO, as well as domain-specific ontologies such as AGROVOC or Gene Ontology bring useful information for fine-grained entity recognition. And, knowledge graph embedding is an effective way that exploits such a knowledge base to generate knowledge representations of entities. It projects entities and relations in a knowledge graph into a continuous vector space. One can take advantages of the knowledge graph embedding for entity recognition using some methods surveyed in [4] and [14].

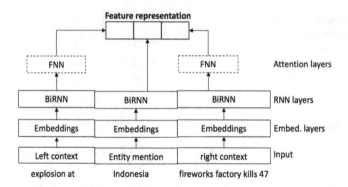

Fig. 4. An illustration of feature representation for mention "Indonesia" in the sentence "explosion at Indonesia fireworks factory kills 47" in which the left context is "explosion at" and the right context is "fireworks factory kills 47".

However, until now not much work in literature has taken the knowledge graph embedding into consideration for entity recognition. Indeed, [5] was a rarely successful work that exploited knowledge graph embedding for fine-grained entity typing.

3   Datasets

CoNLL 2003 is the most popular benchmark dataset for evaluating the performance of CGER systems. It consists of newswire from the Reuters RCV1 corpus tagged with four entity types of named entities: location (LOC), organization (ORG), person (PER), and miscellaneous (MISC). In particular, CoNLL 2003 includes 204,567 training tokens and 46,666 testing tokens. F_1 score is mostly used to evaluate the performance of CGER systems. Wiki, OntoNotes, BBN and NYT are four popular benchmark datasets for evaluating the performance of FGER systems. Wiki dataset consists of 1.5M sentences sampled from Wikipedia articles which are manually annotated using 112 types. OntoNotes consists of 13,109 news documents where 77 test documents manually annotated using 89 types. BBN consists of 2,311 Wall Street Journal articles which are manually annotated using 47 types. The NYT dataset has more than 1.8 million news articles (from 1987 to 2007) which are manually annotated using 446 types. Statistics of these datasets in detail is presented in [1] and [25]. Evaluation metrics are presented in [26].

4   Coarse-Grained Entity Recognition

A general architecture, namely BiRNN-CRF, for entity recognition presented in Fig. 5 that was used in [9,16–19,21,23] and [31] with some variants. Combination of BiLSTM and CRF in the places of BiRNN and Inference layers respectively, in this BiRNN-CRF architecture, was first proposed in [31]. Previously, in [42], the authors proposed the same architecture in which FNN was used instead of BiRNN. Then studies in [9, 19, 21] and [23] conducted different experiments to find optimal parameters of several different settings. [18] improves the studies in [9, 19, 21] and [23] by a neural reranking model.

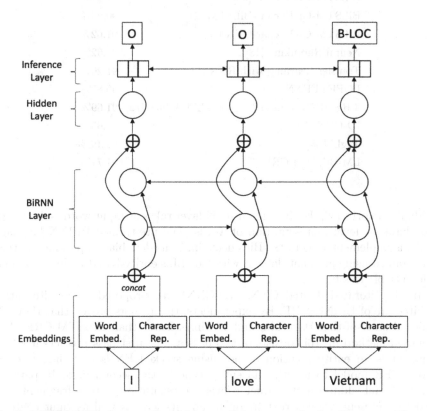

Fig. 5. BiRNN-CRF: A general architecture for coarse-grained entity recognition, in which the inference layer employs CRF model.

In [17], the authors proposed a transfer learning approach to multi-task joint training of several sequence modeling tasks, *i.e.*, POS, Chunk, and CGER. In particular, they proposed three variants of the BiRNN-CRF architecture for the settings of cross-domain, cross-application, and cross-lingual transfer, in which GRUs is used in the places of RNNs. The method proposed in [8] extends BiRNN-CRF architecture by a pre-trained BiRNN language model. In particular, each output vector of the BiRNN layer is concatenated with the corresponding output vector of the pre-trained BiRNN language model before being fed into the next layer. This method is extended in [7] by highway units for language models. [12] extends the proposed method in [8] by combining stacked residual RNNs with language models and bias decoding for CGER. Residual RNN is inspired by the

Table 1. Reprinted Performance (F1 score) comparison of state-of-the-art systems on English Entity Recognition datasets: CoNLL 2003

Method	F1
BiLSTM-CRF [31]	90.10%
BiLSTM-CNN-CRF [23]	91.21%
BiLSTM-CRF, char-BiLSTM [21]	90.94%
BiLSTM-CRF, char-CNN [19]	91.62%
Neural Reranking [18]	91.62%
Transfer Learning with RNN [17]	91.26%
FOFE+FFNN [16]	90.85%
3 Res-RNN + foreLM + backLM + bias [12]	91.69%
ID-CNN [10]	90.65%
TagLM [8]	**91.93%**
LM-BiLSTM-CRF [7]	91.71%

residual networks [20]. Each residual RNN layer takes as input word embeddings and character-level representation of words. The pre-trained BiRNN language model is implemented as in [8]. Bias decoding considers bias as parameters that were trained using gradient descent with CoNLL's entity-based F_1 loss − a non-differentiable loss function.

In [10], iterated dilated CNNs (ID-CNN) was proposed as an alternative architecture of BiRNN-CRF. By experiments, the authors proved that ID-CNN achieved state-of-the-art results and 14 times faster than BiLSTM-CRF. The proposed method in [16] consists of two passes. The first pass predicts if a text fragment is an entity mention or not using stacked FNNs and then feeds it into a CFR model to identify its entity type in the second pass. In particular, word embeddings and character-level representation of text fragments are used to represent a given text fragment and its left, as well as right context. Since FNN requires the inputs having a fixed-size length, the fixed-size ordinally forgetting encoding method proposed in [34] is used to transform a variable-length sequence, *i.e.*, a text fragment or its context, into a fixed-size one. Table 1 presents state-of-the-art results of coarse-grained entity recognition.

Table 2. Reprinted results of state-of-the-art fine-grained entity recognition methods on the three datasets

Typing methods	Wiki/FIGER(GOLD)			OntoNotes			BBN		
	Acc	Ma-F1	Mi-F1	Acc	Ma-F1	Mi-F1	Acc	Ma-F1	Mi-F1
Feature representation [1]	0.658	0.812	0.774	0.522	0.685	0.633	0.604	0.741	0.757
Attention [11]	0.545	0.747	0.715	0.503	0.679	0.616	−	−	−
AFET [25]	0.553	0.693	0.664	0.551	0.711	0.647	0.670	0.727	0.735

5   Fine-Grained Entity Recognition

Since distant supervision generates training data with noise, in [25] and [26], the authors proposed different methods that jointly learn embedding of entity mentions, text features, and type labels (type-paths) to reduce label noise for fine-grained entity typing systems. [33] and [24] also focus on learning embeddings of features and type labels. However, these methods did not use any deep learning model. The common idea in [1,6,11,30] is jointly learning feature representations and embeddings of type labels. The general model for automatically learning feature representations of an entity mention and its context is presented in Fig. 4, of which [6] and [11] used attention mechanism.

Since fine-grained types tend to form a hierarchical structure (*e.g.*, *actor* is a subtype of *artist*, which in turn is a subtype of *person*), state-of-the-art methods could utilize the type hierarchy to encode type labels [1,11]. Moreover, due to label noise in training data, state-of-the-art methods divide the training set into two parts. One consists of *clean* entity mentions and another consists of *noisy* entity mentions. Then, two different loss functions are used to model *clean* and *noisy* entity mentions. In such a case, one can collectively minimize the objectives of two kinds of loss functions by solving joint optimization problem that is a linear combination of the sum loss of each kind. Table 2 presents state-of-the-art results of fine-grained entity recognition.

6   Conclusion

In this paper, we present a short review on entity recognition consisting of coarse- and fine-grained entity recognition. For coarse-grained entity recognition, a generally state-of-the-art approach is presented in Fig. 5 with some variants in different studies. This approach includes three main modules: (i) the first one takes word embeddings or character-level representation of words as inputs, (ii) the second one transforms outputs of the first module by LSTMs or GRUs, and (iii) the third one employs a classical machine learning model, *e.g.*, CRF or SVM. For fine-grained entity recognition, state-of-the-art methods focus on learning feature representations based on deep neural networks in combination with knowledge graph embedding. Overall, those state-of-the-art methods proposed for entity recognition are based on neural networks taking as input word embeddings, character embeddings, or context embeddings, *i.e.*, dense distributed representation. However, they did not exploit linguistic knowledge. In addition, recent research in the field of neuroscience has shown that one of principles of neocortical function is sparse distributed representation. Besides, some deep neural networks such as capsule networks [13] or generative adversarial networks [37] have been proved superior to CNN in the field of computer vision. One can put these lines of research for entity recognition in the future.

References

1. Abhishek, A.A., Awekar, A.: Fine-grained entity type classification by jointly learning representations and label embeddings. In: Proceedings of the 15th Conference of the European Chapter of the Association for Computational Linguistics (EACL), pp. 797–807 (2017)
2. Arik, S.O., et al.: Deep Voice: Real-time Neural Text-to-Speech. arXiv preprint arXiv:1702.07825 (2017)
3. Cui, K.Y., Ren, P.J., Chen, Z.M., Lian, T., Ma, J.: Relation enhanced neural model for type classification of entity mentions with a fine-grained taxonomy. J. Comput. Sci. Technol. **32**(4), 814–827 (2017)
4. Cai, H., Zheng, V.W., Chang, K.C.C.: A Comprehensive Survey of Graph Embedding: Problems, Techniques and Applications. arXiv preprint arXiv:1709.07604 (2017)
5. Huang, L., May, J., Pan, X., Ji, H., Ren, X., Han, J., Zhao, L., Hendler, J.A.: Liberal entity extraction: rapid construction of fine-grained entity typing systems. Big Data **5**(1), 19–31 (2017)
6. Karn, S.K., Waltinger, U., and Schütze, H.: End-to-end trainable attentive decoder for hierarchical entity classification. In: EACL, pp. 752–758 (2017)
7. Liu, L., et al.: Empower Sequence Labeling with Task-Aware Neural Language Model. https://arxiv.org/pdf/1709.04109v3.pdf (2017)
8. Peters, M.E., Ammar, W., Bhagavatula, C., Power, R.: Semi-supervised sequence tagging with bidirectional language models. In: Proceedings of the 55th Annual Meeting of the Association for Computational Linguistics (ACL), pp. 1756–1765 (2017)
9. Reimers, N., Gurevych, I.: Optimal Hyperparameters for Deep LSTM-Networks for Sequence Labeling Tasks. arXiv preprint arXiv:1707.06799 (2017)
10. Strubell, E., Verga, P., Belanger, D., McCallum, A.: Fast and accurate sequence labeling with iterated dilated convolutions. In: Proceedings of the Conference on Empirical Methods on Natural Language Processing (EMNLP) (2017)
11. Shimaoka, S., Stenetorp, P., Inui, K., Riedel, S.: Neural architectures for fine-grained entity type classification. In: EACL, pp. 1271–1280 (2017)
12. Tran, Q., MacKinlay, A., Yepes, A.J.: Named entity recognition with stack residual LSTM and trainable bias decoding. In: Proceedings of the 8th International Joint Conference on Natural Language Processing (IJCNLP) (2017)
13. Sabour, S., Frosst, N., Hinton, G.E.: Dynamic routing between capsules. In: Advances in Neural Information Processing Systems (NIPS) (2017)
14. Wang, Q., Mao, Z., Wang, B., Guo, L.: Knowledge graph embedding: a survey of approaches and applications. IEEE Trans. Knowl. Data Eng. **29**, 2724–2743 (2017)
15. Wang, Y., Kosinski, M.: Deep neural networks are more accurate than humans at detecting sexual orientation from facial images. J. Pers. Soc. Psychol. (2017)
16. Xu, M., Jiang, H., Watcharawittayakul, S.: A local detection approach for named entity recognition and mention detection. In: ACL, pp. 1237–1247 (2017)
17. Yang, Z., Salakhutdinov, R., Cohen, W.W.: Transfer learning for sequence tagging with hierarchical recurrent networks. In: ICLR (2017)
18. Yang, J., Zhang, Y., Dong, F.: Neural reranking for named entity recognition. In: Proceedings of RANLP (2017)
19. Chiu, J.P., Nichols, E.: Named entity recognition with bidirectional LSTM-CNNs. TACL **4**(1), 357–370 (2016)

20. He, K., Zhang, X., Ren, S., Sun, J.: Deep residual learning for image recognition. In: Proceedings of the IEEE Conference on Computer Vision and Pattern Recognition (CVPR), pp. 770–778 (2016)
21. Lample, G., Ballesteros, M., Subramanian, S., Kawakami, K., Dyer, C.: Neural architectures for named entity recognition. In: The 15th Annual Conference of the North American Chapter of the Association for Computational Linguistics: Human Language Technologies (NAACL-HLT) (2016)
22. Oord, A.V.D. et al.: Wavenet: A Generative Model for Raw Audio. arXiv preprint arXiv:1609.03499 (2016)
23. Ma, X., Hovy, E.: End-to-end sequence labeling via bi-directional LSTM-CNNs-CRF. In: ACL, pp. 1064–1074 (2016)
24. Ma, Y., Cambria, E., Gao, S.: Label embedding for zero-shot fine-grained named entity typing. In: COLING, pp. 171–180 (2016)
25. Ren, X., He, W., Qu, M., Huang, L., Ji, H., Han, J.: AFET: automatic fine-grained entity typing by hierarchical partial-label embedding. In: EMNLP (2016)
26. Ren, X., He, W., Qu, M., Voss, C.R., Ji, H., Han, J.: Label noise reduction in entity typing by heterogeneous partial-label embedding. In: Proceedings of the 22nd ACM SIGKDD Conference on Knowledge Discovery and Data Mining (KDD) (2016)
27. Wu, Y., et al.: Google's Neural Machine Translation System: Bridging the gap between Human and Machine Translation. arXiv preprint arXiv:1609.08144 (2016)
28. Xiong, C., Merity, S., Socher, R.: Dynamic memory networks for visual and textual question answering. In: International Conference on Machine Learning (ICML), pp. 2397–2406 (2016)
29. Corro, L.D., Abujabal, A., Gemulla, R., Weikum, G.: FINET: context-aware fine-grained named entity typing. In: EMNLP, pp. 868–878 (2015)
30. Dong, L., Wei, F., Sun, H., Zhou, M., Xu, K.: A hybrid neural model for type classification of entity mentions. In: IJCAI, pp. 1243–1249 (2015)
31. Huang, Z.H., Xu, W., Yu, K.: Bidirectional LSTM-CRF Models for Sequence Tagging. arXiv:1508.01991 (2015)
32. LeCun, Y., Bengio, Y., Hinton, G.: Deep learning. Nature **521**(7553), 436–444 (2015)
33. Yogatama, D., Gillick, D., Lazic, N.: Embedding methods for fine grained entity type classification. In: Proceedings of the 53rd Annual Meeting of the Association for Computational Linguistics and the 7th International Joint Conference on Natural Language Processing (ACL-IJCNLP), pp. 291–296 (2015)
34. Zhang, S., Jiang, H., Xu, M., Hou, J., Dai, L.: The fixed-size ordinally forgetting encoding method for neural network language lodels. In: Proceedings of the 53rd Annual Meeting of the Association for Computational Linguistics and the 7th International Joint Conference on Natural Language Processing (ACL-IJCNLP) (2015)
35. Chen, D., Manning, C.: A fast and accurate dependency parser using neural networks. In: EMNLP, pp. 740–750 (2014)
36. Cho, K., et al.: Learning phrase representations using RNN encoder-decoder for statistical machine translation. In: EMNLP, pp. 1724–1734 (2014)
37. Goodfellow, I., et al.: Generative adversarial nets. In: NIPS, pp. 2672–2680 (2014)
38. Pennington, J., Socher, R., Manning, C.D.: Glove: global vectors for word representation. In: EMNLP, vol. 14, pp. 1532–1543 (2014)
39. Mikolov, T., Sutskever, I., Chen, K., Corrado, G., and Dean, J.: Distributed representations of words and phrases and their compositionality. In: NIPS (2013)
40. Ling, X., Weld, D.S.: Fine-grained entity recognition. In: AAAI (2012)

41. Yosef, M.A., Bauer, S., Hoffart, J., Spaniol, M., Weikum, G.: HYENA: hierarchical type classification for entity names. In: Proceedings of the 24th International Conference on Computational linguistics (COLING), pp. 1361–1370 (2012)

42. Collobert, R., Weston, J., Bottou, L., Karlen, M., Kavukcuoglu, K., Kuksa, P.: Natural language processing (almost) from scratch. J. Mach. Learn. Res. **12**, 2493–2537 (2011)

43. Turian, J., Ratinov, L., Bengio, Y.: Word representations: a simple and general method for semi-supervised learning. In: ACL, pp. 384–394 (2010)

44. Nadeau, D., Sekine, S.: A survey of named entity recognition and classification. Lingvisticae Investig. **30**(1), 3–26 (2007)

45. Etzioni, O., Cafarella, M., Downey, D., Popescu, A.M., Shaked, T., Soderland, S., Weld, D.S., Yates, A.: Unsupervised named-entity extraction from the web: an experimental study. Artif. Intell. **165**(1), 91–134 (2005)

46. Finkel, J.R., Grenager, T., Manning, C.: Incorporating non-local information into information extraction systems by Gibbs sampling. In: ACL, pp. 363–370 (2005)

47. Tsochantaridis, I., Hofmann, T., Joachims, T., Altun, Y.: Support vector machine learning for interdependent and structured output spaces. In: Proceedings of the Twenty-First International Conference on Machine Learning (ICML) (2004)

48. Tjong Kim Sang, E.F., De Meulder, F.: Introduction to the CoNLL-2003 shared task: language-independent named entity recognition. In: NAACL-HLT, pp. 142–147 (2003)

49. Chieu, H.L., Ng, H.T.: Named entity recognition with a maximum entropy approach. In: Conference on Natural Language Learning (CoNLL), pp. 160–163 (2003)

50. Fleischman, M., Hovy, E.: Fine grained classification of named entities. In: Proceedings of the 19th International Conference on Computational linguistics (COLING) (2002)

51. Collins, M., Singer, Y.: Unsupervised models for named entity classification. In: Joint Conference on Empirical Methods in Natural Language Processing and Very Large Corpora (1999)

52. LeCun, Y., Bottou, L., Bengio, Y., Haffner, P.: Gradient-based learning applied to document recognition. Proc. IEEE **86**(11), 2278–2324 (1998)

53. Hochreiter, S., Schmidhuber, J.: Long short-term memory. Neural Comput. **9**(8), 1735–1780 (1997)

54. Cunningham, H., Wilks, Y., Gaizauskas, R.J.: GATE: a general architecture for text engineering. In: COLING, pp. 1057–1060 (1996)

55. Hinton, G.E., McClelland, J., Rumelhart, D.: Distributed representations. In: Rumelhart, D., McClelland, J. (eds.) Parallel Distributed Processing, vol. 1, pp. 77–109 (1986)

An Analysis of Software Bug Reports Using Random Forest

Ha Manh Tran$^{(\boxtimes)}$, Sinh Van Nguyen, Synh Viet Uyen Ha, and Thanh Quoc Le

Computer Science and Engineering, International University - Vietnam National University, Ho Chi Minh City, Vietnam
{tmha,nvsinh,hvusynh,lqthanh}@hcmiu.edu.vn

Abstract. Bug tracking systems manage bug reports for assuring the quality of software products. A bug report also referred as trouble, problem, ticket or defect contains several features for problem management and resolution purposes. Severity and priority are two essential features of a bug report that define the effect level and fixing order of the bug. Determining these features is challenging and depends heavily on human being, e.g., software developers or system operators, especially for assessing a large number of error and warning events occurring on software products or network services. This study proposes an approach of using random forest for assessing severity and priority for software bug reports automatically. This approach aims at constructing multiple decision trees based on the subsets of the existing bug dataset and features, and then selecting the best decision trees to assess the severity and priority of new bugs. The approach can be applied for detecting and forecasting faults in large, complex communication networks and distributed systems today. We have presented the applicability of random forest for bug report analysis and performed several experiments on software bug datasets obtained from open source bug tracking systems. Random forest yields an average accuracy score of 0.75 that can be sufficient for assisting system operators in determining these features. We have provided some analysis of the experimental results.

Keywords: Random forest · Decision tree
Software bug report · Network fault detection · Fault management

1 Introduction

Fault detection plays an important role in operating computer systems. The more complex computer systems are, the more difficult fault detection is. Several hindrances of operating computer systems and services today focus on service availability, performance unpredictability and failure control [1] that are closely associated with fault detection. A normal fault detecting mechanism usually works with the involvement of system operators and the support of multiple monitoring tools. A running computer system requires monitoring tools running

© Springer Nature Switzerland AG 2018
T. K. Dang et al. (Eds.): FDSE 2018, LNCS 11251, pp. 273–285, 2018.
https://doi.org/10.1007/978-3-030-03192-3_21

along with. These monitoring tools keep reporting the status of the system. System operators observe and analyze abnormal signs on the report and the system, then create and submit a bug report to a bug tracking system (BTS) for resolution. Research activities have dealt with automating some parts of the fault detecting mechanism. One of the recently advanced research activities aims at exploiting monitoring log data and historical bug data to early notify the critical status of a system, or even forecast the forthcoming fault of a system.

Bug tracking systems store bug report data to control the quality of software products. They are frequently used to organize the workflows that produce bug reports and forward to system operators for resolution. A bug report contains many features for problem management and resolution purposes. Two essential features, namely severity and priority, define the effect level and fixing order of the bug, respectively. Determining these features is to a large extent a human driven process. Evaluating a large number of error and warning events occurring on real time software products or network services autonomously is challenging. This study proposes an approach of using random forest for evaluating severity and priority for software bug reports autonomously. The contribution of this study is thus threefold:

1. Investigating bug features extracted from the unified bug schema [2] for evaluating severity and priority
2. Proposing an approach of using random forest for evaluating the priority and severity features of software bug reports
3. Providing the prototyping implementation and experiments for the fault analysis approach on a 100-workstation computing cluster

The rest of the paper is structured as follows: the next section includes some background of fault data analysis techniques applied to software maintenance, system failure and reliability, some related work of random forest applied to failure detection and prediction. Section 3 describes the fundamentals of decision tree and random forest, the applicability of software bug data processing, and several processes of building random forest for bug report datasets. Some mathematical formulas and explanations are referred from the study of Breiman et al. [3]. Several experiments in Sect. 4 report the performance and efficiency of bug data analysis before the paper is concluded in Sect. 5.

2   Related Work

Several studies have used fault case analysis for fault detection and resolution. The study of Tran et al. [2] has proposed a semantic search approach for bug reports. The approach includes crawling bug reports from bug tracking systems, extracting semi-structured bug data, and describing a unified data model to store bug tracking data. This model derived from the analysis of the most popular systems is used for semantic search. The model also facilitates fault feature extraction and analysis using machine learning techniques. Another study of Tran et al. [4] has reduced the computation problem by analysing several types of fault classifications and relationships. This approach exploits package dependency, fault dependency, fault keywords, fault classifications to seek the relationships between fault causes. Evaluating these approaches have performed on software bug datasets obtained from different open source bug tracking systems. Wang et al. [5] have proposed an automatic fault diagnosis method for Web applications in cloud computing. The online incremental clustering method identifies access behavior patterns and models the correlation between workload and the metric of resource utilization. The method detects anomalies by discovering the abrupt change of correlation coefficients, and locates suspicious metrics using the feature selection method. Tran et al. [6] introduced the DisCaRia system that applies a distributed case-based reasoning approach to exploring fault solving resources on peer-to-peer networks. The prototyping system has been deployed and currently measured on the EmanicsLab distributed computing testbed [7]. The recent study of Ferreira et al. [8] has proposed an approach of using machine learning techniques for automated fault detection on solar-powered wireless mesh networks. The approach applies knowledge discovery methodology and a pre-defined dictionary of faults and solutions for classifying new faults. Several classification algorithms include Neural Network, Naive Bayes, Support Vector Machine (SVM), Decision Tree and k-Nearest Neighbors (kNN).

Several other studies have used decision tree and random forest for fault classification and prediction. Sinnamon et al. [9] have applied the binary decision diagram to identify system failure and reliability. Large systems usually produce thousands of events that consume a large amount of processing time. This diagram associated with if-then-else rules and optimized techniques reduces time consuming problem. The study of Reay et al. [10] has proposed an analysis strategy aiming at increasing the likelihood of obtaining a binary decision diagram for any given fault tree while ensuring the associated calculations as efficient as possible. The strategy contains 2 steps: simplifying the fault tree structure and obtaining the associated binary decision diagram. The study also includes quantitative analysis on the set of binary decision diagrams to obtain the probability of top events, the system unconditional failure intensity and the criticality of the basic events. Guo et al. [11] have proposed an approach of using random forests for predicting fault prone modules in software development process. The approach exploits the information of the previous projects including modules, defects, locations, metrics, to predict the current project with an assumption of stable development environment. The approach presents several advantages of running efficiently on large datasets and outperforming the other classifiers in terms of robustness and noise. The recent study of Tran et al. [12] has proposed an approach for evaluating the severity level of events using a classification decision tree. The approach exploits existing fault datasets and features, such as bug reports and log events to construct a decision tree that can be used to classify the severity level of new events. This study includes the prototyping implementation and evaluation of the approach for various bug report and log event datasets. The system operators thus refer to the result of classification to determine proper actions for the suspected events with a high severity level. The study in this paper is motivated by the previous study [13] that provides the comparison of machine learning techniques for bug report datasets and recommends random forest as the most promising technique for analysis performance. While the previous approaches focus more on avoiding, detecting and resolving faults on the monitored systems, i.e., passive approaches rely on correct configurations or solutions for the detected faults, this active approach scrutinizes log events from currently running systems and historical bug reports from bug tracking systems in order to classify potential events with high severity that might cause crucial faults on running systems in the near future.

3   Bug Analysis Approach

Random forest is a classifier consisting of a number of decision trees that depend on the independently sampled values of random vectors with the same distribution. The precision of a random forest relies on the strength of the individual trees in the forest and the correlation between trees. The idea of random forest is to select the best decision trees from an ensemble of decision trees built by the subsets of the training dataset. A decision tree algorithm divides the training dataset into subsets with similar instances and uses entropy to calculate homogeneous values for instances. The decision tree and random forest growing processes are referred by the previous studies [3].

3.1   Entropy Splitting Rule

A decision tree is built top-down from a root node and involves partitioning data into subsets that contain instances with similar values (homogeneous). The decision tree algorithm uses entropy to calculate the homogeneity of a sample.

$$H(S) = -\Sigma_{x \in X} P(x) log P(x) \tag{1}$$

where, S is the current dataset for which entropy is being calculated. X is a set of classes in S. $P(x)$ is a ratio between the number of elements in class x and the number of elements in set S. When $H(S) = 0$, the set S is perfectly classified.

Information gain $IG(A)$ is the measure of difference in entropy from before to after the set S is split on an attribute A. In the other words, how much uncertainty in S is reduced after splitting the set S on the attribute A.

$$IG(A, S) = H(S) - \Sigma_{t \in T} P(t) H(t) \tag{2}$$

where, $H(S)$ is entropy of the set S. T is the subset created from splitting S by A. $P(t)$ is a ratio between the number of elements in t and the number of elements in S. $H(t)$ is entropy of subset t. Information gain can be calculated (instead of entropy) for each remaining attribute. The attribute with the largest information gain is used to split S in the current iteration.

3.2 Decision Tree Growing Process

This process uses a dataset with features as input. A feature value can be ordinal categorical, nominal categorical or continuous. The process use entropy splitting rules to choose the best split among all the possible splits that consist of possible splits of each feature, resulting in two subsets of features. Each split depends on the value of only one feature. The process starts with the root node of the tree and repeatedly runs three steps on each node to grow the tree, as shown in Fig. 1 on the left side.

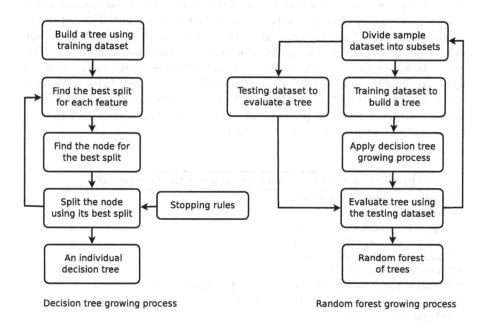

Fig. 1. Processes of growing a decision tree (left) and a random forest (right)

The first step is to find the best split of each feature. Since feature values can be computed and sorted to examine candidate splits, the best split maximizes the defined splitting criterion. The second step is to find the best split of the node among the best splits found in the first step. The best split also maximizes the defined splitting criterion. The third step is to split the node using its best split found in the second step, then the process repeats the first step if the stopping rules are not satisfied. The process generates a decision tree when the stopping rules are satisfied, as follows:

1. If a node becomes pure; that is, all cases in a node have identical values of the dependent variable, the node will not be split.
2. If all cases in a node have identical values for each predictor, the node will not be split.

3. If the current tree depth reaches the user-specified maximum tree depth limit value, the tree growing process will stop.
4. If the size of a node is less than the user-specified minimum node size value, the node will not be split.
5. If the split of a node results in a child node whose node size is less than the user specified minimum child node size value, the node will not be split.

3.3 Random Forest Growing Process

The decision tree growing process is one of the main steps of the random forest growing process that also uses a dataset with features as input. The dataset is randomly partitioned into the training and testing datasets. The training dataset is the input of the decision tree growing process to construct a decision tree. The testing dataset is used to evaluate the constructed tree, as shown in Fig. 1 on the right side. This process repeats to select an sufficient number of trees. We have used Scikit Learn [14], NumPy [15] and SciPy [16] libraries for constructing a random forest. Algorithm 1 presents several steps to build a random forest for the bug dataset with the support of these libraries.

Algorithm 1. Constructing a random forest for the bug dataset

---

  Input : Processed bug dataset
  Output: Random forest
 1 Import random forest and caret libraries;
 2 Load the dataset into data-frame;
 3 Factorize the data-frame to numeric;
 4 Parse factor to class label;
 5 Partition the dataset to the training and testing datasets;
 6 Construct random forest with the dataset;
 7 **return** Random forest;

The first step is to load the random forest and caret libraries to build a random forest and a confusion matrix for validating the accuracy of the random forest. The second step is to load the bug dataset into data-frame that is a special tabular data structure to prepare for training and testing the random forest. The feature values of the bug dataset are factorized by numeric values in the third step because the algorithm cannot deal with text and enumerate values. An integer represents a distinct value, e.g., the priority feature contains 4 values: P1 (urgent), P2 (high), P3 (normal), and P4 (low) corresponding to 1, 2, 3, and 4 after factorization. The fourth step is to parse the class label of the bug dataset to factor variable type, which is a vector of integer values with a corresponding set of character values to use when the factor is displayed. This parsing step also decides the classification response of the random forest. We have used 25% of the bug dataset for evaluation and 75% of the bug dataset for building the random forest. The later dataset includes the testing and training datasets used to classify observations into severity or priority level. The algorithm also relies on a variable that specifies a certain number of random features to grow a single tree.

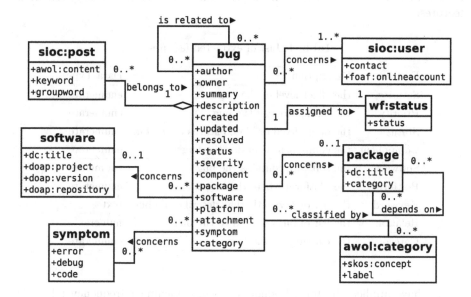

Fig. 2. Unified bug schema represented as a UML diagram

3.4 Bug Data Processing

Bug dataset contains software bug reports obtained from several open source BTSs such as Bugzilla [17], Launchpad [18], Mantis [19], Debian [20]. Bug reports from different BTSs share several common features. Administration features, such as severity, status, platform, content, component, keyword, etc., are represented as field-value pairs. Description features, such as problem description, follow-up discussion are represented as textual attachments. The unified bug schema shown in Fig. 2 aims to support for semantic bug search [2]. This schema contains several features extracted by BTSs and new features to minimize the loss of bug report information, for example, the new relation feature establishes the relationships between bugs, or the new category feature provides for more sophisticated bug classification. We have used a web crawler to get access to BTSs, retrieve and parse the HTML pages of bug reports to extract their content following the unified bug schema that allows various types of bug reports to be stored in one database. Most of bug features can be extracted from bug content. However, BTSs provide very different classifications for some features including severity, priority, status. The model classifies the severity feature into *critical, normal, minor, feature*, and the priority feature into *urgent, high, normal, low*.

Several bug features are concerned with severity and priority. Table 1 presents a list of essential features extracted from the bug schema. These features cause profound impact on determining the severity and priority of a bug report. The keyword feature contains the result of processing the description and discussion features.

Table 1. List of extracted features

Feature	Description	Data Types
Severity	the effect level of the bug	enumerate
Priority	the fixing order of the bug	enumerate
Status	the open, fixed or closed status of the bug	enumerate
Component	the component contains the bug	enumerate
Software	the software contains the bug	enumerate
Platform	the platform where the bug occurs	enumerate
Keyword	the list of keywords that describe the bug	text
Relation	the list of bugs related to the bug	numeric
Category	the category of the bug	enumerate

We have applied the term frequency–inverse document frequency (tf × idf) method to process the description and discussion features that only include the textual data of a bug report. This method measures the significance of keywords to bug reports in a bug dataset by the occurrence frequency of keywords in a bug report over the total number of keywords of the bug report (term frequency) and the occurrence frequency of keywords in other bug reports over the total number of bug reports (inverse document frequency). This process results in a set of keywords that best describe the bug report, i.e., a set of distinct keywords with high significance. Algorithm 2 describes several steps to process the textual data of the bug dataset. The first step is to load the bug dataset as raw keyword set. The next three steps are to remove redundant words, meaningless words, and special characters using stop-word set and regular expression. The last step is to apply the tf × idf method for the remaining keywords.

4   Evaluation

We have used a bug dataset of 300.000 bug reports occurring on Windows platform (Win platform) and 300.000 bug reports occurring on other platforms (All platform), and a computer with Intel Core I7, 3.8 GHZ each core, 8GB RAM and Ubuntu 16.04 LTS 64-bit to build decision trees and random forests. We have verified the priority and severity of bug reports obtained from both the bug tracking systems and the fault detection system operating on a cluster of 100 workstations at International University–Vietnam National University's computing center. This cluster provides computing and storage services based on

Algorithm 2. Filtering keywords for a bug dataset

Input : Raw keyword set
Output: Filtered keyword set
1 Load raw keyword set;
2 Remove duplicated and redundant words by using stop-word set;
3 Remove meaningless words by using regular expression;
4 Remove memory addresses by filtering special characters ;
5 Process tf×idf on the keyword set;
6 **return** Filtered keyword set;

the OpenStack platform [21]. These bug reports have already included priority and severity values assessed by system operators for comparison. We have run experiments for several times and then computed the average evaluation scores with errors.

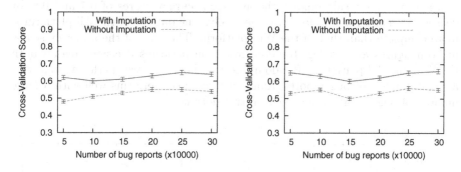

Fig. 3. Cross-validation comparison between using and not using imputation on win platform (left) and all platform (right)

Several bug reports contain incomplete values for the extracted features due to the lack of responses during data collection. There are several methods, such as case deletion, mean imputation, median imputation, k-nearest neighbor, etc., to fill incomplete values for improving accuracy. We have applied the median imputation method for filling incomplete values by the mean of all known values of the features in the class to which the bug report with incomplete values belongs. The method approximately increases the average cross-validation scores of the Win platform and All platform datasets to 0.65, as shown in Fig. 3. The dataset of the All platform suffers less impact by incomplete values than that of the Win platform.

The first experiment compares severity accuracy between decision tree and random forest for Win and All platforms. For the Win platform, Fig. 4 on the left side reports the average accuracy scores of 0.65 and 0.75 for decision tree and random forest, respectively. While decision tree line slightly decreases as the size

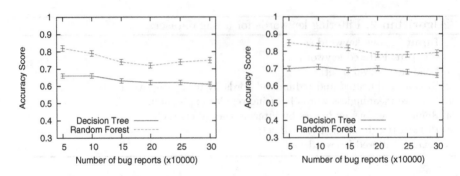

Fig. 4. Severity accuracy comparison between decision tree and random forest on win platform (left) and all platform (right)

of datasets increases, random forest line starts with high scores, reduces and then linearly increases again as the size of datasets increases. For the All platform, Fig. 4 on the right side reports the average accuracy scores of 0.7 and 0.8 for decision tree and random forest, respectively. The dataset of the All platform is more complete than that of the Win platform. The lines of the accuracy scores in both figures are similar. Random forest outperforms decision tree on both the Win and All platforms. Random forest and decision tree performs well with a small number of bug reports. Random forest also performs well with a large number of bug reports while decision tree does not.

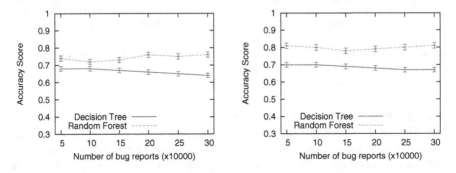

Fig. 5. Priority accuracy comparison between decision tree and random forest on win platform (left) and all platform (right)

The second experiment compares priority accuracy between decision tree and random forest for Win and All platforms. For the Win platform, Fig. 5 on the left side reports the average accuracy scores of 0.66 and 0.73 for decision tree and random forest, respectively. While decision tree line linearly decreases as the size of datasets increases, random forest line remains stable as the size of datasets increases. For the Win platform, Fig. 5 on the right side reports the average accuracy scores of 0.69 and 0.81 for decision tree and random forest, respectively. Priority lines are more stable than severity lines, thus severity to some extent is more difficult to determine than priority. The dataset of All platform is more consistent than that of Win platform because of the big gap between the lines of accuracy scores on All platform. The shape of accuracy scores is also similar.

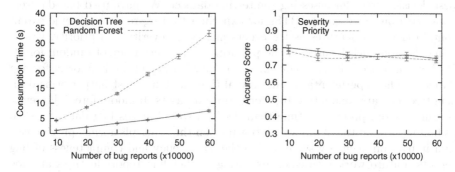

Fig. 6. Time consumption comparison between decision tree and random forest (left) and random forest accuracy comparison between severity and priority classification (right) on the whole dataset

The third experiment presents time consumption for constructing decision tree and random forest for the whole dataset of 600.000 bug reports, as shown in Fig. 6 on the left side. Decision tree line linearly increases and random forest line exponentially increases as the size of datasets increases. It takes approximately 8 s and 33 s to build a decision tree and a random forest for 600.000 bug reports, respectively. Time consumption depends on number of bug reports and features. Time consumption for random forest also depends on constructing multiple decision trees and selecting the best decision trees. Processing large bug datasets consumes much time. Figure 6 on the right side reports the accuracy score of random forest for severity and priority classification using the whole dataset. Severity and priority lines slightly decrease as the size of datasets increases. The accuracy scores of severity and priority are similar on average because they are similar features in terms of data type and determination process. The average accuracy score of 0.75 can be sufficient for assisting system operators in determining these features.

5 Conclusion

We have proposed an approach of using the random forest technique for software bug report analysis. This approach targets to determining the severity and priority of a bug report automatically. Evaluating these features heavily depends on human being. This approach can be applied to evaluating a large number of log events for fault detection and forecast in large, complex communication networks and distributed systems. The log event dataset is so huge that system operators and even supporting tools cannot process quickly, thus resulting in neglecting potentially critical errors or warning events leading to critical errors. Instead of constructing a decision tree for learning from the training dataset and assessing the testing dataset, a random forest constructs a number of decision trees from the subsets of the training dataset and selects the best decision trees for assessing the testing dataset. We have used bug datasets obtained from open source BTSs for experiments. Bug reports to some extent contain the same features as log events including severity and priority. Evaluating the approach focuses on the performance and accuracy of random forest. We have measured the time consumption and accuracy of random forest for bug datasets. The experimental results reveal that random forest outperforms decision tree, i.e., approximately 10% for various datasets. Random forest, however, consumes more processing time than decision tree. Various bug datasets, such as bug reports on the Windows platform or on the other platforms, provide certain impact on accuracy score due to the consistency and completeness of bug reports. Future work focuses on exploring further bug report features and log event datasets to improve the accuracy of random forest and applying the fault detection and forecast tool for the realistic systems.

Acknowledgements. This research activity is funded by Vietnam National University in Ho Chi Minh City (VNU-HCM) under the grant number B2017-28-01.

References

1. Armbrust, M., et al.: A view of cloud computing. ACM Commun. **53**(4), 50–58 (2010)
2. Tran, H.M., Lange, C., Chulkov, G., Schönwälder, J., Kohlhase, M.: Applying semantic techniques to search and analyze bug tracking data. J. Netw. Syst. Manag. **17**(3), 285–308 (2009)
3. Breiman, L.: Random forests. Mach. Learn. **45**(1), 5–32 (2001)
4. Tran, H.M., Le, S.T.: Software bug ontology supporting semantic bug search on peer-to-peer networks. New Gener. Comput. **32**(2), 145–162 (2014)
5. Wang, T., Zhang, W., Wei, J., Zhong, H.: Fault detection for cloud computing systems with correlation analysis. In: Proceedings IFIP/IEEE International Symposium on Integrated Network Management IM 2015, pp. 652–658 (2015)
6. Tran, H.M., Schönwälder, J.: Discaria - distributed case-based reasoning system for fault management. IEEE Trans. Netw. Serv. Manage. **12**(4), 540–553 (2015)
7. Hausheer, D., Morariu, C.: Distributed test-lab: EMANICSLab. In: The 2nd International Summer School on Network and Service Management (ISSNSM 2008), University of Zurich, Switzerland, June 2008
8. Ferreira, V.C., Carrano, R.C., Silva, J.O., Albuquerque, C.V.N., Muchaluat-Saade, D.C., Passos, D.G.: Fault detection and diagnosis for solar-powered wireless mesh networks using machine learning. In: Proceedings IFIP/IEEE Symposium on Integrated Network and Service Management (IM 2017), pp. 456–462 (2017)

9. Sinnamon, R.M., Andrews, J.D.: Fault tree analysis and binary decision diagrams. In: Proceedings in Reliability and Maintainability Annual Symposium, pp. 215–222 (1996)
10. Reay, K.A., Andrews, J.D.: A fault tree analysis strategy using binary decision diagrams. Reliab. Eng. Syst. Saf. **78**(1), 45–56 (2002)
11. Guo, L., Ma, Y., Cukic, B., Singh, H.: Robust prediction of fault-proneness by random forests. In: Proceedings 15th International Symposium on Software Reliability Engineering (ISSRE 2004), pp. 417–428, Washington, IEEE (2004)
12. Tran, H.M., Nguyen, S.V., Le, S.T., Vu, Q.T.: Applying data analytic techniques for fault detection. Trans. Large-Scale Data- Knowl. -Cent. Syst. (TLDKS) **31**, 30–46 (2017)
13. Tran, H.M., Nguyen, S.V., Le, S.T., Ha, S.V.U.: A comparison of machine learning techniques for fault data analysis. Technical report, International University–VNU HCMC (2018)
14. Pedregosa, F., et al.: Scikit-learn: machine learning in Python. J. Mach. Learn. Res. **12**, 2825–2830 (2011)
15. Oliphant, T.: A Guide to NumPy. Trelgol Publishing, USA (2006)
16. Silva, F.B.: Learning SciPy for Numerical and Scientific Computing. Packt Publishing, Birmingham (2013)
17. Mozilla Bug Tracking System. https://bugzilla.mozilla.org/. Accessed Aug 2017
18. Launchpad Bugs. https://bugs.launchpad.net/. Accessed Aug 2017
19. Mantis Bug Tracker. https://www.mantisbt.org/. Accessed Aug 2017
20. Debian Bug Tracking System. https://www.debian.org/Bugs/. Accessed Aug 2017
21. OpenStack Cloud Software 2010. http://www.openstack.org/. Access in Aug 2017

Motorbike Detection in Urban Environment

Chi Kien Huynh$^{(\boxtimes)}$, Tran Khanh Dang, and Thanh Sach Le

Ho Chi Minh City University of Technology, Ho Chi Minh City, Vietnam
{hckien,khanh,ltsach}@hcmut.edu.vn

Abstract. Vehicle detection is one of the key components in vision-based intelligent transportation system (ITS). Having an accurate detector in difficult and varied environments is a prerequisite for a usable ITS. Many existing methods were tested on environments where there are a lot of four-wheelers and the general driving is in order. However, the situation is different in Vietnam and similar countries where motorbike is the major vehicle choice. This makes traffic much more unpredictable and difficult for vision detectors. In this work, we applied different detection methods to analyze their performance on our traffic datasets. The results show that deep neural network approach offers better accuracy than the others. They also prove that in circumstances that are under control, these methods can be readily integrated in practical traffic systems.

Keywords: Convolutional neural network
Motorbike detection · Vision-based traffic system

1 Introduction

In various large-scale metropolises, traffic congestion is a prevalent result caused by fast population growth and inadequate transportation infrastructure. Although better urban planning can help easing this problem in the future, there are immediate workarounds that can be used. Recently, widespread intelligent systems have been deployed in many cities as a mean to manage traffic [1,7,10]. A notable feature among these systems is the use of computer vision methods to detect, track or identify vehicles. Information retrieved from these tasks can be used for urban analysis, to find optimal routes in rush hours and many more purposes. Using such systems can be helpful for any area with intense traffic.

A few overpopulated cities in Vietnam are also facing problems in traffic because facilities couldn't have adapted quickly enough. An intelligent transportation system can be imported to mitigate those problems. However, established ones cannot be readily used in Vietnam urban areas. The main reason is that in Vietnam, along with a few other countries, the main type of vehicle is motorbike. This leads to several different difficulties to existing intelligent systems as they were designed for environments where car or truck are primary. As shown in Fig. 1, when the number of motorbikes is high, the situation can be very

© Springer Nature Switzerland AG 2018
T. K. Dang et al. (Eds.): FDSE 2018, LNCS 11251, pp. 286–295, 2018.
https://doi.org/10.1007/978-3-030-03192-3_22

Fig. 1. Traffic at a 6-way intersection in Ho Chi Minh city, Vietnam.

chaotic. Since motorbikes are very flexible, drivers can create situations which four-wheelers normally cannot. Because of this, there should be a considerable amount of experiments need to be conducted before a vision-based system can work in practice.

This paper will show experimental results of different detection methods on our traffic image datasets. Discussions regarding the advantages and drawbacks of them will be made in Sect. 4. The last section will also mention some challenges for current methods in our problem.

2   Related Works

Many object detection methods usually have three main components: a candidate generator, feature extractor and a classifier. The main purpose of the first component is to find potential patches in the image and forward them to the other two. A common candidate proposal method is sliding window . It searches in a manner close to brute-forcing. Since it is slow and ineffective, many better alternatives have been proposed. In [17], there is an initial sparse search in the framework to locate body parts of a pedestrian. The center of the pedestrian is then estimated using partial least square regression. This task greatly reduces the amount of time needed to find candidates. There are also other alternative searching methods such as those proposed in [12] and [15] for faster detection.

Regarding feature extraction for object detection, all of them can be grouped into 2 main categories. The first one consists of hand-crafted algorithms such as histogram of oriented gradient (HOG) [3] or local binary pattern (LBP) [14]. Other features in the second category are those that can be learned using machine learning models [4,11]. Features come from hand-crafted methods are usually more generic, well-suited for a wide range of object types because their designs are not dependent on a specific scope of images. In contrast, features learned from machine learning models usually carry characteristics of the dataset that was used for training. They usually cannot be directly reused for new object types without transfer learning or learning anew.

In detection, a classifier receives feature vectors computed from any candidate image patch and classifies them into one of the predefined labels. One of the most popular choices is support vector machine (SVM) [2]. A single SVM classifier can only work on binary classification problems, but they can be applied in an one-versus-all or one-versus-one manner to classify more than two labels. Otherwise, a single ensemble classifier can be trained with AdaBoost [13].

The listed components can be combined using deep neural networks. In [5], there is no separate feature extractor or classifier. Instead, a single convolutional neural network was used for both roles. Other works such as [8] and [6], the detection was performed in an end-to-end manner using a single neural network. The last layer of these networks does not only provide classification scores but also the information of the bounding boxes of objects.

3   Detection Methods

This section introduces methods which were used in our experiments. The first subsection presents our recommended method in details. The second subsection lists all baseline detection methods that will be compared to the recommended one.

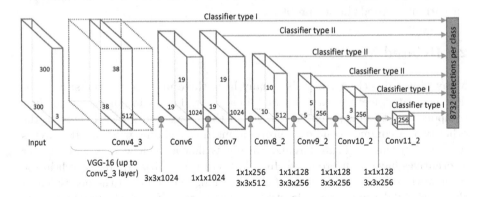

Fig. 2. SSD architecture. Green text denotes sizes of the filters of convolutional layers. The format is $H \times W \times N_c$ where H being height, W being width and N_c being number of channels. Classifier type I and II are convolutional layers with filter size $3 \times 3 \times (4 \times (NumOfClasses + 4))$ and $3 \times 3 \times (6 \times (NumOfClasses + 4))$ respectively. (Color figure online)

3.1   Single Shot Multibox Detector [6]

Single shot multibox detector (SSD) is an end-to-end detection neural network. Unlike older models, it covers all three stages in detection with its intrinsic architecture. The network consists of base convolutional layers and extra layers which serve different purposes. Base layers are reused from classification network called

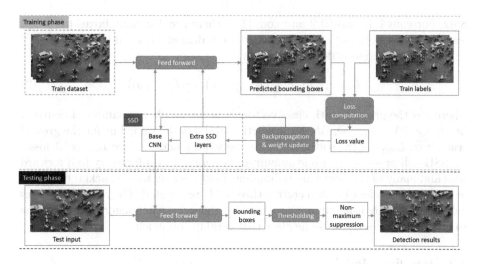

Fig. 3. The training and testing phases of SSD.

VGG [9]. A common practice here is to fix the weights of VGG's layers in training to avoid overfitting, as long as they were already trained for a classification task previously. However, in our case, we found that there is no need in pre-training VGG separately from SSD.

Extra layers from SSD provide extraction of deeper, more abstract features. Additionally, the final layers, denoted as classifiers in Fig. 2 are designed to predict bounding boxes and object types. These classifiers are convolutional layers that produce K vectors per 3×3 input region, each vector follows the format:

$$[p, \Delta x, \Delta y, \Delta w, \Delta h] \tag{1}$$

Here, p is the class probability vector with length equal to the number of classes we want to detect. $\Delta x, \Delta y, \Delta w, \Delta h$ are the difference in x, y, width and height between the predicted box and its corresponding default box. K is the number of default boxes.

The use of default boxes is an additional trick to stabilize the training process. Without them, there is a high possibility that predicted boxes in the first few train iterations will float around. As a consequence, the same prediction might not always be matched to the same ground truth even if it should be so. This makes the loss function hard to converge. To tackle this, every predicted box, no matter where they locate, are always hard-assigned to some predefined default boxes. All ground truth boxes will then be matched to these predefined ones using Jaccard overlap rate. Any pair of prediction and ground truth that share the same default box will contribute in the loss function. Predictions that are not matched to any ground truth are ignored or used as negative samples.

Clearly, the number of default boxes we choose will affect the accuracy of the detection. In our work, we reuse the original values proposed by the authors of

SSD. For conv4_3, conv10_2 and conv11_2, there are 4 default boxes per 3×3 feature map region. For other layers, we use 6 default boxes in total.

The loss function is defined as follow:

$$L(x, c, l, g) = \frac{1}{N}(L_{conf}(p, c) + \alpha L_{loc}(l, g)) \tag{2}$$

where c is the ground truth class vector, N is the number of samples in current batch, $l = [\Delta x, \Delta y, \Delta w, \Delta h]$ for prediction and g is the same but for the ground truth box. L_{conf} is the softmax log loss between p and c where L_{loc} is L1 loss.

SSD will produce the same amount of bounding boxes for every feed forward pass and many of the boxes are redundant. Boxes belong to the background class or has class confident below a certain threshold are rejected. The remaining boxes will be input to non-maximum suppression, a greedy search algorithm that aims to eliminate boxes of the same class that overlap too much.

3.2   Baseline Models

Apart from SSD, we also include experimental results of the following detector for better comparison:

1. This method uses sliding window for candidate proposal, HOG [3] as the main feature and SVM for classification.
2. This one also uses HOG but partial least square regression was incorporated to improve the detection speed [17].
3. Deep convolutional neural network [5]. Different from SSD, this network was trained on small, fixed-sized image patches of motorbikes for classification purpose. In detection phase, the input image is rescaled with different scale factors and fed into the network multiple times. This allow the networks to detect motorbikes of different sizes.

Dataset A   Dataset B   Dataset C

Fig. 4. Examples of the three datasets used in our experiments.

4 Experiment and Results

4.1 Dataset

In our experiment, we used three different datasets in total:

- Dataset A: this consists of images uniformly sampled from [16] with a rate of five frames each second. We use the first 1000 frames for training and the next 200 for testing.
- Dataset B: consists of 200 train images and 200 test images. This dataset is considered more difficult as the video was recorded at a medium height hence there are a lot of occlusions.
- Dataset C: consists of 2000 train images and 1847 test images. This is the most difficult dataset since there was a mild rain at the time the video was captured.

4.2 Configuration

- SSD hyper parameters: We used SSD300 in our experiments. That is, input images must be resized to 300×300 before being fed into SSD. The base layers of SSD are completely the same as those in VGG.
- Training: The learning rate we chose was 0.001, momentum rate was 0.9. We trained SSD on our GTX 670 with a batch size of 2. The training was stopped after 100000 iterations.
- Testing: The rejection threshold for class confidence we chosen to be 0.2 for all vehicles.

4.3 Evaluation Metrics

After receiving predictions from a detector, we compare them with ground truth bounding boxes. A prediction is considered true positive if it satisfies the following condition:

$$IoU = \frac{area(BB_{dt} \cap BB_{gt})}{area(BB_{dt} \cup BB_{gt})} \geq 0.5 \tag{3}$$

Here, BB_{dt} and BB_{gt} respectively denote predicted bounding box and ground truth box. IoU measures the ratio between the area of the intersection and the area of the union of both boxes. $IoU = 1$ means that the prediction is the perfect match, $IoU = 0$ means that two boxes do not intersect. If this value is less than 0.5 than BB_{dt} will be considered false positive. If two detected boxes share the same ground truth and both have $IoU \geq 0.5$, the box with lower confident will be counted as false positive. Any ground truth with no prediction assigned to it is seen as false negative. The number of true positives, false positives and false negatives of the entire test dataset are then used to compute the following metrics:

$$precision = \frac{TP}{TP + FP} \tag{4}$$

$$recall = \frac{TP}{TP + FN} \tag{5}$$

$$F - score = 2 \times \frac{precision \times recall}{precision + recall} \tag{6}$$

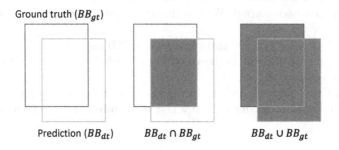

Fig. 5. Visualization of the areas in (3).

The above metrics are computed separately for each class of vehicles if the detector produces more than one.

Table 1. Motorbike detection result on dataset A.

	Precision	Recall	F-score
[3]	0.79	0.70	0.74
[17]	0.80	0.75	0.77
[5]	0.84	0.79	0.81
SSD	**0.92**	**0.90**	**0.91**

4.4   Results and Discussions

Results on dataset A of SSD and different baseline models are shown in Table 1. SSD outperformed the other methods by a large margin. The first two methods are very sensitive to occlusions which cause them to miss many motorbikes on the top half. The third method has better performance because the neural network was trained specifically to classify motorbike. HOG was designed to be generic therefore the first two methods could not work as well. Overall, SSD has the best results because its feature detection layers were trained specifically for the type of vehicle in our problem. Moreover, bounding box information was able to be encoded in classifier layers in SSD, making it much more robust and accurate than [5], which is only able to predict boxes of fixed sizes.

Table 2 shows detection results of SSD on all vehicle classes that appear in dataset A, B and C. Note that the training and testing was done separately

Table 2. Detection results of SSD on all datasets.

Dataset	Class	Precision	Recall	F-score
A	Motorbike	0.92	0.90	0.91
	Car	0.96	0.82	0.88
	Bus	0.88	0.88	0.88
B	Motorbike	0.67	0.66	0.67
	Car	0.69	0.74	0.71
	Bus	0.75	0.76	0.76
C	Motorbike	0.47	0.61	0.53
	Car	0.58	0.62	0.60
	Bus	0.82	0.99	0.90
	Truck	0.75	0.72	0.74

Fig. 6. A few detection results on three datasets. For visibility, only results on motorbikes are drawn. These images are also zoomed in and cropped. Green box: true positive, blue box: false positive, red box: false negative. (Color figure online)

for each dataset. A trend we can see with the results here is that the more occlusions occur, the less accurate the detector becomes. Dataset B and C have fairly low viewpoints, unlike dataset A. Because of this reason, many motorbikes are hidden behind other foreground vehicles. There are cases where the only visible parts of the motorbike are the safety helmets. These are proven to be difficult for SSD to detect, hence the F-score of 0.67 and 0.53 for motorbikes on dataset B and C.

Results on car are similar to those on motorbike albeit better. In contrast, SSD perform consistently well when it comes to bus. This is reasonable because buses and cars have more distinctive visual features. Additionally, buses are less likely to be hidden behind other vehicles.

Dataset C has the worst overall results than the other two. Firstly, it has the lowest capturing point among all three videos. Secondly, the rain has a severe effect on the illumination and causes noisy reflections on the road. The variety of raincoats is a factor as well.

The execution time for SSD was roughly 64.5 frames per second on our GTX 1080 Ti and 10.5 frames per second on our GTX 670. Speed comparison could not be made here as the other three baseline models do not use the same hardware. Moreover, GPU acceleration wasn't used in our experiments for [3] and [17]. Note that the reported speed only takes feed forward time of SSD into account, preprocessing time and non-max suppression time were not included but they are expected to be small.

5 Conclusion

In this paper, we have shown detection results of different methods in some urban traffic datasets coming from Vietnam. The reported figures show that convolutional neural networks have outperformed traditional pipelines in vehicle detection in cluttered environments. SSD, a network that can encode bounding box information in its final layers, has a large performance advantage over the others. The main reason of the better performance is that deep neural networks, after training on a specific dataset, can extract more well-suited features for the classes they know. Additionally, networks such as SSD that was designed for bounding box predictions will be able to produce better, more refined box shapes than the older approaches.

Even so, the network still could not detect vehicles accurately when there are a lot of occlusions caused by low capturing point. This weakness will be addressed in our future work. Additionally, a common downside of using deep neural networks is that we have to train them again every time there is a new dataset that is different from previous ones. This makes it harder to continuously apply them in real application because the process of labeling and training takes a lot of effort and time. This point should be considered in the future as well.

Acknowledgment. This work has been supported by the Advanced Computing Lab, Ho Chi Minh City University of Technology.

References

1. Chen, B.H., Huang, S.C.: Probabilistic neural networks based moving vehicles extraction algorithm for intelligent traffic surveillance systems. Inf. Sci. **299**, 283–295 (2015)
2. Cortes, C., Vapnik, V.: Support-vector networks. Machine learning **20**(3), 273–297 (1995)
3. Dalal, N., Triggs, B.: Histograms of oriented gradients for human detection. In: 2005 IEEE Computer Society Conference on Computer Vision and Pattern Recognition, CVPR 2005, vol. 1, pp. 886–893. IEEE (2005)

4. Girshick, R., Donahue, J., Darrell, T., Malik, J.: Rich feature hierarchies for accurate object detection and semantic segmentation. In: Proceedings of the IEEE Conference on Computer Vision and Pattern Recognition, pp. 580–587 (2014)
5. Huynh, C.K., Le, T.S., Hamamoto, K.: Convolutional neural network for motorbike detection in dense traffic. In: 2016 IEEE Sixth International Conference on Communications and Electronics (ICCE), pp. 369–374. IEEE (2016)
6. Liu, W., et al.: SSD: Single shot multibox detector. In: Leibe, B., Matas, J., Sebe, N., Welling, M. (eds.) ECCV 2016. LNCS, vol. 9905, pp. 21–37. Springer, Cham (2016). https://doi.org/10.1007/978-3-319-46448-0_2
7. Pan, C., Yan, Z., Xu, X., Sun, M., Shao, J., Wu, D.: Vehicle logo recognition based on deep learning architecture in video surveillance for intelligent traffic system (2013)
8. Redmon, J., Divvala, S., Girshick, R., Farhadi, A.: You only look once: unified, real-time object detection. In: Proceedings of the IEEE Conference on Computer Vision and Pattern Recognition, pp. 779–788 (2016)
9. Simonyan, K., Zisserman, A.: Very deep convolutional networks for large-scale image recognition (2014). arXiv preprint arXiv:1409.1556
10. Tang, Y., Zhang, C., Gu, R., Li, P., Yang, B.: Vehicle detection and recognition for intelligent traffic surveillance system. Multimedia Tools Appl. **76**(4), 5817–5832 (2017)
11. Thai, N.D., Le, T.S., Thoai, N., Hamamoto, K.: Learning bag of visual words for motorbike detection. In: 2014 13th International Conference on Control Automation Robotics and Vision (ICARCV), pp. 1045–1050. IEEE (2014)
12. Uijlings, J.R., Van De Sande, K.E., Gevers, T., Smeulders, A.W.: Selective search for object recognition. Int. J. Comput. Vis. **104**(2), 154–171 (2013)
13. Viola, P., Jones, M.: Rapid object detection using a boosted cascade of simple features. In: 2001 Proceedings of the 2001 IEEE Computer Society Conference on Computer Vision and Pattern Recognition, 2001 CVPR , vol. 1, p. I. IEEE (2001)
14. Wang, X., Han, T.X., Yan, S.: An HOG-LBP human detector with partial occlusion handling. In: 2009 IEEE 12th International Conference on Computer Vision, pp. 32–39. IEEE (2009)
15. Wang, X., Yang, M., Zhu, S., Lin, Y.: Regionlets for generic object detection. In: 2013 IEEE International Conference on Computer Vision (ICCV), pp. 17–24. IEEE (2013)
16. Whitworth, R.: Ho Chi Minh City (Saigon), Vietnam Rush Hour Traffic in Real Time (2013). http://www.robwhitworth.co.uk/ Accessed 12 Mar 2018
17. Wu, J., Chen, W., Huang, K., Tan, T.: Partial least squares based subwindow search for pedestrian detection. In: 2011 18th IEEE International Conference on Image Processing (ICIP), pp. 3565–3568. IEEE (2011)

Data Analytics and Recommendation Systems

Comprehensive Review of Classification Algorithms for Medical Information System

Anna Kasperczuk and Agnieszka Dardzinska[⊠]

Department of Mechanical Engineering, Division of Biocybernetics and Biomedical Engineering, Bialystok University of Technology, ul. Wiejska 45c, 15-351 Bialystok, Poland
{a.kasperczukuk,a.dardzinska}@pb.edu.pl

Abstract. Nowadays, the Internet and information systems become an integral part of everyday life. The trend of using advanced recommendation systems is still growing in various areas, also in medicine. Two of the diseases where diagnosis is a big problem for specialists are colon disease and Crohn's disease. The course of the disease strongly resembles other diseases in the large intestine, so it became extremely important to help doctors and find symptoms that would clearly indicate the colon disease, excluding others. In order to find rules that distinguish these two diseases, together data mining and statistical methods were mixed and used.

Keywords: Classification · Decision tree · Decision system
Information system

1 Introduction

Machine Learning algorithms have been widely used to solve various kinds of data classification problems also in medicine. Ulcerative colitis is a disease that causes long-term inflammation of the colon, which creates irritation or ulcers. This can lead to debilitating abdominal pain and potentially life-threatening complications. It affects only the colon or rectum and destroys the innermost part of the mucosa, not passing through the mouth. Ulcerative colitis causes inflammation and ulcers in the large intestine, which can cause a frequent feeling of need for bowel movement. Exact causes of the disease are not known, therefore their search is extremely important.

2 Main Assumptions

We work on data presented in form of a decision table $S = (X, A, V)$, where:

- X is a nonempty, finite set of objects,
- A is a nonempty, finite set of attributes,
- $V = \{V_a : a \in A\}$ is a set of all attributes values.

Additionally, $a : X \to V_a$ is a function for any $a \in A$, that returns the value of the attribute of a given object [4]. The attributes are divided into different categories: set of

T. K. Dang et al. (Eds.): FDSE 2018, LNCS 11251, pp. 299–309, 2018.
https://doi.org/10.1007/978-3-030-03192-3_23

stable attributes A_{St} (e.g. date of birth, place of birth, color of skin), set of flexible attributes A_{Fl} (blood pressure, weight, sugar level) and set of decision attributes D (e.g. method of treatment, class of illness) such that $A = A_{St} \cup A_{Fl} \cup D$. In this paper we analyze information systems with only one decision attribute d. The example of an information system S is represented as Table 1 [4, 8].

Table 1. Information system S

X	a	b	c	d
x_1	a_1	b_2	c_2	d_1
x_2	a_1	b_1	c_1	d_1
x_3	a_2	b_1	c_1	d_1
x_4	a_2	b_2	c_1	d_2
x_5	a_2	b_2	c_2	d_2
x_6	a_2	b_1	c_1	d_1
x_7	a_2	b_2	c_1	d_2
x_8	a_2	b_1	c_2	d_2

Information system is represented by eight objects, one stable attribute a (its value cannot be changed), two flexible attributes b, c (their values can change under some conditions) and one decision attribute d.

3   Classification

The classifier is an algorithm that implements classification, especially in a concrete implementation. There are many different classifiers and many different types of classification results. Moreover it is difficult, especially working with medical data, to decide which classifier is the most effective one for the given set of data. It is already widely known that some classifiers perform better than others on different datasets. Having medical data and decide which classifier gives better results there are two options. First is to put all the trust in an expert's opinion based on his knowledge and experience. Second is to run through each possible classifier that could work on the dataset, and identify rationally the one which performs the best [2, 3]. We use the classification method, where both data mining techniques and statistical methods divide objects into different decision classes.

Mixture of data mining algorithms [6] with statistical methods [2] is an algorithm that creates a step-by-step guide how to determine the output of a new data instance. It is the process of finding a set of models that differentiate data classes and concepts. We use it to predict group memberships for data instances [7]. In first step we describe a set of predetermined classes on the basis of logical regression. Each tuple is assumed to belong to a predefined class as determined by classification attribute, the set of tuples are used for model construction, called training sets. The model can be represented as classification rules, decision trees or mathematical formulas. It is used then for prediction of future data trends, or eventually reclassification of objects. It estimates the accuracy of the constructed model by using certain test cases. Test sets are always independent of the training sets [3, 6].

3.1 Decision Trees

Among the classification methods, one of the most popular method is decision tree. It is particularly attractive because of the intuitive way of knowledge representation understood by people [10, 11]. Initially decision trees appeared in the 1960s in the areas of research on psychology and sociology. In computer science, for the first time they found their application in the works in the 80's [1, 13].

Compared to other methods of classification, decision trees can be constructed relatively quickly. Their main advantage is the clear representation of knowledge, the possibility of using multidimensional data, and scalability with the use of large data sets. Additionally, the accuracy of this method is comparable to the accuracy of other classification methods. However, the main disadvantage of the discussed method is the high sensitivity to the missing values of attributes, because at their bases there is an explicitly expressed assumption of full availability of information gathered in the database. The disadvantages also include the inability to capture the correlation between attributes [13]. Therefore we can use ERID algorithm first, which help us to reduce some missing values in dataset with high accuracy.

Classification trees are used to determine the affiliation of objects to the quality class of a dependent variable. This is done based on measurements of one or more prediction variables. The classification tree presents the process of dividing the set of objects into homogeneous classes. The division is based on the values of the features of the objects, the leaves correspond to the classes to which the objects belong, while the edges of the tree represent the values of the features on the basis of which the division was made [13].

The process of creating a decision tree is based on the recursive division of the teaching set into subsets, which takes place to achieve their homogeneity due to the belonging the objects to classes. The goal is to create a tree with the fewest number of nodes, and as a consequence, the simplest classification rules [1].

The decision tree creation algorithm can be written as follows [7, 10]:

1. For a given set of objects, using ERID algorithm we find all missing values of attributes, compose the containment relation, and make more complete new information system;
2. For more complete set of attributes values corresponding to the set of objects we check whether they belong to the same class (if they belong - end the process, if they do not belong - consider all possible divisions of a given set into all possible homogeneous subsets);
3. Evaluate the quality of each of these subsets according to the previously accepted criterion and select the best one;
4. Split the set of objects on the basis of step 3;
5. Repeat above steps for each of the subsets.

3.2   Support Vector Machine (SVM)

Vector transport machine (SVM), which Vladimir Vapnik and Corinna Cortes [15] made for the first time when removing the cover on the floors and/or in the car. SVM is a version of a binary classifier that gives a set of input data and then classifies one device. The goal is to map the n-dimensional entrance space to a higher space. Thanks to the new ticket is classified by constructing a linear class. In SVM, a sample of data is viewed as a p-dimensional vector that SVM separates with a hyperplane of sets $(p - 1)$. The SVM algorithm has the advantage that it does not affect the minimum minima [14]. We modified this method, and the constraint is softened. Therefore these hyperplanes are built more independently. The main procedure starts with partitioning all negative objects into dense clusters. The same step is repeated for all positive objects also dividing them into dense clusters. To learn a negative rule, we take all objects in one of this negative clusters jointly with all positive objects. The algorithm [12] constructs a minimal number of hyperplanes needed to build classification part of a rule describing this negative cluster. The same procedure is repeated for all the remaining negative clusters. Rules describing positive clusters are constructed the same way. Taking the medical data with 152 instances affected by ulcerative colitis, as an example, we show that the overall support and confidence of rules, extracted from that database, using our strategy [11, 12] is much higher than the confidence and support of rules obtained using methods described in Fig. 1 shows that there are many possible hyper-planes that can perfectly separate the two classes [15]. However, we need to find the best hyperplane that represents the largest distance between the two classes. SVM maximizes the margin between the hyperplane and two classes.

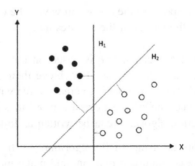

Fig. 1. Two separated classes described by the method [15]

In two dimension space, two groups can be separated by a line, using the equation $ax + by \geq c$ for the first group and $ax + by \leq c$ for the second group.

To choose the best possible hyperplane and minimize the risk of overfitting, it is very important to find the one with the maximum margin between the two classes. This is a typical optimization problem that can be solved using the Lagrangian formula. After finding the optimal hyperplane, only the data points closest to the hyperplane will have a positive weight, while others will take zero. Points regarding data in which distances are closest to the decision surface are called support vectors and are the most critical elements of training data. The position of the hyperplane is shifted when the support vectors are removed.

The distance between the data point (x_0, y_0) and the straight $ax + by + c = 0$ can be measured using the formula below:

$$\frac{|ax_0 + by_0 + c|}{\sqrt{(a^2 + b^2)}}$$

We have L training data, where each instance of X_i has D attributes and two classes: -1 and 1. We assume that the training data can be separated in a linear way, therefore we can draw a hyperplane that separates two classes. This hyperplane can be described as $x.w - b = 0$, where w is normal for the hyperplane. H_1 is a hyperplane for the first class, and H_2 is a hyperplane for the second one. $H_1 : x_i.w - b = 1$ and $H_2 : x_i.w - b = -1$. The perpendicular distance from the hyperplane is $\frac{b}{\|w\|}$. All points that are closest to H_1 and H_2 are auxiliary vectors.

We define d_1 as the distance from H_1 to the hyper-plane and d_2 as the distance from H_2 to the above-mentioned hyperplane. The SVM margin is the distance from H_1 to H_2 and is expressed as $d_1 + d_2$.

The distance between H_0 and H_1 expresses the following formula:

$$\frac{|w.x + b|}{\|w\|} = \frac{1}{\|w\|},$$

The distance between H_1 and H_2 is equal $\frac{2}{\|w\|}$.

The distance between two hyperplanes (H_1 and H_2) can be maximized by minimizing the value of $\|w\|$. The margin is $\frac{1}{\|w\|}$ and can be maximized using the following formula:

$$\min\|w\| = y_i(x_i.w + b) - 1 \geq 0, \forall_i$$

Minimizing $\|w\|$ is equivalent to minimizing $\frac{1}{2}\|w\|^2$ then using QP optimization (Quadratic Programming). In the next step, find $\frac{1}{2}\|w\|^2$ such that $y_i(x_i.w + b) - 1 \geq 0, \forall_i$. Minimization can be continued through the use of Lagrange multipliers α, where $\alpha_i \geq 0, \forall_i$.

$$L = \frac{1}{2}\|w\|^2 - \alpha[y_i(x_i.w + b) - 1 \geq 0\forall_i]$$

$$L = \frac{1}{2}\|w\|^2 - \sum_{i=1}^{L} \alpha_i[y_i(x_i.w + b) - 1]$$

$$L = \frac{1}{2}\|w\|^2 - \sum_{i=1}^{L} \alpha_i y_i(x_i.w + b) + \sum_{i=1}^{L} \alpha_i$$

For derivatives of 0, we get:

$$w = \sum_{i=1}^{l} \alpha_i y_i x_i, \quad \sum_{i=1}^{l} \alpha_i y_i = 0$$

3.3   Rating of the Classifier

Each built-in classifier should be evaluated in terms of its quality. For this purpose, two sets of data are necessary. The first, so-called training set is intended for learning the classifier. The second - validation test is used to test the classifier. In both sets, it is necessary to know how the samples belong to the classes. In many cases, the division of data into a teaching and testing set is not given from above. Then, a random division into two disjoint sets can be repeatedly made, usually in such way that the test set is smaller than the teaching one. In such case, we deal with simple validation. Another type of validation is k-fold validation called the k-fold cross-check [9] (called k-fold cross validation). In this method, the input set is divided into k subsets. Then, each of the subsets is a test set, and the classifier is taught on the $k-1$ of the other subsets. In this way, the validation is repeated k times, and the final result is usually the average of all repetitions [9].

Various metrics are used to evaluate the classifier [5]. In order to present the metrics used in the work, the designations as in Table 1 for different cases of classifier response were adopted depending on the class value for the sample. In the field of machine learning, specifically the problem of statistical classification, the confusion matrix (Table 2), also known as the error matrix [8, 14] is a specific table layout of usually supervised learning (in unprotected learning mode it is usually called matching matrix). Each row in the matrix represents occurrences in the projected class, while each column represents occurrences in the actual class or vice versa [12].

Table 2. Confusion matrix if a is taken to be the positive class (e.g. patient has the provided disease)

	a	b
Actual $a = 0$	TP	FN
Actual $b = 1$	FP	TN

In order to evaluate the quality of a binary classifier, a group of additional metrics should be considered. A true positive (TP) example is the one whose true label is 1 and the classifier has returned such label. The concepts of genuinely negative, false positive and false negative examples (which are denoted as follows: TN, FP, FN) are analogously defined.

Sensitivity – (TPR, hit rate, recall) the probability that the classification will be correct, provided that the case is positive. For a medical case, it may be the probability that the test performed by a sick patient will show that he has the predicted illness. Sensitivity can be described by the following formula:

$$TPR = \frac{TP}{TP + FN}$$

Specificity – (TNR) the probability that the classification will be correct, provided the case is negative. An example is the probability that a healthy person will not be diagnosed with a test. Specificity is defined by the following formula:

$$TNR = \frac{TN}{TN + FP} = 1 - FPR$$

False positive rate – (FPR) the coefficient of instances falsely classified as a given class, which we write with the following formula:

$$FPR = \frac{FP}{FP + TN} = 1 - TNR$$

False discovery rate – (FDR) error factor type I. The FDR aims to control the expected proportion of "discoveries" that are false (incorrect rejections):

$$FDR = \frac{FP}{FP + TP}$$

Positive predictive value – (PPV, precision) this indicator answers the example question: If the test result is positive, what is the probability that the patient has the illness? We can express the measure using the following formula:

$$PPV = \frac{TP}{TP + FP}$$

Negative predictive value – (NPV) the indicator answers the question: If the test result is negative, what is the probability that the patient is healthy?

$$NPV = \frac{TN}{TN + FN}$$

F1-score – the harmonic mean of precision and sensitivity, and its set of values is the interval [0, 1]. The measure assesses the relationship between sensitivity and precision. However, it does not include true negative results:

$$F_1 = 2 \cdot \frac{PPV \cdot TPR}{PPV + TPR} = \frac{2 \cdot TP}{2 \cdot TP + FP + FN}$$

4 Experiments

Our dataset contains clinical data of 152 patients affected by ulcerative colitis. Patients are characterized by 117 attributes and classified into two groups: patients with ulcerative colitis (UC) and patients with Crohn disease (CD). Our goal was to find classification rules. The study group consisted of patients with inflammatory bowel diseases. In the first group, ulcerative colitis was diagnosed (N = 86, women N = 32, men N = 54), and the second group were patients with Crohn disease (N = 66, women N = 32, men N = 34).

Too many variables can negatively impact the performance of the model. As a consequence, the first stages of the study, during which initial data processing is performed, are important. The data can be subjected to selection, transformation, or delete unwanted variables.

After completing the data using ERID and removing variables where the percentage of missing data exceeded 60%, the number of attributes decreased. There are 73 attributes left. Subsequently, all the attributes associated with treatment were excluded from the analysis, since predicates describing the treatment cannot determine the occurrence of the disease. Then, the attributes were selected using significance tests. Finally, a set of attributes was obtained that significantly differed in the two analyzed groups. The next stages of the analysis were carried out using data mining methods. Classification algorithms such as J48, SVM and Random Forest were used. Finally, the best algorithm was selected by analyzing the quality of classification measures (Tables 3, 4, 5, 6 and 7).

Table 3. Confusion matrix for J48 algorithm

Observed effects	Expected effects	
	UC	CD
UC	81	5
CD	6	60

After using logical regression model connected with ERID algorithm, the highest values of sensitivity and high specificity were obtained in the case of the Random Forest algorithm. For the aforementioned classifier, the sensitivity value was 100%, which proves the ideal ability to detect patients with CD. The specificity value determining the ability to detect people with UC within 98.48%. After applying the J48 algorithm, sensitivity of 94.19% and specificity of 90.91% were achieved. In the case of SVM, the sensitivity reached 93.02%, and the specificity was 84.85%.

The frequency of false alarms in the case of the J48 algorithm was at the level of 0.09, while the frequency of false discoveries was 0.07. For the SVM and Random Forest algorithms, these values were 0.15 and 0.11 and 0.02 and 0.01 respectively.

In the next step, the predictive properties of the constructed model were determined. The positive precision indicator in the case of the J48 algorithm was at the level

Table 4. Confusion matrix for SVM

Observed effects	Expected effects	
	UC	CD
UC	78	8
CD	17	49

Table 5. Confusion matrix for Random Forest algorithm

Observed effects	Expected effects	
	UC	CD
UC	86	0
CD	1	65

Table 6. The values of the measures

	FPR	FDR	PPV	NPV	F-score
J48	0.09	0.07	0.93	0.92	0.94
SVM	0.15	0.11	0.89	0.90	0.91
Random Forest	0.02	0.01	0.99	1.00	0.99

Table 7. Sensitivity and specificity

	Sensitivity	Specificity
J48	94.19%	90.91%
SVM	93.02%	84.85%
Random Forest	100.00%	98.48%

of 0.93, while the other two methods were respectively: 0, 89 and 0.99. The negative precision value was J48: 0.92, SVM: 0.9 and Random Forest: 1, respectively.

In addition, the value of F1-score, which is a balanced measure, which to a certain extent describes the model as a whole, was calculated. In the first discussed algorithm F1 = 0.92, for the other two F1 = 0.91 for SVM and F1 = 0.99, for Random Forest.

The proposed method was compared with currently used methods. All variables were introduced to the classifier and three algorithms were compared: J48, SVM and Random Forest. The results are shown (Tables 8 and 9).

Table 8. The values of the measures

	FPR	FDR	PPV	NPV	F-score
J48	0.11	0.08	0.92	0.87	0.91
SVM	0.26	0.18	0.82	0.86	0.86
Random Forest	0.06	0.05	0.95	0.97	0.97

Table 9. Sensitivity and specificity

	Sensitivity	Specificity
J48	89.53%	89.39%
SVM	90.70%	74.24%
Random Forest	97.67%	93.94%

Sensitivity in the case of the J48 algorithm was 89.53% and reached a value lower by more than 5 percentage points, comparing with the classifier discussed earlier. At the same time, it was the lowest value among the three compared algorithms. For a classifier built using the SVM method, the value discussed was 90.70%, while for Random Forest it was 97.67%. These values, in both cases, were lower compared to the model built on the basis of the developed methodology.

Similar results were obtained for specificity. The measure in question in the case of J48 reached the value of 89.39%, SVM - 74.24%, and for Random Forest - 93.94%. In the case of three algorithms, the level of specificity was lower compared to the classifier discussed earlier.

The instance rate falsely classified as a given class (FPR) has reached the following values for three algorithms respectively: 0.11 (J48), 0.26 (SVM), 0.06 (Random Forest). The type I error rate (FDR) assumed the following levels: 0.08, 0.18, 0.05.

The positive precision value was 0.92 (J48), 0.82 (SVM), 0.95 (Random Forest). The negative pretension for J48 was 0.87, SVM 0.86, Random Tree 0.97.

The harmonic mean of precision and sensitivity, i.e. the measure of F1, achieved high, but less satisfactory values, comparing with the classifier built by using the developed methodology. This value reached the following levels: 0.91 (J48), 0.86 (SVM), 0.97 (Random Forest).

5 Conclusion and Future Work

In this work we dealt with the data of patients suffering from ulcerative colitis and Crohn's disease. In order to find rules that distinguish these two diseases, classification methods were used. Three popular methods were compared: methods of decision trees (J48 and Random Forest) and SVM. Patients' data were selected using statistical methods. The proposed method gives better results than the method consisting in the introduction of all attributes to the model. In the future, the obtained classification models will be used to build the rules of action from classification rules to reclassify patients from one class to another (more desirable one).

Acknowledgements. This work was supported by MB/WM/8/2016 and financed with use of funds for science of MNiSW. The Bioethical Commission gave the permission for the analysis and publication of our results.

References

1. Breiman, L., Friedman, J.H., Olshen, R.A., Stone, C.J.: Classification and Regression Trees. Wadsworth International Group, Belmont (1984)
2. Cheng, J., Greiner, R.: Learning Bayesian belief network classifiers: algorithms and system. In: Stroulia, E., Matwin, S. (eds.) AI 2001. LNCS (LNAI), vol. 2056, pp. 141–151. Springer, Heidelberg (2001). https://doi.org/10.1007/3-540-45153-6_14
3. Dardzinska, A.: Action Rules Mining. Springer, Heidelberg (2013). https://doi.org/10.1007/978-3-642-35650-6
4. Fawcett, T.: An introduction to ROC analysis. Pattern Recogn. Lett. **27**, 861–874 (2006)
5. Frawley, W., Piatetsky-Shapiro, G., Matheus, C.: Knowledge discovery in databases, an overview. Knowl. Disc. Databases 1–27 (1991)
6. Hand, D., Mannila, H., Smyth, P.: Eksploracja danych. Wydawnictwa Naukowo – Techniczne, Warszawa, 35–61, 91–127, 181–201 (2005)
7. Kasperczuk, A., Dardzinska, A.: Comparative evaluation of the different data mining techniques used for the medical database. Acta Mechanica et Automatica **10**(3), 233–238 (2016)
8. Kohavi, R.: A study of cross-validation and bootstrap for accuracy estimation and model selection. In: Proceedings of the International Joint Conference on Artificial Intelligence, vol. 2, pp. 1137–1143 (1995)
9. Powers, D.M.W.: Evaluation: from precision, recall and F-measure to ROC, informedness, markedness and correlation. J. Mach. Learn. Technol. **2**(1), 37–63 (2011)
10. Quinlan, J.R.: Introduction of decision trees. In: Machine Learning, pp. 81–106. Kluwer Academic Publishers (1986)
11. Ras, Z.W., Dardzinska, A., Liu, X.: Rule discovery by axes-driven hyperplanes construction. In: Kłopotek, M.A., Wierzchoń, S.T., Trojanowski, K. (eds.) Intelligent Information Processing and Web Mining. Advances in Soft Computing, vol. 25. Springer, Heidelberg (2004). https://doi.org/10.1007/978-3-540-39985-8_62
12. Ras, Z.W., Dardzinska, A., Liu, X.: System ADReD for discovering rules based on hyperplanes, special issue on selected problems in knowledge representation. Int. J. Eng. Appl. Artif. Intell. **17**(4), 401–406 (2004)
13. Raś, Z.W., Dardzińska, A.: Data security and null value imputation in distributed information systems. In: Raś, Z.W., Dardzińska, A. (eds.) Monitoring, Security, and Rescue Techniques in Multiagent Systems. Advances in Soft Computing, vol. 28. Springer, Heidelberg (2005). https://doi.org/10.1007/3-540-32370-8_9
14. Stehman, S.V.: Selecting and interpreting measures of thematic classification accuracy. Remote Sens. Environ. **62**(1), 77–89 (1997)
15. Vapnik, V.: The Nature of Statistical Learning Theory. Springer, New York (1995). https://doi.org/10.1007/978-1-4757-2440-0

New Method of Medical Incomplete Information System Optimization Based on Action Queries

Katarzyna Ignatiuk[1], Agnieszka Dardzinska[1(✉)],
Małgorzata Zdrodowska[1], and Monika Chorazy[2]

[1] Department of Mechanical Engineering,
Department of Biocybernetics and Biomedical Engineering,
Bialystok University of Technology, ul. Wiejska 45c, 15-351 Bialystok, Poland
ignatiukkatarzyna@gmail.com,
{a.dardzinska,m.zdrodowska}@pb.edu.pl
[2] Department of Neurology, Medical University of Bialystok,
ul. M. Skłodowskiej - Curie 24A, 15-276 Bialystok, Poland
neurosek@umb.edu.pl

Abstract. In this paper we assume there is a group of connected distributed information systems (*DIS*). They work under the same ontology. Each information system has its own knowledgebase. Values of attributes in incomplete information system *IS* form atomic expressions of a language used for communication with others. Collaboration among systems is initiated when one of them is asked to resolve a query containing nonlocal attributes for *IS*. When query fails, then the query answering system (*QAS*) is trying to replace values in a query by new values from their corresponding neighborhoods. *QAS* for *IS* can also collaborate and exchange knowledge with other information systems. In all such cases, it is called intelligent. As the result of its request, knowledge is extracted locally in each information system and sent back to the client. The outcome of this step is collective knowledgebase. In this paper we present a method of identifying which information system is semantically the closest to *IS*. We propose a new measure supporting choice of closest pair of systems, which determines the distance between the two systems. The proposed method was tested and verified in medical systems with randomly selected data. The satisfying initial results were obtained and based on them, the proposed measure can be successfully used in medical systems to support the work of doctors and the treatment of patients.

Keywords: Query answering system · Knowledge extraction · Action query
Incomplete medical information system

1 Introduction

Modern query answering systems area of research is related to enhancements of query answering systems into intelligent systems. The emphasis is on problems in users posing queries and systems producing answers. This becomes more and more relevant

© Springer Nature Switzerland AG 2018
T. K. Dang et al. (Eds.): FDSE 2018, LNCS 11251, pp. 310–322, 2018.
https://doi.org/10.1007/978-3-030-03192-3_24

as the amount of information available from local or distributed information sources increases. We need systems not only easy to use but also intelligent in answering the users' needs, especially when we work on medical data. A query answering system often replaces human with expertise in the domain of interest, thus it is important, from the user's point of view, to compare the system and the human expert as alternative means for accessing information [17]. Especially medical datasets have a huge amount of information about each patient and provide great opportunity for knowledge discovery algorithms. One of the goals is to find a way to process the enormity of accumulated knowledge. In this paper we assume that there is a group of collaborating medical information systems, which are incomplete, coupled with a Query Answering System (QAS) and a knowledgebase which is initially empty. By incompleteness we mean a property which allows us to use attributes and their values with corresponding weights [3, 4]. Additionally, we assume that the sum of these weights for one particular attribute value for one object has to be equal 1. The definition of an information system of type λ given in this paper was initially proposed in [14]. The type λ was introduced with a purpose to check the weights assigned to values of attributes by Chase algorithm [4, 9]. If a weight is less than λ, then the corresponding attribute value is ruled out and weights assigned to the remaining attribute values are equally adjusted so its sum is equal one again. Semantic inconsistencies are due to different interpretations of attributes and their values among sites (for instance one site can interpret the concept *healthy* or *high* differently than other sites). Different interpretations are also implied by the fact that each site may differently handle incompleteness [15].

In this paper, special attention will be paid to the incompleteness of medical information systems and action queries which should be answered correctly. This is an important issue affects to doctors decision and patient treatment.

2 Main Assumptions

In reality, different data are often collected and stored in information systems residing at many different locations, built independently, instead of collecting and storing them at a single location. In this case we talk about distributed information systems. It is very possible that some attributes are missing in one of these systems and occur in others. More formally, the information system S is a triplet $S = (X, A, V)$ where [2]:

- X is a nonempty, finite set of objects,
- A is a nonempty, finite set of attributes,
- $V = \cup \{V\_a : a \in A\}$ is a set of attribute values, where V_a is a set of attribute values a for any a $\in$ A.
  We assume, that [4]:
- $V_a \cap V_b = \emptyset$ for any a, b $\in$ A such that a $\neq$ b,
- a : X $\rightarrow$ V_a is a partial function for every a $\in$ A.

2.1   Incomplete Information Systems

We call information system S as an incomplete information system IS of type λ, if S is an incomplete information system introduced by Pawlak in [10] and the following four conditions hold:

1. X is a set of objects, A is a set of attributes, and $V = \cup\{V_a : a \in A\}$ is a set of values of attribute a
2. $(\forall x \in X)\,(\forall a \in A)[a_S(x) \in V_a \text{ or } a_S(x) = \{(a_i, p_i) : a_i \in V_a \wedge p_i \in [0,1] \wedge 1 \le i \le m\}]$
3. $(\forall x \in X)\,(\forall a \in A)\big[(a_S(x) = \{(a_i, p_i) : 1 \le i \le m\}) \Rightarrow \sum_{i=1}^{m} p_i = 1\big]$
4. $(\forall x \in X)\,(\forall a \in A)\big[(a_S(x) = \{(a_i, p_i) : 1 \le i \le m\}) \Rightarrow (\forall i)(p_i \ge \lambda)\big]$

Table 1 presents the example of incomplete information system.

Table 1. Part of the incomplete information system S based on blood database

Patient	RBC [$10^6/\mu l$]	WBC [$10^3/mm^3$]	HGB [g/dl]	MCH [pg]	PLT [$10^3/\mu l$]	Blood disease
x_1	4.1	6.2	9.0	23.2		Anemia
x_2	2.8	41.8		23.4	87.1	Leukemia
x_3	3.2	3.2	9.5		90.3	Pancytopenia
x_4	7.4		19.2	36.1	350.7	Polycythemia vera

Objects in Table 1 are four patients. The attributes of these objects are the characteristics of patients: {RBC, WBC, HGB, MCH, PLT, Blood disease} and numerical values of attributes create values of these attributes. Attributes in information system can be divided into stable (they cannot change their values), semi-stable (they can change their values under some particular assumptions) and flexible (they can change their values). Gender is an example of stable attribute, patient's age is semi-stable attribute, blood parameters are flexible ones. Also the decision attribute can be treated as a flexible attribute, and the patients can be classified for different classes of blood diseases.

Single patient data informs about his health condition. To change this state, the doctor indicates treatment based on his knowledge. He can also use knowledge in the form of patterns extracted from the medical databases. One of the methods of presenting patterns are classification rules or action rules, whose general form is represented by the dependence $A \to C$ [2]. The first part is created by the values of the classification attributes (stable, semi-stable, flexible). The implication part is formed by the values of the decision attribute. Example rule for the object x_4 from Table 1. has the form $(RBC = 7.4) * (HGB = 19.2) * (MCH = 36.1) * (PLT = 350.7) \to (blood\ disease = polycythemia\ vera)$. The value of the blood disease is a consequent and the rest of the rule is the antecedent.

To verify extracted rules forming knowledgebase, two known statistical measures are generally used: confidence (*conf*) and support (*sup*). They are calculated in a standard form [14, 15].

2.2   Types of Incomplete Information Systems

One of the main problems of medical information systems is their incompleteness. By incompleteness we mean null values among selected attributes because of different reasons. In medical information system, the cause of these incompleteness may be unreliable supplemented patient documentation, loss of part of documentation, errors during transferring information from paper to electronic documentation or intentional encrypting of sensitive patient data. Table 1 shows an example of an incomplete information system based on a medical database. The values $PLT(x1), HGB(x2), MCH(x3)$ and $WBC(x4)$ are not defined. In medical information systems we can observe four main types of incompleteness [2].

Type 1
Incompleteness of the first type is defined by the assumption that at least one attribute $a \in A$ is a partial function [2]:

$$(\exists x \in X)(\exists a \in A)[a(x) = Null] \tag{1}$$

Table 1 is an example of such *IS*. Null value is interpreted as "undefined" value. In the system of this type, an undefined value can take different values, not necessarily the value which already exists in the system.

Type 2
By the incompleteness of this type we understand the situation where all attributes in $S = (X, A, V)$ are functions of the type $a : X \rightarrow 2^{V_a} - \{\emptyset\}$ [2].

If $a(x) = \{a_1, a_2, \ldots, a_n\} \subseteq V_a$ then we can say that the value of attribute a is one from $a_1, a_2, \ldots, a_n$.

If $a(x) = V_a$, then all values of the attribute a are equally probable and $a(x) = null$ corresponds to "blank".

Table 2. Incomplete information system of Type 2

Patient	Name	Last name	Blood disease
x_1	Ann, Emily, John	Brown	Anemia, pancytopenia
x_2	Kate	Johnson	
x_3	John, Kate		Polycythemia vera
x_4	Lucy, John	Smith, Taylor	Anemia, leukemia

Table 2 represents such *IS*. Null value for *Last name*(x_3) means, that *Last name*$(x_3) = Brown$ OR *Last name*$(x_3) = $ *Johnson* OR *Last name*$(x_3) = Taylor$ OR *Last name*$(x_3) = Smith$, all with the same probability.

Null value corresponding with Blood *disease*$(x2)$ means, that *Blood disease*$(x_2) = $ *anemia* OR *Blood disease*$(x_2) = leukemia$ OR *Blood disease*$(x_2) = $ *pancytopenia* OR *Blood disease*$(x_2) = polycythemia vera$.

Type 3

For the incompleteness of type 3 we assume, that all attributes in $S = (X, A, V)$ are functions of type $a : X \to 2^{V_a}$ [2]. This type differs from the previous one, because we allow having the empty set as the value of some attributes in S. When $a(x) = \emptyset$, then the value of attribute a for the object x does not exist (Table 3).

Table 3. Incomplete information system of Type 3

Patient	Name	Last name	Blood disease
x_1	Ann, Emily, John	Brown	Anemia, leukemia
x_2	Kate	Johnson	
x_3	John, Kate	$\emptyset$	Polycythemia vera
x_4	Lucy, John	Smith, Taylor	$\emptyset$

Last name$(x3) = \emptyset$ is interpreted as „ x_3 doesn't have last name". It means that the value doesn't exist for x_3. *Blood disease*$(x_4) = \emptyset$ is interpreted that the object x_4 doesn't have any blood disease.

Type 4

For this type of incompleteness, we assume that all attributes in $S = (X, A, V)$ are functions of the type: $a : X \to 2^{V_a \times R}$. When we assume that $a(x) = \{(a_1, p_1), (a_2, p_2), \ldots (a_n, p_n)\}$ and p_i is a confidence for a_i, then [2] (Table 4):

$$\sum_{i=1}^{n} p_i = 1 \tag{2}$$

Table 4. Incomplete information system of Type 4

Patient	Name	Last name	Blood disease
x_1	Ann, Emily, John	Brown	(Anemia, $\frac{3}{4}$), (leukemia, $\frac{1}{4}$)
x_2	Kate	Johnson	$\emptyset$
x_3	John, Kate		Polycythemia vera
x_4	Lucy, John	Smith, Taylor	(Anemia, $\frac{1}{2}$), (pancytopenia, $\frac{1}{2}$)

In this case *Blood disease*$(x1) = \{(anemia, \frac{3}{4}), (leukemia, \frac{1}{4})\}$ is interpreted as "the confidence that x_1 has an anemia is $\frac{3}{4}$ or that he has a leukemia is $\frac{1}{4}$. The object x_2 has not got any blood disease. The object x_3 has polycythemia vera with the confidence equal to 1. For the object x_4, the anemia and pancytopenia are equally likely with the confidence $\frac{1}{2}$.

2.3   Distributed Information Systems

Assume that *IS* is an incomplete information system working as an element in distributed information system, and the query q is submitted to this system. The syntax of the query q contains values unknown to *IS*. For example the value of MCH for object $x3$ from Table 1. when the query is $q(RBC, HGB, MCH) = (RBC = 3.2) * (HGB = 9.5) * (MCH = ?)$. Missing values can be replaced by statistical or rule-based methods suggested values, for example, by the rules extracted in Chase algorithm [4]. Another approach is to create *QAS* [13, 15] that uses the knowledge collected from other, connected information systems. It is concerned with identifying all objects in the system satisfying a given description. For example an information system might contain information about patents and classify them using four attributes of "blood pressure", "weight", "gender" and "muscle tension". A simple query might be to find all patients with blood pressure above 140/90 mmHg which are women. When information system is incomplete, patients which are women and unknown blood pressure can be handled by either including or excluding them from the answer to the query. Therefore the correct answer cannot be made. But we can discover rules for blood pressure level in terms of the attributes weight, gender and muscle tension. These rules could then be applied to patients with unknown blood pressure level to generate values that could be used in answering the query. However the rules have to work under the same ontology [9]. It [6, 7, 16] is used to build a semantical connection between systems helping in communication to each other. This is important for semantical inconsistencies caused by different interpretation of attributes and their values by different systems. For instance, one medical system can interpret the concept weight differently than other one. The definition of a distributed information system was initially introduced in [15] and next applied in [1, 10–13].

By an incomplete distributed information system we mean a pair DIS $= (\{ S_i \}_{i \in I}, L)$ where [2]:

- $S_i = (X_i, A_i, V_i)$ is an information system for any $i \in I$, and $V_i = \cup \{V_{ia} : a \in A_i\}$,
- $\exists i \in I\ S_i$ is incomplete,
- L is a symmetric, binary relations on the set I,
- I is a set of sites.

Two systems S_i, S_j are called neighbors in distributed information system if $(i, j) \in L$. Distributed information system is object–consistent if the following condition holds [2]:

$$(\forall i)(\forall j)(\forall x \in X_i \cap X_j)(\forall a \in A_i \cap A_j)$$

$$\left[\left(a_{[S_i]}(x) \subseteq a_{[S_j]}(x) \right) \ or \ (a_{[S_j]}(x) \subseteq a_{[S_i]}(x)) \right],$$

where a_s denotes that a is an attribute in S. The inclusion $(a_{[S_i]}(x) \subseteq a_{[S_j]}(x))$ means that the system S_i has more precise information about the attribute a for object x than system S_j. Object–consistency means that information about objects in one of the systems is either the same or more general than in the other. Saying other words, two consistent systems cannot have conflicting information about any object x which is stored in both of them. System in which the above condition does not hold is called object–inconsistent. The result of collaboration between the systems is creation of the knowledgebase which collects rules defined as expressions written in predicate calculus and originates from various information systems. We are interested in how to build intelligent QAS using action queries for incomplete information systems. If query is submitted to IS, the first step is to make IS as complete as possible. We use not only functional dependencies to chase IS but also use rules discovered from a complete subsystem to do the chasing. In the first step, intelligent QAS identifies all incomplete attributes used in a query. The values of all incomplete attributes are treated as concepts to be learned. Incomplete information is replaced by new data based on these rules. When the process of removing incompleteness in the local information system is completed, QAS changes the query into action query and start to work in a usual way. In this paper, we present the new method which helps to decide whether the selected information system is the closest one (in a semantical sense) to the system which has to answer given query q. First we use ERID algorithm [5] to extract rules from each information system independently. Confidences and supports of rules are used to construct the measure of the distance between pair of systems. In this paper we propose measure which is the modification of the work from [3].

3 Searching the Closest Information System

Ontologies and inter-ontology relationships between information systems can represent a particular point of view of the global information system by describing customized domains. To allow intelligent query processing, it is often assumed that an information system is coupled with some ontology. Inter-ontology relationships can be seen as semantical bridges between ontologies built for each of the autonomous information systems so they can collaborate and understand each other. In [5, 11], the notion of optimal rough semantics and the method of its construction have been proposed. Rough semantics can be used to model semantic inconsistencies among sites due to different interpretations of incomplete values of attributes.

3.1 The Closest Information System

Assume now that user submits a query to one of the information system, which cannot be answered because some of the attributes used in a query are unknown or hidden in the information system representing system S and knowledgebase K. Assume, we have a set of collaborating distributed information systems (DIS) working under the same ontology, and the user asks a query $q(Q)$ for an information system (S, K) from DIS, where $S = (X, A, V), K = \emptyset$), Q is the set of attributes used in $q(Q)$, and $A \cap B \neq \emptyset$ [3]. All attributes in $Q \backslash [A \cap Q]$ are called foreign for (S, K). Since (S, K) can collaborate with other information systems in DIS, values of hidden or missed attributes for (S, K) can be extracted from other information systems in DIS. Assume now that we have three, object–consistent and incomplete collaborating information systems: (S, K), (S_1, K_1), (S_2, K_2) where: $S = (X, A, V), S_1 = (X_1, A_1, V_1)$, $S_2 = (X_2, A_2, V_2)$ and $K = K_1 = K_2 = \emptyset$ [1, 2]. If the consensus between (S, K) and (S_1, K_1), based on the knowledge extracted from $S(A \cap A_1)$ and $S_1(A \cap A_1)$, is chosen by (S, K) as closer information system than consensus (S, K) and (S_2, K_2), the knowledge from first pair becomes more helpful in recreate given query. Rules defining hidden attribute values for S are then extracted at S_1 and stored in K. Assuming that systems S_1 and S_2 store the same sets of objects and use the same attributes describing them, how can we choose the closer and better system to the solve the query?

First, the attributes common to the two systems should be indicated. Next, the ERID algorithm extracts rules for each system forming knowledgebase. From the set of rules we choose rules that exist in both systems and for each pair we calculate their confidence and support. On the basis of these measures the factor of fitting two systems is calculated:

$$d(S_i, S_j) = \frac{\sum_r d_r (S_i \rightarrow S_j)}{max(\sum sup_r S_i, \sum sup_r S_j)}$$

where:

$$d_r(S_i \rightarrow S_j) = \sqrt{(conf_r S_i \cdot sup_r S_i)^2 + (conf_r S_j \cdot sup_r S_j)^2}$$

We say that system S_i is closer to Sj than S_k when d(Si, Sj) is closer to 1 than $d(S_i, S_k)$. From all the distributed systems we choose the one with maximum value of $d(S_i, S_j)$, which corresponds to the closest information system to the client.

Example
Let us assume we have three medical information systems: S_1, S_2 and S_3. They are presented in Tables 5, 6 and 7. The systems are incomplete, object-consistent, created in different locations and they create Query Answering System.

Table 5. Information system S_1

X	a	b	c	g
x_1	1	2		3
x_2	2	2		3
x_3	3	3		1
x_4	1	1		2
x_5	2	3		1
x_6	3	2		2

Table 6. Information system S_2

X	a	b	c	d	e	g
x_7	1	1	3		1	3
x_8	2	1	2		2	2
x_9	1	2	2		1	3
x_{10}	3	2	1		2	2
x_{11}	1	3	3		1	1
x_{12}	3	3	1		2	1

Table 7. Information system S_3

X	a	b	c	d	e	g
x_{13}	2	3	1	2		1
x_{14}	1	2	2	2		3
x_{15}	2	1	1	3		3
x_{16}	3	2	2	3		2
x_{17}	1	2	3	1		3
x_{18}	2	3	1	2		1
x_{19}	3	2	3	3		2
x_{20}	1	1	3	1		3

Information system S_1 received a query $q(a,c,g) = a3 * c? * g2$ and has no information about hidden attribute c, which appears in other systems such as S_2 and S_3. Our goal is to choose one of them, from which we will be able to predict the values of attribute c in system S_1 and to answer query $q(a,c,g)$. Because attributes a,b,g are common for all the systems, first we extract the rules describing them in terms of other attributes. If the system is incomplete, we use ERID algorithm. For each rule we calculate support and confidence in a standard way [1]. Next, we pair the systems: S_1 with S_2 and S_1 with S_3. For each pair we select rules the same way for the two systems. Tables 8 and 9 present joint rules for paired systems with the confidence and support for each rule and in each system.

Next the factor of fitting two systems: S_1 and S_2 is calculated:

$$d(S_1, S_2) = \frac{21.21}{max(21, 22)} = \frac{21.21}{22} = 0.964.$$

And the same for systems S_1 and S_3:

$$d(S_1, S_3) = \frac{37.197}{max(40, 23)} = \frac{37.197}{40} = 0.929.$$

Since the factor between S_1 and S_2 is closer to 1 than the factor between S_1 and S_3, we choose S_2 as the closer information system to the communication with the

Table 8. The common rules for systems S_1 and S_2 with their confidences and supports

	S_1		S_2	
	conf	sup	conf	sup
$b_2 \rightarrow a_1$	0.(3)	1	0.5	1
$b_2 \rightarrow a_3$	0.(3)	1	0.5	1
$b_1 \rightarrow a_1$	1	1	0.5	1
$b_3 \rightarrow a_3$	0.5	1	0.5	1
$g_3 \rightarrow a_1$	0.5	1	0.5	2
$g_1 \rightarrow a_3$	0.5	1	0.5	1
$g_2 \rightarrow a_3$	0.5	1	0.5	1
$b_2 * g_3 \rightarrow a_1$	0.5	1	0.5	1
$b_2 * g_2 \rightarrow a_3$	0.5	1	0.5	1
$a_1 \rightarrow g_3$	1	1	0.(6)	2
$a_3 \rightarrow g_1$	0.5	1	0.5	1
$a_3 \rightarrow g_2$	0.5	1	0.5	1
$b_1 \rightarrow g_2$	1	1	0.5	1
$b_2 \rightarrow g_3$	0.(6)	2	0.5	1
$b_2 \rightarrow g_2$	0.(3)	1	0.5	1
$b_3 \rightarrow g_1$	1	2	1	2
$a_1 * b_2 \rightarrow g_3$	1	1	1	1
$a_3 * b_2 \rightarrow g_2$	1	1	1	1
$a_3 * b_3 \rightarrow g_1$	1	1	1	1

incomplete system S_1. From all rules describing attribute c in terms of a, b and g, we choose the rules by which the system S_1 can answer the query $q(a, c, g) = a3 * c1 * g2$. Based on the system S_1, attribute c has a value equal to 1. The rules from knowledgebase K allow us to answer other questions in the system S_1.

3.2   Experiments and the Results

As the testing data set we have taken 1,000 tuples randomly selected from a database of some neurological disease. The sample table, containing 60 attributes, was randomly partitioned into five subtables of equal size containing 200 tuples each. Next, from each of these subtables 10 attributes (columns) have been randomly removed leaving four data tables of the size 200 × 50 each. One of these tables is called a client and the other 4 are servers. Now, for all objects at the client site, values of one of the attributes, e.g. c, which was chosen randomly, have been hidden. At each server site, if attribute c was listed in its domain schema, descriptions of h using WEKA software have been learned. All these descriptions, in the form of rules, have been stored in the knowledge base of the client. ERID and distributed chase was applied to predict what is the real value of the hidden attribute for each object x at the client site. The threshold value $\lambda = 0.115$ was used to rule out all values predicted by distributed Chase with confidence below

Table 9. The common rules for systems S_1 and S_3 with their confidences and supports

	S_1		S_3	
	conf	sup	conf	sup
$b_1 \rightarrow a_1$	0.5	1	1	1
$b_1 \rightarrow a_2$	0.5	2	0.(3)	1
$b_2 \rightarrow a_3$	0.5	2	0.(3)	1
$b_3 \rightarrow a_2$	1	2	0.(3)	1
$g_1 \rightarrow a_2$	1	2	0.5	1
$g_2 \rightarrow a_3$	1	2	0.5	1
$g_3 \rightarrow a_1$	0.75	3	0.5	1
$g_3 \rightarrow a_2$	0.25	1	0.5	1
$b_2 * g_2 \rightarrow a_3$	1	2	1	1
$b_2 * g_3 \rightarrow a_1$	1	2	0.5	1
$b_3 * g_1 \rightarrow a_2$	1	2	0.5	1
$a_1 \rightarrow g_3$	1	3	0.5	1
$a_2 \rightarrow g_1$	0.(6)	2	0.5	1
$a_3 \rightarrow g_2$	1	2	0.5	1
$a_2 \rightarrow g_3$	0.(3)	1	0.5	1
$b_2 \rightarrow g_3$	0.5	2	0.(6)	2
$b_2 \rightarrow g_2$	0.5	2	0.(3)	1
$b_3 \rightarrow g_1$	1	2	1	2
$a_1 * b_2 \rightarrow g_3$	1	2	1	1
$a_2 * b_3 \rightarrow g_1$	1	1	1	1
$a_3 * b_2 \rightarrow g_2$	1	2	1	1

that threshold. Almost all hidden values (189 out of 200) have been discovered correctly. For each site we calculated the distance to client, and we chose the closest one. The results were promising with high accuracy (above 90%).

4   Conclusion

One of the main problems of medical information systems is their incompleteness. It has a significant impact on the discovered knowledge from medical databases. To help the decision process in the incomplete system, a method of discovering rules based on knowledge gathered in distributed information systems was proposed.

In this study, we proposed the factor of fitting two systems which can help to find the closest information systems. On the basis of this measure, it is possible to build more precise knowledgebase about patients and answer the query asked for system without valuable information.

Our method has been analyzed based on the medical information systems with missing data and allowed to ascertain which system integration gives better results.

Acknowledgements. Research was performed as a part of project no. MB/WM/6/2017 and financed with use of funds for science of MNiSW.

References

1. Glebocka, A.D.: Null values and chase in distributed information systems. In: Negoita, Mircea Gh., Howlett, Robert J., Jain, Lakhmi C. (eds.) KES 2004. LNCS (LNAI), vol. 3214, pp. 1143–1149. Springer, Heidelberg (2004). https://doi.org/10.1007/978-3-540-30133-2_152

2. Dardzinska, A.: Action Rules Mining, pp. 5–19. Springer, Berlin (2013). https://doi.org/10.1007/978-3-642-35650-6

3. Dardzinska, A., Ignatiuk, K., Zdrodowska, M.: Query answering system as a tool in incomplete distributed information system optimization process. In: Dang, T.K., Wagner, R., Küng, J., Thoai, N., Takizawa, M., Neuhold, Erich J. (eds.) FDSE 2017. LNCS, vol. 10646, pp. 101–109. Springer, Cham (2017). https://doi.org/10.1007/978-3-319-70004-5_7

4. Dardzińska, A., Raś, Zbigniew W.: $CHASE_2$ – rule based chase algorithm for information systems of type λ. In: Tsumoto, S., Yamaguchi, T., Numao, M., Motoda, H. (eds.) AM 2003. LNCS (LNAI), vol. 3430, pp. 255–267. Springer, Heidelberg (2005). https://doi.org/10.1007/11423270_14

5. Dardzinska, A., Ras, Z.: Extracting rules from incomplete decision systems: system ERID. In: Young Lin, T., Ohsuga, S., Liau, C.J., Hu, X. (eds.) Foundations and Novel Approaches in Data Mining, pp. 143–153. Springer, Heidelberg (2006). https://doi.org/10.1007/11539827_8

6. Guarino, N.: Formal ontology in information systems. In: Proceedings of FOIS 1998, pp. 3–15, Trento, Italy (1998)

7. Guarino, N., Giaretta, P.: Ontologies and knowledge bases, towards a terminological clarification. Towards Very Large Knowledge Bases: Knowledge Building and Knowledge Sharing, pp. 25–32 (1995)

8. Laudon K., Laudon J.: Management Information System: Managing the Digital Firm, pp. 14–16. Prentice Hall, New Jersey (2012)

9. Mizoguchi, R.: Tutorial on ontological engineering—Part 1: introduction to ontological engineering. New Gener. Comput. **21**(4), 365–384 (2003)

10. Pawlak, Z.: Information systems—theoretical foundations. Inf. Syst. J. **6**(1981), 205–218 (1991)

11. Ras, Z.: Collaboration control in distributed knowledge-based system. Inf. Sci. **96**(3), 193–205 (1997)

12. Ras, Z.: Query answering based on distributed knowledge mining. In: Proceedings of the 2nd Asia-Pacific Conference on Intelligent Agent Technology: Research and Development, pp. 17–27, Maebashi City, Japan (2001)

13. Ras, Z.: Reducts-driven query answering for distributed knowledge systems. Int. J. Intell. Syst. **17**(2), 113–124 (2002)

14. Ras, Z., Dardzinska, A.: Solving failing queries through cooperation and collaboration. World Wide Web J. **9**(2), 173–186 (2006)

15. Ras, Z., Dardzinska, A.: Cooperative multi-hierarchical query answering systems. In: Meyers, R. (ed.) Encyclopedia of Complexity and Systems Science, pp. 1532–1537. Springer, New York (2009). https://doi.org/10.1007/978-0-387-30440-3_100

16. Ras, Z., Joshi, S.: Query approximate answering system for an incomplete DKBS. Foundamenta Informaticae J. **30**(3), 313–324 (1997)
17. Van Heijst, G., Schreiber, A., Wielinga, B.: Using explicit ontologies in KBS development. Int. J. Hum. Comput. Stud. **46**(2), 183–292 (1997)
18. Yoo, I., Alafaireet, P., Marinov, M., Pena-Hernandez, K., Gopidi, R., Chang, J., Hua, L.: Data mining in healthcare and biomedicine: a survey of the literature. J. Med. Syst. **36**(4), 2431–2448 (2012)

Cloud Media DJ Platform: Functional Perspective

Joohyun Lee, Jinwoong Jung, Sanggil Yeoum, Junghyun Bum,
Thien-Binh Dang, and Hyunseung Choo[✉]

Department of Electrical and Computer Engineering, Sungkyunkwan University, Suwon, Korea
{joohyun7,sjud325,sanggil12,bumjh,dtbinh,choo}@skku.edu

Abstract. Content services have been provided to people in a variety of ways. Jukebox provides an automated music-playing service. User inserts a coin and presses a music button desired, the jukebox automatically selects and plays the record. DJs in Korean cafes receive the contents they want and play it through the speakers in the store. In this paper, we propose a service platform that reinvents the Korean cafe DJ in an integrated environment of IoT and cloud computing. In addition, we analyze the functional aspects of the services provided by the proposed platform. The user in a store requests contents (music, video, message) through the service platform. The contents are provided through the public screen and speaker in the store where the user is located. This allows people in the same location store to enjoy the contents together. The user information and the usage history are collected and managed in the public cloud. Therefore, users can receive customized services regardless of stores. Based on the implementation results of the platform, it is shown that the proposed platform provides more functions and advantages than other streaming services. Also proposed platform can provide contents efficiently to concurrent users.

Keywords: Internet of Things · Cloud computing · Streaming service

1 Introduction

Users have been served contents in various offline ways for a long time. Examples of typical services are Jukebox and DJ. Jukebox existed a lot from 1940s to 1960s, especially in the 1950s. Jukebox has records so that when users insert a coin, they can listen to the music. Users can listen music with higher volume and better sound quality through Jukebox. In Korea, DJs in music cafe stores called "Dabang" were popular during the 1970s and 1980s. The cafe DJ shares music and messages to the users in the store. The music is played with the public speaker of the store. People in the same store could enjoy it together through these services.

Advances in cloud computing become possible to provide personalized online streaming services based on cloud [1]. However, users have difficulty in selecting the contents that suits them from a lot of contents. A lot of research has been proposed on algorithms and systems that recommend user preferences as a solution to these problems [2–5]. In particular, a cloud-based recommendation system recommends contents based on an age, location, usage history, and preferences [6, 7]. However, such a system is only a recommendation and does not guarantee the satisfaction of the user. It is also hard

© Springer Nature Switzerland AG 2018
T. K. Dang et al. (Eds.): FDSE 2018, LNCS 11251, pp. 323–334, 2018.
https://doi.org/10.1007/978-3-030-03192-3_25

to offer a personal online streaming service that user can enjoy with other users in real-time.

This paper proposes and implements a Cloud Media DJ Platform that reinvent a DJ of a music cafe. The platform is based on an integrated environment of Internet of Things (IoT) and cloud computing. Based on a single public cloud, services provided by this platform allow users to request content services directly. User information and usage history are collected and analyzed in the public cloud. Even when the users visit other stores, they can receive customized content services based on their information and previous usage history.

We analyze and evaluate the services of the proposed platform, focusing on the functional aspects. The platform provides four main services. Types of services are music, video, message, and combined service. User can share contents with other users in real time through the platform and receive scalability and convenience services through public speakers and screens in the store.

The remainder of this paper is organized as follows. Related work is discussed in Sect. 2. The proposed platform is discussed in Sect. 3, and the implementation and evaluation is discussed in Sect. 4. Conclusions are presented in Sect. 5.

2   Related Work

2.1   Recent Internet of Things and Cloud Computing

Internet of Things (IoT) is a technology that allows communication between objects by connecting to the Internet with sensor and communication functions. The kind of object include a user device, home appliance, and the like things. Objects connected by the Internet exchange and analyze data. Cloud computing provides an environment where IT services such as storing data and providing contents are available anytime, anywhere. It also provides large scale capacity and processing by using servers on the Internet. IoT technology has limited processing capacity and storage problems. This leads to performance and security problems.

A solution to these problems is presented to integrate IoT with cloud computing. The technology that combines IoT and cloud computing is called CloudIoT [8]. This new paradigm provides many services and applications to users. In recent years, the two themes have been popular and researched. Integrated research and application programs are also actively being proposed. In general, the infinite capabilities and resources of the cloud can solve the problems of the IoT: storage, processing, and communication. CloudIoT paradigm can lead to big advances in life and activity of human [9]. Multimedia applications based on this paradigm provide efficient services and build new businesses [10].

2.2   Offline DJ and Online DJ

Offline DJ (Disk Jockey) is a person who plays the music recorded in real time to users located in the same store. Typical DJ types are digital radio DJ and club DJ. Radio DJ plays music on a digital radio station. Club DJ plays a variety of music in clubs and encourages people to dance. They act as a mediator for the music recorded to the users and recreate the music. Online DJ is a person who broadcasts online. It is usually called broadcaster, and it is called BJ (Broadcasting Jockey) in Korea. They get popularity from users and attempt to build new businesses.

2.3   Streaming Services

Currently, many companies around the world provide streaming services. A typical music streaming service is Apple music, and is a global music streaming service released by Apple in the U.S. [11]. A typical video streaming service is YouTube [12]. It is Google free video sharing site, a global platform that allows users to create and share video contents. YouTube helps and provides customized contents for over 1 billion users [13]. In recent years, cloud-based streaming services have been one of the core innovations in digital streaming service. Many music providers like Apple, Amazon, Google and Microsoft are currently offering cloud-based streaming services. The advantage of cloud-based streaming services is that users can store and easily access the contents [14]. Cloud-based streaming services are actively researched, and applications that provide contents to users are also developed. Users receive a variety of contents and services. However, users have difficulty in selecting their preferred services in a lot of contents. Content providers need a way to effectively inform users a lot of contents. The competitiveness of content providers is determined by providing contents that users prefer. Algorithm and applications for providing customized contents to users are being studied [15, 16].

3   Proposed Platform

3.1   Platform Overview

Cloud Media DJ Platform provides customized content services for mobile users in an integrated environment of IoT and cloud computing. The platform consists of user device, public cloud, streaming client, public screens and speakers. Users select for contents want to receive in stores where they are located, and provided it as public screens and speakers. The user receives one or more of the music, the video, and the messaging services. The user device is a communication device that use applications in the networks environment. Public cloud consists of operation server and streaming servers. The operation server collects user information and performs analysis. The detailed functions of the operation server are described in the following paragraphs. Streaming server is an external affiliated server with contents data. For example, it could be Spotify or YouTube which hold and provide music and video. Depending on the type of contents requested by the user, the streaming server provides the contents data. The streaming server that provides the music and the streaming server that provides the video may differ and include multiple affiliate streaming servers. The streaming client is located in the store, receives contents data, and transmits it to the public screens or speakers. The screens and the speakers connect with the streaming client to play the contents received by the streaming client. Figure 1 shows the overall structure of platform performed by a public cloud.

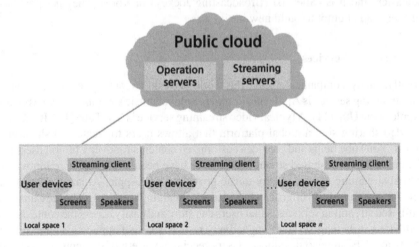

Fig. 1. Structure of Cloud Media DJ Platform

When a user requests contents desired through the application, the public cloud collects the contents data requested. The collected data is sent to a streaming server, and the contents data is transmitted to a streaming client at a store where the user is located. The streaming client transmits the contents to the public screens and speakers, so that the contents to be played. Streaming client can be implemented with minimum specifications so it can be used lightly in stores. Users in the same store receive content services through public screens and speakers.

The operation server of platform mainly performs user information collection and analysis functions. Figure 2 shows the structure of the operation server. The operation server consists of device communication, user data storage, information analysis, and

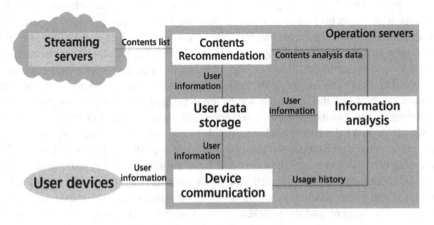

Fig. 2. Structure of operation server

contents recommendation. Device communication is an interface with a user device. The interface serves as the communication module that the user device can access. User data storage stores user information and usage history. Information analysis part analyzes and integrates user data in operation servers. Contents recommendation transmits the contents list to the streaming server so that the contents are played.

3.2   Service Design

The proposed platform provides four main services. The types of services are music, video, messaging, and combined service, and perform the following operations. Figure 3 shows the operation of the service, and the numbers are to explain the process.

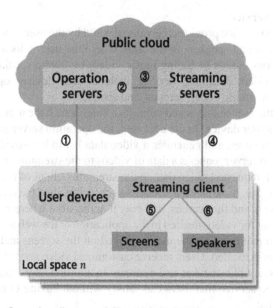

Fig. 3.   Operation diagram of Cloud Media DJ Platform (step ①–step ⑤)

Music Streaming Service

Music streaming service provides the music that users want to listen in the store where the user is located. Users request services by searching and selecting the music they want through the application. The components for music streaming service are the user device, operation server, streaming server, streaming client, public screens, and speakers. The operation server and the streaming server can provide services to the store. The streaming client, screen, and speaker exist in the store. Operation server and streaming server serve all stores. Streaming client, public screens, and speakers are in each store.

In step ①, the user searches for a music desired through the mobile device and requests it. The operation server of public cloud collects the music requested and generates a list to be provided to the store where the user is located in step ②. In step ③, the operation server requests contents data from the streaming server based on the generated list. In the next step, the streaming server sends the corresponding data to the streaming client in the store. In the last step, the streaming server sends the music data to the speaker and the screen so that it can be play in store.

The user can view music related information on user device or public screen. In addition, the user searches for a music desired and directly select for it. The selected music is provided to the public speakers of the store where the user is located, and the music information is provided to the screens. The music information includes a title, a singer, and an album photo. The user is provided with a music directly selected on the speakers. It allows users to provide customized music and share the music with other users in real time. Also, the store manager automatically provides customized music services without having to manage the music playing lists.

Video Streaming Service

The video streaming service provides the video desired by the user. The video is played on the public screens and speakers in the store where the user is located. The service provides with various videos including music video, by public speakers and screens. Components for video streaming service include user device, operation server, streaming server, streaming client, screens, and speakers.

In first step ①, the user searches and requests a video which the user wants to receive in a store through a user device. In the next step, the operation server collects the video data requested by the user, and generates a video data list to be provided to a store. In step ③, the operation server requests a data of videos to the streaming server. Streaming server transmits the requested video data to the streaming client of the store where the user is located in step ④. In step ⑤, the streaming server transfers the received video data to the public speakers and the screens, and play it at the store where user is located.

The user searches for videos desired in the application or the web at the store, request directly. The video requested by the user is played on the screens and speakers of the store where the user is located. Users receive customized video service by receiving the video requested as a public speaker and a screen. In addition, store manager automatically provides video service that users can satisfy without having to manually control the video playing list.

Messaging Service

The messaging service is a service that allows users to request a message and receive it on a public screen. Users request to a message that they want to leave or share with the people in the store. This message is displayed on the screen. Messaging service provides users to share information about simple messages or special events with people who is in the same store. Messaging services display simultaneously with music or video if user wants.

Unlike other services, the messaging service does not require a streaming server and is simple. It operates as follows. In the first step, the user requests a message to be provided as a public screen in the store where the user is located through the mobile device. In step ②, the operation server processes the message requested of the user and the location of the store. In step ③, the processed message data is transmitted to the streaming client of the store. In the last step, the streaming client sends the data received from the operation server to the public screen for playing.

When a user enters a store, the user requests for the message desired through the mobile application or web. This allows the user to use it for the purpose of the event or for sharing with people in the same store. In addition, the store manager can automatically customized messaging service to satisfy the user and view comments that people leave in the store.

Combined Service

Combined service provides two services at the same time. The three services are synchronized and users can receive services synchronized if they request for more than one service. For example, if a user requests a video and a message, they are automatically synchronized and the video to be played on the screen and the message to be viewed. The message is shown at the top of the screen.

When a user requests to a music and a message at the same time, the music is output to the speaker and the message is output to the screen. The length of the message and the time when the message is output can be adjusted by the user. When a user requests a video and a message at the same time, the sound of the video is output to the public speaker, and the video and the message are output to the public screen. At this time, the message is displayed at the top of the screen.

Combined services provide more than one service at the same time. If the user requests more than one service, they to be synchronized and provided services together. Combined services provide more than three main services, it can extend the service area and the effectiveness of the platform.

4   Function Discussion

We develop platform based on the Google cloud platform using an API that provides music and video. The API used is provided by Spotipy and YouTube. Cloud Media DJ Platform developed with a 9798 lines of code, of which the public screen developed with 1725 lines of code. As a result of the implementation, the user requests a desired music, video, and message. The data requested is provided to the public screen and speaker of the store through the operation server and the streaming server. Figure 4 shows the implementation of the four services displayed on the public screen. The upper left is the music streaming service, and the upper right is the video streaming service. The lower left is messaging service, and the lower right is combined service.

Fig. 4. Public screen according to services

4.1 Functional Evaluation

Based on the results of the implementation, the platform compares and evaluates functions according to three kinds. Table 1 compares the Cloud Media DJ Platform with the international famous streaming services. Most popular international streaming services are Apple Music and YouTube. Users can search for contents and see the ranking of popular contents through all three services. In addition, the user receives contents preferred through the service. Apple Music requires a monthly fee, but proposed platform and YouTube are free. Unlike the other two services, the proposed platform provides services based on the location and enables real-time sharing with people in the same store. Apple Music and YouTube provide music and video services, but the platform provides not only music and video, but also messages and combined services. In

Table 1. Comparison with popular streaming services

Function	Apple Music	YouTube	Cloud Media DJ Platform
Contents searching, view ranking	O	O	O
Providing user-preferred contents	O	O	O
Free of charge	×	O	O
Location-based service	×	×	O
Real-time sharing service	×	×	O
Service types	Music, Video	Music, Video	Music, Video, Message
Number of users: number of service recipients	1:1	1:1	1:m

Apple Music and YouTube, personal service users and recipients of services are matched. This platform allows a large number of users in the same store to receive services even if there is only one user.

The second is a comparison based on the type of streaming service. Table 2 compares Cloud Media DJ Platform and type of streaming service. A streaming service is classified into a personal streaming service, a café streaming service, and a proposed platform as a service. A personal streaming service is a method of using a streaming service through a device of user alone. Cafe streaming service is a service provided to users in café stores. In the case of personal streaming service and proposed platform, a user searches contents and view a list of popular contents. In addition, these two service cases provide the user with the contents preferred, and the user provides the service based on the usage history. Therefore, the user can be satisfied. However, users of personal streaming service can not share it with others in real time. The proposed platform provides not only music and video, but also combined services and messages. Finally if a user uses a personal streaming service, the service recipient is only one. On the other hand, the proposed platform and the cafe streaming service can have many service recipients regardless of the number of users.

Table 2. Comparison based on streaming service type

Function		Personal streaming service	Café streaming service	Cloud Media DJ Platform
Contents searching, view ranking		O	×	O
Providing user-preferred contents		O	×	O
User	History based service	O	×	O
	Satisfaction guarantee	O	×	O
Real-time sharing service		×	O	O
Service types		Music, Video	Music, Video	Music, Video, Message
Number of service recipients		1	m	m

The third functional evaluation compares proposed platform and DJ services. DJ types are classified as Online and Offline. Online DJ is a person who broadcasts and services on the Internet. Offline DJ is a person who provides services in the same local place. The proposed platform and Online DJ allows users to browse contents and view popular content lists. Online and Offline DJ provide service to users from one DJ. The content lists to be provided is determined by a DJ. Therefore, it is difficult to provide a service considering the satisfaction of the user and the usage history. On the other hand, the proposed platform allows users to create playlists considering satisfaction of user and usage history. Also, it is possible to provide a combined service of two contents which are not provided by other DJ services. The proposed platform can provide various services considering user satisfaction compared to other DJ services (Table 3).

Table 3. Comparison according to DJ type

Function		Online DJ	Offline DJ	Cloud Media DJ Platform
Contents searching, view ranking		O	×	O
Playlists	Popular ranking	Δ	Δ	O
	Contents preferred	Δ	Δ	O
User	History based service	×	O	O
	Satisfaction guarantee	×	×	O
	Location based service	×	O	O
Service types		Music, Video	Music, Video	Music, Video, Message

4.2 Performance Evaluation

Finally, we discuss the performance evaluation of the platform. Main issue of this platform is collecting and responding to a lot of data.

Figure 5 shows the response times for the number of users requesting to contents at the same time. The response time is the time until the streaming server transmits the contents in response to the request of user. The specifications of public cloud for implementation are 4 vCPUs, 16 GB. We measured the number of stores at 100, 200, and 300. At the same time, even if the number of users requesting contents is 1000, the public cloud of platform can process within 3.5 s.

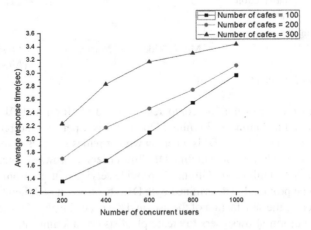

Fig. 5. Average response time by number of users

5   Conclusion and Future Work

We develop a cloud-based platform for automatic customized contents. We also analyzed the service platform based on the platform implementation results. The proposed platform provides automatic customized streaming service and convenient service through public screen and speaker. The user can conveniently receive the service and can share with the users in the same store. Cloud Media DJ Platform also showed that it can provide more convenient interface by comparing features with other streaming services. Finally, we show that Cloud Media DJ Platform process numerous concurrent users quickly based on the public cloud.

In the future, we will add functions to provide music and video according to detailed user information. Only the content services requested by the user is played on public screen and speaker. We will study recommendation algorithms to play when users do not request contents. We will research and develop this platform to enable customized music and video based on user information.

Acknowledgements. This research was supported in part by Korean government, under GITRC support program (IITP-2018-2015-0-00742) supervised by the IITP, Priority Research Centers Program (NRF-2010-0020210), and Human-plus research program (2018M3C1B8023550) respectively.

References

1. Aznoli, F., Navimipour, N.J.: Cloud services recommendation: reviewing the recent advances and suggesting the future research directions. J. Netw. Comput. Appl. **77**, 73–86 (2017)
2. Bobadilla, J., Ortega, F., Hernando, A., Gutiérrez, A.: Recommender systems survey. Knowl.-Based Syst. **46**, 109–132 (2013)
3. Lu, J., Wu, D., Mao, M., Wang, W., Zhang, G.: Recommender system application developments: a survey. Decis. Support Syst. **74**, 12–32 (2015)
4. Chen, H.C., Chen, A.L.: A music recommendation system based on music and user grouping. J. Intell. Inf. Syst. **24**(2–3), 113–132 (2005)
5. Chen, H.C., Chen, A.L.: A music recommendation system based on music data grouping and user interests. In: Proceedings of the Tenth International Conference on Information and Knowledge Management, pp. 231–238 (2001)
6. Alhamid, M.F., Rawashdeh, M., Dong, H., Hossain, M.A., Alelaiwi, A., El Saddik, A.: RecAm: a collaborative context-aware framework for multimedia recommendations in an ambient intelligence environment. Multimed. Syst. **22**(5), 587–601 (2016)
7. Lin, P.J., Chen, S.C., Yeh, C.H., Chang, W.C.: Implementation of a smartphone sensing system with social networks: a location-aware mobile application. Multimed. Tools Appl. **74**(19), 8313–8324 (2015)
8. Babu, S.M., Lakshmi, A.J., Rao, B.T.: A study on cloud based Internet of Things: CloudIoT. In: IEEE Global Conference Communication Technologies (GCCT), pp. 60–65 (2015)
9. Botta, A., De Donato, W., Persico, V., Pescapé, A.: Integration of cloud computing and internet of things: a survey. Future Gener. Comput. Syst. **56**, 684–700 (2016)
10. Zhu, W., Luo, C., Wang, J., Li, S.: Multimedia cloud computing. IEEE Sig. Process. Mag. **28**(3), 59–69 (2011)

334   J. Lee et al.

11. Tae Hyun, K., Jae Ik, L.: Music streaming service UI case study. Korea Sci. Technol. Forum **24**, 159–171 (2016)
12. Pires, K., Simon, G.: YouTube live and Twitch: a tour of user-generated live streaming systems. In: Proceedings of the 6th ACM Multimedia Systems Conference, pp. 225–230 (2015)
13. Chatzopoulou, G., Sheng, C., Faloutsos, M.: A first step towards understanding popularity in YouTube. In: INFOCOM IEEE Conference on Computer Communications Workshops, pp. 1–6 (2010)
14. Lee, J.H., Wishkoski, R., Aase, L., Meas, P., Hubbles, C.: Understanding users of cloud music services: selection factors, management and access behavior, and perceptions. J. Assoc. Inf. Sci. Technol. **68**(5), 1186–1200 (2017)
15. Lee, W.P., Tseng, Y.G.: Incorporating contextual information and collaborative filtering methods for multimedia recommendation in a mobile environment. Multimed. Tools Appl. **75**(24), 16719–16739 (2016)
16. Yang, J., et al.: Multimedia recommendation and transmission system based on cloud platform. Future Gener. Comput. Syst. **70**, 94–103 (2017)

Cloud Media DJ Platform: Performance Perspective

Jinwoong Jung, Joohyun Lee, Sanggil Yeoum, Junghyun Bum,
Thien Binh Dang, and Hyunseung Choo[✉]

Department of Electrical and Computer Engineering, Sungkyunkwan University, Suwon, Korea
{sjud325,joohyun7,sanggil12,bumjh,dtbinh,choo}@skku.edu

Abstract. With the development of mobile devices and the cloud, many kinds of offline services that people have been handling is replaced by online ones. In the past, Korean music cafe DJ performed the role of playing music requested and reciting the story of customers. We replace this offline service with an online one consisting of a public cloud, a streaming server, a streaming client, and a mobile device. The user requests music, video, and messages using the mobile, and receives services through a public screen and speakers in the space where the user is located. We also design and implement Cloud Media DJ platform. This paper focuses on the performance of this platform. We use Docker container distribution technology to accommodate large-scale traffic. In experimental evaluation results, the proposed platform reduces the response time from 25% (minimum) to 75% (maximum) according to the number of concurrent users compared to the existing one.

Keywords: Server · Client · Cloud · Docker · Container

1 Introduction

There were human DJs in Korean music cafes that were popular in 1970s. The music cafes sold drinks like the current cafe, and it was a place where people meet each other. The DJ of music cafes played the music requested by the cafe customers, sang songs, and recited the stories of customers. Nowadays, due to the development of personal music devices, many of the human DJs of music cafes have disappeared and only the music cafes has been left. However still many people want to share their favorite music with friends. How do we implement the offline service that the DJs in music cafes provided for today?

In Cloud Media DJ service, the most important factor is how to replace the music cafe DJ. This is because DJ had led the music cafe. We use a cloud server instead of the DJ. The song of DJ is replaced by the streaming API of the music and videos service provider. The behavior of users who requested the song by letter or speech directly to the DJ is replaced by the application such as the Android and web application. The voice of DJ is replaced by speakers in the cafe and the stories of customer is seen through the screen in the cafe.

Cloud Media DJ service require stable and reasonable server platform. The cafe owner should not be burdened with the payment for the service, and the customer should

© Springer Nature Switzerland AG 2018
T. K. Dang et al. (Eds.): FDSE 2018, LNCS 11251, pp. 335–348, 2018.
https://doi.org/10.1007/978-3-030-03192-3_26

not feel any security threat to use the service. The proposed platform is operated within Google Compute Engine [1], and managed separately for each cafe by Docker [2] container for stable operation of the proposed service. Cafe owner does not require physical server, and can provide new service to customers without difficult management and high cost. The servers managed with Docker container show higher stability and lower latency than a single server structure.

In this paper, we describe Cloud Media DJ service and the proposed platform where the service is installed. First, we will start with the introduction of the proposal service and explain how the role of human DJs of music cafes is implemented in terms of server and client. Next, we propose a low-cost, high-efficiency platform that works well for large-scale traffic and demonstrate that this platform has stability and low response time compared to a single server structure. Finally, we describe conclusions and future work.

2   Related Work

2.1   Mobile and Web Application

The greatest innovation in the 21st century IT era is the birth of mobile devices [3]. The ability to access the Internet anytime and anywhere with a smartphone rather than a desktop is one of the greatest achievements of recent times. We can communicate with external services at any time via a smartphone and keep our data unlimited in external space. In addition, the quality of wireless network communications is evolving very quickly, making the advance of mobile devices faster [4]. Application services evolved together as mobile devices advanced. Now, most of the online services can be run as smartphone applications. Application services are expected to take a large portion in the IT era in the future.

The applications that run mostly on mobile devices are native and web applications [5]. Native applications run directly on mobile operating systems without going through another execution environment. These features make native applications perform faster and can gain most of permissions on mobile devices. However, there is a disadvantage that native applications developed for the Android platform cannot be executed on other operating systems such as iOS. Web applications run through web view in web browsers. It has lower performance than native applications because it runs through the web view. However, the HTML5 [6] standard used on the web is compliant with all of the web browsers. If a developer create a web application, users can run it on any platform without any operating system limitations. An application developer should choose the right application style, keeping in mind the scenarios in which the applications they want to develop, how much their cost, and how much performance they need.

2.2   Cloud and Virtualization

Cloud computing [7] is one of the trends that are very interested today. This means processing information to another computer connected to the Internet rather than to our own physical computer. IT conglomerates are building their own data centers around the world, and many companies are using cloud computing services provided by IT conglomerates without having their servers physically [8]. It does not require high cost and physical management. It also provides high security. Cloud computing includes virtual machine technology that has been in used since the beginning of cloud computing, and container technology that has existed from the past, but has recently emerged.

Virtual machine [9] is software implementation of computer environment. Simply put, emulating a single computer. It features a virtualization of the operating system using hypervisor on existing physical hardware. It is separated from the host operating system, because the operating system is virtualized. Therefore, it is possible to use an operating system different from that of the host operating system, and the host and other VMs can be safely maintained even when one VM is attacked due to the structure separated from the host operating system. However, there are disadvantages as well. The amount of disk space required to virtualize an operating system is very large. And once it is a virtualized operating system, there is a performance penalty compared to the host operating system.

Container is to virtualize a process in contrast with virtual machine. Virtualization of the process does not degrade performance, nor does it require large disk space because the process uses only the resources it needs [10]. However, since it is not an operating system virtualization, it is impossible to use another operating system. It is also more vulnerable to security threats than a virtual machine because it is connected with the host operating system. Nonetheless, the reason for this recent popularity of containers comes from the birth of the Docker. The Docker helps create and manage containers and makes it easy to publish images created. Docker users convert their packages into Docker images and publish them to the public space. Other users can download the image at any time and use that package in an optimal state. For example, we had to set up a web server and a DB server separately in the past. But now we just need to download and run the image that has the perfect installation and configuration using the Docker. In addition, the performance of the Docker engine itself, which is a virtualized process, is very high and stable, which can follow to 100% performance of the host computer.

As mentioned above, there are advantages and disadvantages of virtual machine and Docker container [11]. What virtualization technologies are used by developers to manage their servers is based on the service or platform characteristics. Our proposed platform is to increase stability by placing different servers in each cafe. For example, suppose that all the cafes are managed on a single server. If a cafe receives a bad packet, the entire server will crash and the service will not be available in all cafes. But using cloud computing instance for every cafe requires astronomical costs. In addition, since it is assumed that the number of cafes using our service is large, provision of the virtual machine for each cafe is impossible because of the feature of the virtual machine that uses a lot of computing resources. In this case, if we use container to separate servers for each cafe and manage each container using Docker, it can be more stable and efficient server environment.

3   Platform Design

3.1   Cloud Media DJ Service

Cloud Media DJ service is an online service in which when a cafe customer requests music, videos, or messages, the requested contents are played in the cafe where the customer is located currently. The customer can request the contents using the Android or web application, and the contents are played on the screen and the speakers in the cafe. Cafe customers can listen to music they want instead of music played at random in the cafes, and owners of cafes can make the foundation to change normal customers to their regular customers. Our service does not have a process of cafe selection because it automatically recognizes locations through our applications. After running the application, users can request the contents they want in the cafe with only two or three touches. Also, since the request history of customers is always saved, the music that has been requested in the other cafes appears on the recommendation list or is automatically played in the currently located cafe. Furthermore, by using the service, music and stories can be shared with the people who came together, not by themselves. Since the cloud web server has a web application to be used in cafes, a cafe owner can have a content playback environment on a mobile or a laptop without having a separate device (Fig. 1).

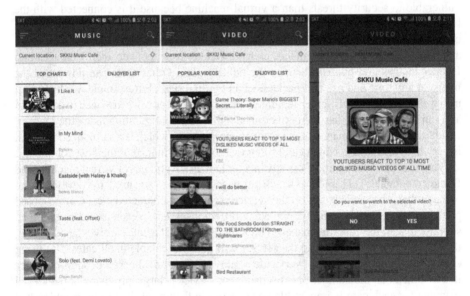

Fig. 1. The screens of our proposed service

The proposed service provides music streaming function and videos streaming function that allow the customer wants to play music and video requested in the cafe. When a cafe customer selects music or videos from popular charts or search that in the application, streaming will be done in real time in the cafe. This is the core of the proposed service, and replaces the biggest role of the DJ of music cafe, such as music and videos playback. Cafe customers can request their messages to show on the cafe screen. Simply put, our service has the function of a message board. When a cafe customer writes some sentences through the application, the sentences appears on the cafe screen. It replace the role of music cafe DJ who recite the story of customers. With this feature, customers can share their stories with other customers in cafes. Political and sensational words within the messages of customers are automatically excluded by words filtering.

The screens in the cafe display of the music list when the music is playing, and the video as full screen when the video is playing. Customers may want to request music and show their messages at the same time, or want to include their messages in the video. The proposed service is able to accept for both music and messages or videos and messages at the same time. If there are music list and messages at the same time in the screen, the messages take up most of the screen, instead of a music list. If there are video and messages at the same time in the screen, the message is displayed as a large caption on the top of the video. Customers can express their stories with appropriate contents through the service. The messages displayed appear for a few seconds depending on their length and then disappear.

3.2 Platform Organization and Service Scenario

Cloud Media DJ service runs on the proposed platform with Public Cloud, Streaming Server, Streaming Client, and User Devices. Public Cloud is the most important component that connects cafes (Streaming Client) to customers and Streaming Server. Streaming Server includes music and videos service providers. It provides streaming contents to Streaming Client. Streaming Client plays the contents received from Streaming Server. Cafe customers can request contents what they want through an Android or a web application on their User Devices. Contents information requested from an application on a smartphone is transmitted to Streaming Server via Public Cloud, and Streaming Server streams contents to Streaming Client. Therefore, there are User Devices and Streaming Client in the cafe space, and Public Cloud and Streaming Server exist outside the cafe space.

Users can freely choose between Android and web applications in the cafe. When requesting contents from Android application, it becomes TCP socket client role, and when requesting contents from web application, it becomes web client role. This is covered in detail in Sect. 3.3. Public Cloud has TCP socket server, web server, and DB server. TCP socket server is used to communicate with Android applications. Web server is used to provide web application of contents request to a user device and to provide web application of contents streaming to Streaming Client. Finally, DB Server is responsible for storing all the information of users, contents request history and the cafe information. Streaming Server is a contents providing company and we selected two companies. We selected Spotify [12] as a music provider and Youtube [13] as a video provider. Streaming Client starts from running a web application. When owners of cafes accesses web server in Public Cloud, the corresponding web application is executed and the contents is streamed through the web page. Any device can open this web page without limitations (Fig. 2).

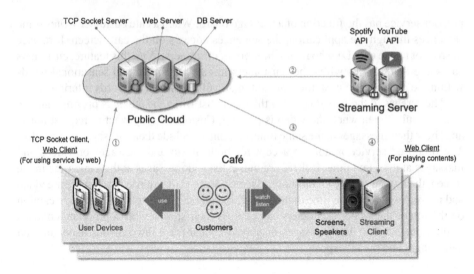

Fig. 2. Main scenario of Cloud Media DJ service

When customers launch an Android or web application via User Devices, it is immediately connected to Public Cloud through socket or web socket [14] communication in the background. ① When users select the contents in a list of popular contents that can be requested through an application or search the contents what they want, the information of the contents, current cafe information, and customer information are transmitted to Public Cloud. The data is transmitted as a UTF-8 [15] string through socket or web socket communication, and is JSON [16] type. The protocol is the most important data that determines which method in Public Cloud to execute. The remaining data contains the contents information, current cafe information, and the user information in order. The data is received in JSON type and can be distinguished into each role in Public Cloud (Fig. 3).

Protocol	Contents Type	Contents ID	Title	Artist	Album Art	Phone Number	Cafe

Fig. 3. The structure of the contents request data

Public Cloud is always running as a role to manage all the components of proposed platform. ② Public Cloud receives the data in JSON type and confirms what data was transmitted through the protocol data. The protocol exists in a variety of categories, such as request for new contents, deleting the contents requested, searching for contents by title or artist names, changing the user information, and collecting previous contents requested list. It is now assumed that this is a new contents request protocol. Each data except protocol is divided into contents information, current cafe information, and the user information and stored in temporary variables. Public Cloud stores these in the contents request history table in DB. This history can be viewed through the previous contents protocol and used for customized recommendations.

When DB storage is completed, the REST API provided by Streaming Server is used. Music that matches the contents information is searched using 'Get a Track Endpoint' of Spotify Web API v1, and the videos are searched using 'Search: list Endpoint' of YouTube Data API v3. Since Public Cloud is implemented as Node.js [17] platform, both Spotify and YouTube APIs use JavaScript libraries. Using the API, Public Cloud can get the search results back in JSON type. Public Cloud collects the contents streaming address from the returned JSON data.

③ Public Cloud transmits the streaming address and the user information to Streaming Client corresponding to current cafe information received from the customer. Public Cloud transmits data in JSON type, and includes the protocol data received when collecting data from User Devices. Streaming Client distinguishes functions by the protocol and executes appropriate functions. Since Public Cloud received new contents request protocol that we assumed earlier, Streaming Client play the contents using the streaming address on the web page (Fig. 4).

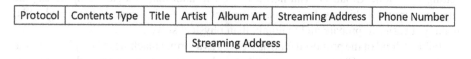

| Protocol | Contents Type | Title | Artist | Album Art | Streaming Address | Phone Number |

| Streaming Address |

Fig. 4. The structure of the contents playback data

④ The contents playback using streaming address on the web page in Streaming Client is divided into two methods. The first is music playback using the HTML5 'video' tag. The HTML5 'video' tag has a 'src' attribute. If 'src' attribute is given streaming address received from Public Cloud, music is streamed. It is important to make real-time response when the contents request come in. Since the 'video' tag of HTML5 has a source playback completion event, when the current music is ended, Streaming Client sequentially plays the next contents collected by Streaming Client.

The second is YouTube video playback using the YouTube IFrame Player API. This is an useful player API to load and play YouTube videos using HTML5 'iframe' tags. Unlike the video tag, there is no 'src' attribute and there is a 'videoId' attribute. Users of API can input the unique ID of the YouTube video what users want to see. Using the API, developers can implement the ability to play and stop YouTube videos in real time. It also allows developers to set the size of the video to be output, and collects the 'onStateChange' event when the playback is complete. When the current video playback ends, the next video is played sequentially, because of this event.

Streaming Client must be able to output messages contents in addition to music and videos contents. The output of messages contents is relatively simple because it is not streaming like music and videos. Streaming Client print the sentences using the JQuery [18] library and CSS. The JQuery library coordinates output times of messages contents. And CSS makes to display messages as full screen without impacting existing other web components through 'position: absolute' and 'z-index' attributes.

When music or videos contents is being played, the messages contents request may arrive. Streaming Client detects the music or a video being played and outputs the messages to the appropriate location. If the music is playing, messages will show in full screen, but if a video is playing, messages will show on top of the video. Streaming Client use flag variables of JavaScript to determine the playback status of the contents, and output the messages to the appropriate location via CSS.

3.3 Client-Server Communication

Public Cloud runs as a single instance within the Google Compute Engine. Google Compute Engine is a core service of Google Cloud Platform. When the user pays the money, it provides the virtual machine of the desired specification. The virtual machine specification the user set once can be changed at any time and this is more secure than the physical server of normal users. Because it is provided by Google which is the IT conglomerate. Regional settings of Google Compute Engine allow users to set the geographical location where their proposed service will work. Using appropriate regional settings, users can use the virtual machine faster. Note that, the current regional setting for Public Cloud is 'asia-northeast1-c'. In addition, Google Compute Engine supports firewalls, history functions, self-monitoring tools, security scan, and SSH, making it easier to prepare and maintain than physical servers.

Public Cloud of the proposed platform exists as a virtual machine in Google Compute Engine, and Public Cloud consists of TCP socket server, web server and DB Server. TCP socket server and web server run on the Node.js platform. Node.js is a server process execution environment platform suitable for interactive programs. DB Server uses MySQL database management system.

TCP socket server running in Node.js is implemented as 'net' module which is the basic module of Node.js. The 'net' module provides TCP socket server and TCP socket client functions. In the proposed platform, only the TCP socket server function of 'net' module is used to communicate with the Socket API of the Android application. If developers create a TCP server by setting the port numbers in the 'net' module, users can connect to the TCP socket server through the IP address and port numbers of Public Cloud by using the Socket API of the Android application as a client (Fig. 5).

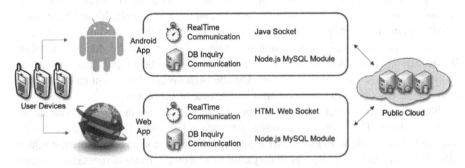

Fig. 5. Communication library between User Devices and Public Cloud

Web server is implemented using 'express' module. The 'express' module contains various convenient modules in a package so that developers can easily create web server. If developers create a web server by setting the port numbers with the 'express' module, users can connect to the web application using the web browser. Cloud Media DJ service requires a real time communication method in order to receive the contents request of users. In this case, the communication used is a web socket. The web socket can be implemented using the 'Web Socket' module in Node.js. Like the 'net' module, it provides web socket server and client functions. Only web socket server is used in our proposed platform, because web applications used by User Devices and Streaming Client assume the role of web socket client. Since web socket server can be created based on web server by default, it follows port numbers of web server. Web applications can be web socket clients using the 'ws' protocol of web socket library. In Node.js, clients can connect through the IP address and port numbers of web server created using 'express' module.

DB Server can query, insert, modify and delete the data by using the 'mysql' module of Node.js when MySQL database management system is installed. Android applications cannot be directly connected to MySQL for some security reasons. Therefore, most Android applications connect to the DB through the back-end web, but Node.js has a 'mysql' module that acts as a back-end. User Devices receive the request of DB related through socket communication, executes the SQL statement with the 'mysql' module in Public Cloud, and returns results to User Devices (Fig. 6).

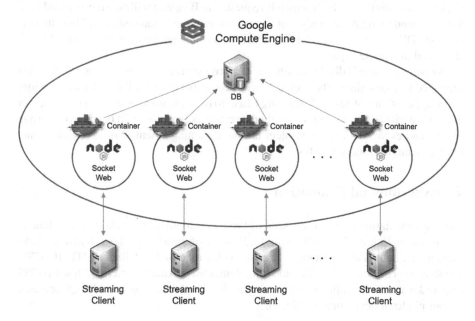

Fig. 6. The structure of Public Cloud

We propose the container structure as a way to minimize the virtual computing costs used by Google Compute Engine, rationalize performance, and increase server stability. If the server is divided into as many containers as the number of cafes, even if the error occurs in a cafe, it does not affect the other cafes. It also combines the event driven features of Node.js, which only work when an event occurs and the features of a container that uses just as much computing resources as needed with process virtualization, so that just one instance of Google Compute Engine can operate the proposed platform stably. For example, we assume that five users in A cafe request the contents. The contents request data is transmitted to a container divided into A cafe. There are 199 containers in Public Cloud in addition to A cafe container, but the contents are only processed in A cafe. And the other 199 containers do not work and do not use computing resources. As a result, A cafe container has the same performance as a server state without dividing into the container. This is demonstrated in detail in Sect. 4.

The Node.js server process, which is implemented all the functions of the proposed service, is packaged as a Docker image. Using that image, hundreds or even thousands of containers can be generated in a single line of code, depending on virtual machine specifications. Thus, in Public Cloud, Node.js server processes including TCP socket server, web server, and DB communication back-end are divided into containers, and the number of this containers corresponding to the number of cafes exists. The reason the DB Server itself is not included in the container is related to the contents request of customers. The customer should be able to see all the contents that they have requested so far from their current location no matter which cafe they request their contents. For example, when a user request the contents history of the user from A cafe, if the history does not contains the contents which is requested in B cafe, it will be a recommendation that is dependent on A cafe only, not a user customized recommendation. Thus, there is only one DB server in Public Cloud, which means that all the contents request history are stored in a unified space.

Streaming Client is displayed when a web browser is connected to a cafe web server separated by a container. The each cafe web server that exists in Public Cloud provides a web application that acts as Streaming Client to the web browser. Since the web server exists for each cafe, there is no worry that the contents requested will be played in wrong places, and even if there is a problem in one Streaming Client, it does not affect any other Streaming Clients of cafes.

4 Experimental Evaluation

Our implementation platform consists of three development projects. Public Cloud is developed in Eclipse [19] IDE with 1,038 lines of JavaScript, Java, and JSP code. Streaming Client is also written in Eclipse, with a total of 1,725 lines of HTML, CSS, and JS code. The Android application is implemented in Android Studio [20] with 6,785 lines of Java code. In addition, many commands related to Docker and MySQL are used for our platform development (Fig. 7).

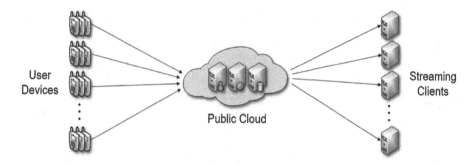

Fig. 7. The structure of single server

We implemented a single server structure, which is the most common server structure, to measure the performance of our proposed platform. A single server structure is one in which a TCP socket server, a web server, and a DB server are existed in a single instance of Google Compute Engine. A TCP socket server and a web server communicate with all the cafes, and a DB Server stores all the information. The instance specification of Google Compute Engine set to 4 vCPUs and 16 GB memory (Fig. 8).

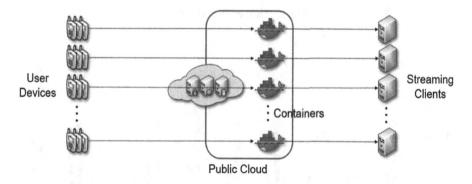

Fig. 8. The structure of using container

The proposed platform uses container technology on one instance of Google Compute Engine to provides communication servers as many as the number of cafes. As mentioned in Sect. 3.3, the TCP socket server and web server are included in containers that exist as many as the number of cafes. One TCP socket server and one web server communicate with only one Streaming Client. One DB server stores all the information, just like a single server structure. The instance specification is same with the single server structure as 4 vCPUs and 16 GB memory (Figs. 9 and 10).

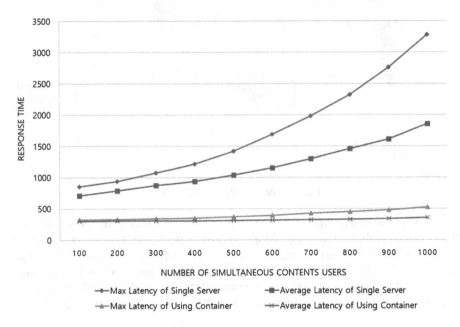

Fig. 9. Latency in 200 cafes

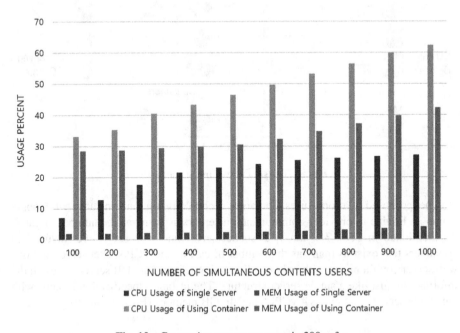

Fig. 10. Computing resources usage in 200 cafes

We assumes that there are 200 cafes which use the proposed service, and a large number of customers each request one content. The above graph shows the response time which when completing the request process for the content in Public Cloud, and shows the computing resources usage rate of Public Cloud. The response time of the content request is used to determine if the customer who use the proposed service will feel uncomfortable when large traffic occurs. The usage rate of computing resources is used to determine if Public Cloud can use the instance of Google Compute Engine as efficiently as possible. Since Spotify API and Youtube API cannot be used for performance measurement, we decided to implement Spotify sample data in external space and use this sample data instead.

As a result of measuring the performance of the two structures, there are advantages and disadvantages of each structure. The latency graph shows that the using container structure has an overwhelmingly shorter response time than the single server structure. This can reduce latency by as less as 25% to as much as 75% in large traffic. The graph of computing resource usage shows that the using container structure efficiently uses CPU and memory by as less as two times to as much as fourteen times compared to a single server structure. Contrary, this indicates that a single server structure does not require high specification. The single server structure sometimes causes data loss in large-scale traffic because a large amount of contents request data is received by one server and data collision occurs. This is the reason why the response time is high. The proposed platform is a platform suitable for the using container structure because stability, high efficiency and short response time are the top priority in the service for the customers of cafes.

5   Conclusion and Future Work

Through the proposed platform, customers can use the contents request service stably, and the cafe owner can provide the service with low cost and high efficiency. In addition, cafe owners will be able to easily attract regular customers without difficult management through the automated service. Using the Docker container increases the stability of the server, so it can provide commercial service without problems even for large-scale traffic. This structure can reduce latency from 25% (minimum) to 75% (maximum) in various situations. The current recommendation function only recommends previously requested contents. Therefore, we will develop a practical customized recommendation function based on the age ranges of customers and the contents genre. Furthermore, we plan to develop a function to grasp the precise position of the customer by using the inaudible frequency instead of the GPS information which is inaccurate in indoor environments.

Acknowledgements. This research was supported in part by Korean government, under GITRC support program (IITP-2018-2015-0-00742) supervised by the IITP, Priority Research Centers Program (NRF-2010-0020210), and Human-plus research program (2018M3C1B8023550) respectively.

References

1. Krishnan, S.P.T., Gonzalez, J.L.U.: Google compute engine. In: Building Your Next Big Thing with Google Cloud Platform, pp. 53–81, Apress, Berkeley (2015)
2. Merkel, D.: Docker: lightweight Linux containers for consistent development and deployment. Linux J. **2014**(239), 2 (2014)
3. Bi, Q., Zysman, G.L., Menkes, H.: Wireless mobile communications at the start of the 21st century. IEEE Commun. Mag. **39**(1), 110–116 (2001)
4. Kumar, S.: Mobile communications: global trends in the 21st century. Int. J. Mob. Commun. **2**(1), 67–86 (2004)
5. Charland, A., Leroux, B.: Mobile application development: web vs. native. Queue **9**(4), 20 (2011)
6. Anthes, G.: HTML5 leads a web revolution. Commun. ACM **55**(7), 16–17 (2012)
7. Armbrust, M., et al.: A view of cloud computing. Commun. ACM **53**(4), 50–58 (2010)
8. Gupta, P., Seetharaman, A., Raj, J.R.: The usage and adoption of cloud computing by small and medium businesses. Int. J. Inf. Manag. **33**(5), 861–874 (2013)
9. Li, Y., Li, W., Jiang, C.: A survey of virtual machine system: current technology and future trends. In: 2010 Third International Symposium on Electronic Commerce and Security (ISECS), pp. 332–336. IEEE (2010)
10. Xavier, M.G., Neves, M.V., Rossi, F.D., Ferreto, T.C., Lange, T., De Rose, C.A.: Performance evaluation of container-based virtualization for high performance computing environments. In: 2013 21st Euromicro International Conference on Parallel, Distributed and Network-Based Processing (PDP), pp. 233–240. IEEE (2013)
11. Felter, W., Ferreira, A., Rajamony, R., Rubio, J.: An updated performance comparison of virtual machines and Linux containers. In: 2015 IEEE International Symposium on Performance Analysis of Systems and Software (ISPASS), pp. 171–172. IEEE (2015)
12. Kreitz, G., Niemela, F.: Spotify–large scale, low latency, P2P music-on-demand streaming. In: 2010 IEEE Tenth International Conference on Peer-to-Peer Computing (P2P), pp. 1–10. IEEE (2010)
13. Burgess, J., Green, J.: YouTube: Online Video and Participatory Culture. Wiley, Hoboken (2013)
14. Fette, I., Melnikov, A.: The websocket protocol, (No. RFC 6455) (2011)
15. Yergeau, F.: UTF-8, a transformation format of ISO 10646, (No. RFC 3629) (2003)
16. Crockford, D.: The application/json media type for JavaScript object notation (JSON), (No. RFC 4627) (2006)
17. Tilkov, S., Vinoski, S.: Node.js: using JavaScript to build high-performance network programs. IEEE Internet Comput. **14**(6), 80–83 (2010)
18. Volder, K.: JQuery: a generic code browser with a declarative configuration language. In: Van Hentenryck, P. (ed.) PADL 2006. LNCS, vol. 3819, pp. 88–102. Springer, Heidelberg (2005). https://doi.org/10.1007/11603023_7
19. Shavor, S., D'Anjou, J., Fairbrother, S., Kehn, D., Kellerman, J., McCarthy, P.: The Java Developer's Guide to Eclipse. Addison-Wesley Longman Publishing Co., Inc., Boston (2003)
20. Zapata, B.C.: Android Studio Application Development. Packt Publishing Ltd., Birmingham (2013)

Analyzing and Visualizing Web Server Access Log File

Minh-Tri Nguyen[1]([✉]), Thanh-Dang Diep[1], Tran Hoang Vinh[1],
Takuma Nakajima[2], and Nam Thoai[1]

[1] Faculty of Computer Science and Engineering,
Ho Chi Minh City University of Technology, VNUHCM, Ho Chi Minh City, Vietnam
{nmtribk,dang,1414700,namthoai}@hcmut.edu.vn
[2] Graduate School of Information Systems,
The University of Electro-Communications, Chofu-shi, Tokyo, Japan
tnakajima@comp.is.uec.ac.jp

Abstract. Websites have endlessly multiplied during the recent decades and the number of visitors to the websites keeps the pace with them simultaneously, which leads to the process of huge data creation. The data are believed to consist of hidden knowledge well worth considering in various activities related to e-Business, e-CRM, e-Services, e-Newspapers, e-Government, Digital Libraries, and so on. In order to extract knowledge from the web data efficiently, a process called web usage mining is applied to such data. In this literature, we use the process to uncover interesting patterns in web server access log file gathered from Ho Chi Minh City University of Technology (HCMUT) in Vietnam. Moreover, we propose a novel model to construct and add new attributes encompassing country, province (or city), Internet Service Provider (ISP) from the existing attribute IP. The model belongs to attribute construction (or feature construction) which is one of strategies of data transformation being a data pre-processing technique. By utilizing the aforementioned mining process, we have wide knowledge about user access patterns for every country, province and ISP. Such knowledge can be leveraged for optimizing system performance as well as enhancing personalization. Furthermore, the valuable knowledge can be useful for deciding reasonable caching policies for web proxies.

Keywords: Web usage mining · Data transformation · Server log file
Access log · Extended common log file format

1 Introduction

It is widely acknowledged that Internet plays a vital role in our daily lives. With the advent of the Internet, we can straightforwardly reap much information by means of surfing websites. The number of websites and their visitors has proliferated every day. In particular, the number of Internet users by December 31, 2017, was 4,156,932,140 [4] which is around 55% of the world's population. The

© Springer Nature Switzerland AG 2018
T. K. Dang et al. (Eds.): FDSE 2018, LNCS 11251, pp. 349–367, 2018.
https://doi.org/10.1007/978-3-030-03192-3_27

number of active websites is nearly 300 million [2]. The huge data generated from websites are supposed to contain hidden knowledge concerning the business or patterns characterizing customer profile and behavior. With the rapid growth of the World Wide Web, the study of knowledge discovery in web, modeling and predicting user access patterns on a website has become very important [8]. Since the web technology largely feeds on ideas and knowledge rather than being dependent on fixed assets, it gave birth to new companies, such as Yahoo, Google, Netscape, e-Bay, e-Trade, Expedia, Amazon, and so on [7]. With the large number of companies using the Internet to distribute and collect information, how to extract knowledge from web data has gained a lot of attention from researchers as well as practitioners and it has formed a specialized research area called web data mining. To extract interesting patterns from web data, data mining techniques can be applied. However, due to web data's unstructured and semi-structured features, the data mining techniques cannot be applied to web data directly, which is specifically addressed by web data mining techniques [18].

Based on the primary kinds of data which are mined in the mining process, web data mining can be categorized into three types: web structure mining, web content mining and web usage mining [13].

- Web structure mining: Web structure mining focuses on mining knowledge from relationships between hyperlinks. For example, we can discover websites which are interested by almost users or common interests shared by communities of users.
- Web content mining: The purpose of web content mining is to extract knowledge from web page contents. For example, we can analyze the users' opinions about products based on their reviews and postings.
- Web usage mining: Web usage mining refers to the discovery of user access patterns based on log files. For example, the website performance can be optimized based on frequent user access patterns.

In web data mining, web usage mining is high flying due to effective use in numerous web related applications [10]. A previous study [7] summarizes a wide range of applications. However, this paper focuses mainly on the following applications:

1. Personalization for a user can be achieved by keeping track of previously accessed pages. These pages can be used to identify the typical browsing behavior of a user and subsequently to predict desired pages.
2. By determining frequent access behavior for users, needed links can be identified to improve the overall performance of future accesses.
3. In addition to modifications to the linkage structure, identifying common access behaviors can be used to improve the actual design of web pages and to make other modifications to the site.
4. Web usage mining of patterns provides a key to understanding web traffic behavior, which can be used to deal with policies on web caching, network transmission, load balancing, or data distribution.

5. Web usage mining is also useful for detecting attempted break-ins to the system.

In this paper, we apply the web usage mining process to log file gathered from HCMUT's website [3]. The process encompasses three key stages: data pre-processing, pattern discovery, and pattern analysis. In the data pre-processing, we use data cleaning to eliminate unnecessary log entries containing references to style files, graphics, or sound files, and data reduction to remove attributes which do not provide useful information for our analysis. In addition, pageview identification is applied to the log file for determining real single user actions (such as a click-through) while data transformation is utilized for application-oriented analysis. In terms of the data transformation, we propose a novel model related to attribute construction to add new attributes including country, province, ISP from the existing attribute IP. The overall progress on the stage is easily implemented by taking advantage of Apache Spark [1]. We use statistics to mine patterns from pre-processed data in the pattern discovery and extract interesting patterns (knowledge) in the pattern analysis after that. The pattern discovery and analysis are both realized by making use of Tableau [5]. The attained knowledge helps not only in improving the website performance and users' experience but also in being potential to improve more reasonable caching policies [15,16].

The remainder of the paper is organized as follows. The next section describes the structure of the web log file. Section 3 shows how to mine our web log file as well as the details of implementation, and how to use the domain knowledge with the aim of adding new attributes which are essential for our analysis. Section 4 depicts the findings, together with extracted knowledge about user preferences in different geographical regions while Sect. 5 surveys the state of the art and related work. Section 6 discusses other techniques which are able to apply to the kind of web log file for various purposes. Finally, Sect. 7 gives some conclusions and the further research direction of the study.

2   Web Log File Structure

While users surf websites, their interactions are recorded in web log file. There are three main sources of the raw web log file encompassing client log file, proxy log file and server log file. The client log file stores the most genuine behavior of users [14]. However, the log file is difficult to get due to the requirement of the collaboration from users. The proxy log file stores pages or objects cached from many websites and browsed by numerous users. Therefore, unleashing the true picture of user behavior is very difficult [10]. In the scope of the literature, we only take account of the server log file which is considered as the most reliable and accurate for web usage mining process by most researchers [10].

There are 4 common types of the server log file including access log, agent log, error log and referrer log [23,24]. The referrer log contains information concerning referrer. When you jump from any website to another, the server will capture the former website and store it in the column referrer. The log file is always exploited by Google [14]. The error log consists of error messages or codes, such as "Error

404 File Not Found". The error log file is more helpful for the website designer to optimize the website links [10]. Agent log file records the information about the website users' browser, browser's version and operating system [24]. The information is very useful for website designer and administrator for developing the website compatible with various operating systems and web browsers. The access log file is the major log of web server which records all the clicks, hits and accesses made by any website user [10]. There is a little useful information which can be extracted from the log, such as user behavior and interest, network traffic, the number of pageviews, and so on. In the scope of this paper, we only consider the access log.

There are three main types of web server log file formats available to capture the activities of users on websites [25]. They are Common Log File Format (NCSA), Extended Log Format (W3C) and IIS Log Format (Microsoft). Only the Common Log File Format is taken into consideration in the article. Table 1 gives an example of attributes together with their description in the Common Log File Format.

3   Web Usage Mining

Web mining is considered as a fairly large research area, however, according to Kosala and Blockeel [12], it can be categorized into three sub-areas which depends on which part of the web to mine: web content mining, web structure mining, and web usage mining. In this article, the data that we collected are the HCMUT access log, then we mainly focus on using several web usage mining techniques in order to extract useful knowledge.

Techniques in web usage mining are usually used to discover and analyze click streams, user utilization and other associated data collected or generated when users interact with web server system. The results of these processes may contain information about the user or group of users, which may be more clearly expressed through some statistics or visualization methods.

According to Liu [13], the standard data mining process [20] can also be applied for web usage mining, which contains three inter-dependent stages (Fig. 1): data pre-processing, pattern discovery, and pattern analysis. This section describes in detail how we apply these processes to our current dataset.

3.1   Data Pre-processing

Data pre-processing is an important task in any data mining application. It takes a responsibility for creating suitable target data sets which statistics and data mining algorithms can be applied on. According to Pabarskaite [17], data pre-processing may take 80% of the mining process. It is especially an important part in web usage mining because one of the characteristics of the web log data is that it contains a lot of meaningless data and noise. In some cases, web logs are also collected from multiple sources, across multiple channels or events formatted differently. Actually, data pre-processing is time-consuming, but it is

Table 1. Example of Common Log File Format

127.0.0.1	This is the IP address of the client (remote host) which made the request to the server
-	The "hyphen" in the output indicates that the requested piece of information is not available. In this case, the information that is not available is the RFC 1413 identity of the client determined by identd on the clients machine
frank	This is the userID of the person requesting the document as determined by HTTP authentication
[10/Oct/2000:13:55:36 -0700]	The time that the server finished processing the request
"GET /apache_pb.gif HTTP/1.0"	The request line from the client is given in double quotes. The request line contains a great deal of useful information. First, the method used by the client is GET. Second, the client requested the resource /apache_pb.gif, and third, the client used the protocol HTTP/1.0
200	This is the status code that the server sends back to the client
2326	The entry indicates the size of the object returned to the client, not including the response headers
"http://www.example.com/start.html"	The "Referer" (sic) HTTP request header. This gives the site that the client reports having been referred from
"Mozilla/4.08 [en] (Win98; I; Nav)"	The User-Agent HTTP request header. This is the identifying information that the client browser reports about itself

usually the main task which contains most computationally intensive step in web usage mining, and often requires applying some special algorithms or heuristics which are commonly not used in other research fields. In addition, this data preparation may involve other existing knowledge and data through integrating and transforming to get a suitable dataset which depends on mining purposes.

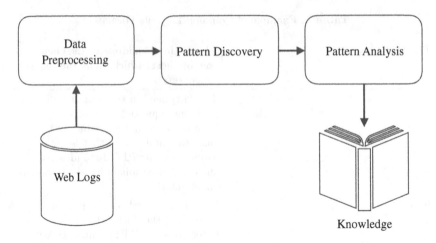

Fig. 1. Web usage mining process

Many studies and practices in web usage mining have focused on preprocessing data. However, depending on the web content, web architecture, and mining purpose, challenges presented in pre-processing process are manifold, which leads to vast proposed algorithms and heuristic techniques, such as data fusion and cleaning, data reduction, user identification, pageview identification, data transformation, sessionization, path completion, data integration, and so on. Besides, the successful process mining web usage is highly dependent on human factor [12]. Analysts' observation would be an important determinant of the quality of mined knowledge. Based on the defined purposes, we mainly use 4 major techniques depicted in Fig. 2 in order to pre-process our log file, including data cleaning, data reduction, pageview identification, and data transformation.

In this study, we focus on analyzing the log files collected from our web server which is classified as web usage data, and stored in the Extended Common Log Format. This is a semi-structured log that is usually stored as text files. Therefore, we transform these data into a structured table, which can be handled more conveniently.

Data Cleaning is usually the first step in the data pre-processing in general. This step significantly reduces the size of the log file so that it actually delivers an effective performance because of reducing a huge amount of computation for following steps. Experimental results on our log data have shown that the number of meaningless records removed is about 46%. Table 2 shows an example of removing meaningless records in our log.

In this example, we only present two attributes of the log file which are *Request* and *Referrer*. Actually, the contents requested can be style or image files (PNG, JPG, GIF, ...) or files defining the structure of the web page which may not provide useful information for analysis or mining task. Therefore, we remove all records in the same format as lines 4 and 5. In addition, the entries that

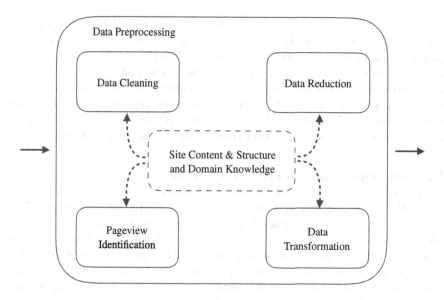

Fig. 2. Data pre-processing

Table 2. Data cleaning

Line	Request	Referrer
1	www.hcmut.edu.vn/	http://www.google.com
2	www.hcmut.edu.vn/vi/	www.hcmut.edu.vn/
3	/includes/css/hcmut/welcome.css	www.hcmut.edu.vn/vi
4	/includes/css/hcmut/images /cell_a_bg.png	www.hcmut.edu.vn/includes /css/hcmut/welcome.css
5	/includes/css/hcmut/images /tintuc/top_menu_bg.png	www.hcmut.edu.vn/includes /css/hcmut/welcome.css
6	/includes/css/hcmut/nivo_slider.css	www.hcmut.edu.vn/vi/
7	/vi/newsletter/view/su-kien	www.hcmut.edu.vn/vi/newsletter/
...		

have status "error" or "failure" are also removed. Moreover, records generated by requests for some resources, such as audio, video, are retained at this stage for later mining purposes.

Data Reduction is similar to data cleaning. This step can substantially reduce the size of a dataset in storage. However, it does not reduce the number of records but only reduces the number of attributes of each log entry by keeping the interesting fields. For the purpose of evaluating a caching algorithm, we only remain several fields which may contain useful information required, including: *Client IP, Date Time, Referrer*. In our case, most of the contents on the website are public, which leads to most logged requests do not contain any authentication information. Client IP addresses, alone, are not effective in identifying a specific user over a period of time. It is not uncommon to find many log entries corresponding to a limited number of proxy server IP addresses from large Internet Service Providers. Therefore, two occurrences of the same IP address in fact, might correspond to two different users [13]. However, we are actually interested in the geographic location where the request was sent, then only using the Clients IP field is enough. Besides, retaining Referrer and Date Time fields would also help determine users' preference over time which may be used for further mining.

Pageview Identification is a process that heavily depends on the site contents, the internal structure of the site, as well as the underlying site domain knowledge. As we all know, theoretically, pageview can be considered as a collection of web objects and resources representing a specific user event such as clicking on a link, opening an option panel, sending a message. Therefore, each pageview may be represented by a combination of static templates and content generated by servers' application based on a set of parameters or by one/many HTML files. It leads to the fact that many requests are generated corresponding to only one user event but those requests may refer to similar resource paths. In fact, a Uniform Resource Identifier (URI) includes some components: scheme, authority, path, query, fragment (Fig. 3). By removing two fields: query and fragment, and retaining the path that users refer to, we obtain the dataset in the same format as the data described in Table 3.

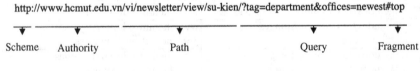

Fig. 3. URI structure

In order to eliminate duplicate entries, we filter these entries by IP and then combine some heuristic techniques of sessionization. The goal of this stage is to reconstruct a pageview. Thereafter, we are able to count the number of times that user hits each page. More specifically, a request r is considered to constructed pageview S if all three following conditions are satisfied: (1) the *Referrer* of r was previously invoked in S, (2) r has the same *Client IP* with the initial element of S called R, (3) $t_1 - t_2 < \theta$, where t_1 is timestamp of r, t_2 is timestamp of R and θ is a threshold determined by experiment. Otherwise, r is used as an initial element of a new constructed pageview. As described in Table 3, red entries will be discarded.

Data Transformation is actually a process that converts data from one format or structure into another one. It is usually used as a fundamental component of

Table 3. Apply pageview identification on log file HCMUT

Line	Client IP	Date Time	Referrer
1	112.197.177.55	[06/Nov/2016:19:50:31 +0700]	http://www.hcmut.edu.vn/vi
2	112.197.177.55	[06/Nov/2016:19:50:31 +0700]	http://www.hcmut.edu.vn/vi
3	112.197.177.55	[06/Nov/2016:19:50:32 +0700]	http://www.hcmut.edu.vn/vi
4	112.197.177.55	[06/Nov/2016:19:50:32 +0700]	http://www.hcmut.edu.vn/vi /newsletter/view/su-kien
5	172.28.2.3	[06/Nov/2016:19:50:32 +0700]	http://www.hcmut.edu.vn/vi
6	172.28.2.3	[06/Nov/2016:19:50:32 +0700]	http://www.hcmut.edu.vn/vi
7	112.197.177.55	[06/Nov/2016:19:50:32 +0700]	http://www.hcmut.edu.vn/vi /newsletter/view/su-kien
...			
8	112.197.177.55	[06/Nov/2016:19:52:04 +0700]	http://www.hcmut.edu.vn/vi
...			

most data integration and data management techniques. The goal of data transformation is to achieve suitable data for subsequent data mining algorithms. Thus, data transformation can be simple or complex, which heavily depends on the difference between source data and the required data of mining algorithms. Some common strategies for data transformation include the following [9]: Smoothing, Attribute construction, Aggregation, Normalization, Discretization, and Concept hierarchy generation for nominal data.

To gather more information about the geographic location of users accessing our web system, we generate some additional attributes based on the client's IP address. In order to achieve this goal, we propose a Client IP Location Lookup model as depicted in Fig. 4.

At the initial stage, all Client IPs are considered as unknown IPs, each IP address would be looked up in our offline IP database. If a fail status is returned, which means this IP address does not exist in the database, then it would be moved to waiting queues of online lookup block. Conversely, the record containing that IP address would be updated with three additional attributes representing location information: country, province, ISP.

In the online lookup block, the IP addresses would be pop out of the queues and searched for location information through several public APIs. At the moment, we mainly use API of https://extreme-ip-lookup.com/json/ and http://ip-api.com/json/. In case the IP address is found, this IP and its location information would be added to the offline database, and the record containing this IP is also expanded as mentioned above. For some private IP addresses and the IP addresses that cannot be found through the two APIs will be considered as unknown IPs and stored separately in another database so that they may be updated manually later.

Besides, using public APIs can save the cost for information lookup, but it also raises some issues like higher response time, unreliable response, and so on. However, in web usage mining, the processing time as well as the reliability of

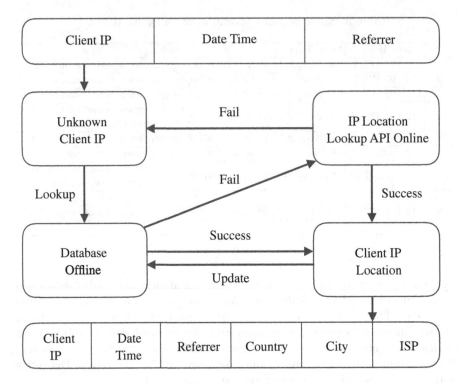

Fig. 4. Client IP lookup model

Table 4. Country correction

Line	Country	Standardized
1	Iran	Iran
2	IRN	Iran
3	Islamic Republic of Iran	Iran
4	ISR	Israel
...		

Table 5. City correction

Line	City	Standardized
1	An Hoa	Ha Noi
2	An Nhon	Dong Thap
3	Angiang	An Giang
4	Anh Son	Ha Giang
...		

information retrieved are important aspects. The reason is that the data used in web usage mining are usually web logs containing millions of entries, and the quality of the output of the pre-processing step plays a vital role which governs the mining results. Nonetheless, the lookup processes completely take place independently, they can effectively be parallelized by some high-level function of Apache Spark [1] for both online and offline lookup. Considering the experimental results, we realize that, issues related to the reliability of data mainly are attributable to countries, cities, and ISPs which have different names being consistent among used APIs or even on the same API. The remaining issues are derived from the confusion between the name of countries and cities as well as the name of cities and towns or villages. In order to overcome these limitations, some of the data normalization and standardized techniques are directly applied in the lookup process. Tables 4, 5 and 6 show some examples for standardizing the data obtained from public APIs. These data would be updated to the offline database so that it can be reused for later lookup.

Table 6. ISP correction

Line	ISP	Standardized
1	Vietnam Post and Telecom Corporation	VNPT-VN
2	Vietnam Post and Telecom Corporation Company	VNPT-VN
3	Vietnam Posts And Telecommunications Group	VNPT-VN
4	Vietnamnet-No 4 Lang Ha Ha Noi	VIETNAMNET-VN
5	Vietnamobile Telecommunications Joint Stock Compan	VIETNAMOBILE-VN
6	Vietnamobile Telecommunications Joint Stock Company	VIETNAMOBILE-VN
...		

In summary, by processing the web log containing more than 74 million records, we not only realize that there are nearly 10 million visits to our website over a year, but also aggregate information about the location of more than 780 thousand different Client IP addresses stored in the offline database.

3.2 Pattern Discovery and Analysis

After pre-processing the log file, we apply fundamental statistics to the log file to determine the number of pageviews based on different geographical regions (country or province) or distinct ISPs. Besides, we determine the number of each kind of contents during one year. After the pattern discovery, we uncover several interesting patterns, such as the most visiting countries or provinces to HCMUT website or which ISPs contributes the number of pageviews to the website most. Furthermore, we attain other knowledge like which contents are the most interested ones per month over one year. Experimental findings are presented in the next section.

4 Experimental Results

In this paper, we concentrate on analyzing HCMUT server log file of size 21 GB approximately. We present normal and interesting patterns by using various maps or pie charts.

Figure 5 shows the number of pageviews corresponding with each nation around the world while the proportion of interesting nations to the globe is depicted in Fig. 6. Vietnam is the most visiting country to HCMUT website

Fig. 5. The number of pageviews to HCMUT website for every nation

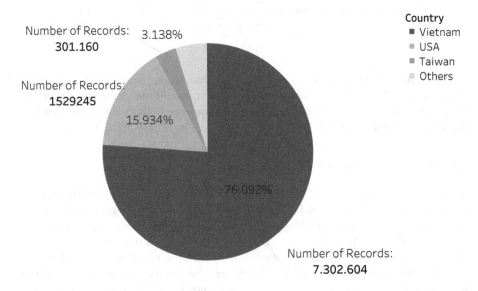

Fig. 6. The proportion of pageviews to HCMUT website for interesting nations

in the world with 76.092%. The next two positions are USA and Taiwan with 15.934% and 3.138% respectively.

Figure 7 elaborates the number of pageviews for all provinces in Vietnam and Fig. 8 describes the proportion of interesting provinces in Vietnam. Users from Ho Chi Minh City visit HCMUT website the most with 73.497%. Hanoi follows Ho Chi Minh City with 13.943%.

Figure 9 outlines the number of pageviews contributed from every ISP in Vietnam while the proportion of pageviews to HCMUT website for interesting ISPs in Ho Chi Minh City and Hanoi is shown in Fig. 10. VNPT-VN is the ISP

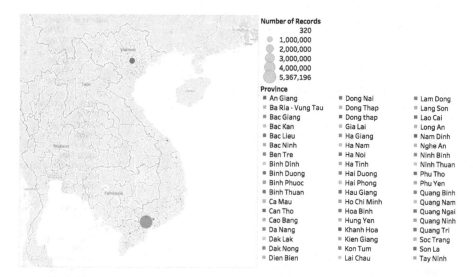

Fig. 7. The number of pageviews to HCMUT website for every province in Vietnam

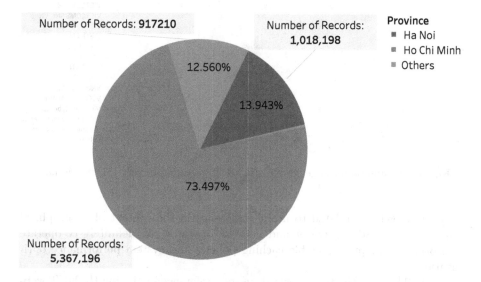

Fig. 8. The proportion of pageviews to HCMUT website for interesting provinces in Vietnam

which contributes to the total pageviews the most for both Ho Chi Minh City and Hanoi with around 48%. VIETTEL-VN occupies 20.563 % for Ho Chi Minh City and 19.673% for Hanoi while FPT-VN takes up 9.629% and 23.14% for Ho Chi Minh City and Hanoi respectively. SPT-VN accounts for 14.487% in Ho Chi Minh City and does not have its branch in Hanoi.

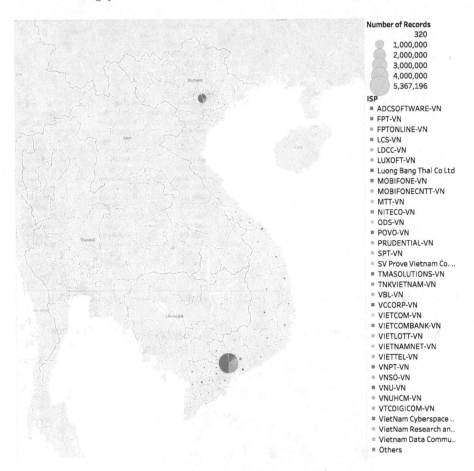

Fig. 9. The number of pageviews to HCMUT website for every ISP in Vietnam

With knowledge related to users' preference in the context of geographical regions, website administrator and network operator can definitely co-operate each other to adopt reasonable caching policies and system performance optimization.

Figure 11 shows the trend towards users' preference in visiting HCMUT website per month. In this figure, we can obviously discover the most visited contents and the considerable change of access patterns every month. The knowledge is extremely helpful in the context of predicting future access patterns, which assists significantly improve system performance as well as make more reasonable caching decisions compared with using current access patterns.

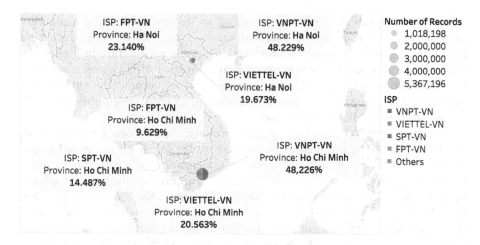

Fig. 10. The proportion of pageviews to HCMUT website for interesting ISPs in Ho Chi Minh City and Hanoi

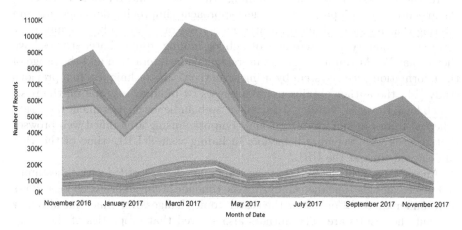

Fig. 11. The trend towards users' preference in visiting HCMUT website per month

5 Related Work

Mining web logs is an enormous research field that has attracted a lot of practitioners. As we mentioned earlier, it may be categorized into web content mining, web structure mining and web usage mining (WUM). There is a little considerable work concerning the field. Because different websites have different web log structures, most of these studies focus on analyzing a particular log and propose particular approaches. In this section, we only list some typical studies related to our work.

In terms of web content mining, there are many studies of user access patterns through web logs. For example, in prior researches [21,22], a mining system called Web Utilization Miner was proposed to discover interesting navigation patterns. At that time, it was satisfied the expert's criteria by exploiting an innovative aggregated storage representation for the information in logs of the real web server using its own mining language (called MINT) which supports the specification of statistical, structural and textual nature. The authors also emphasized the importance of data preparation, or data pre-processing, in filtering meaningless data from datasets.

In the scope of web structure mining, by analyzing web logs, Perkowitz et al. [19] investigated the problem of index page synthesis, which is the automatic creation of pages that facilitate a visitor's navigation of a website. In order to create an adaptive website, they propose an approach called cluster mining to find out collections of cohesive pages and then gather them to the same topic by using their PageGather algorithm to help the website itself improve the organization and presentation. In this work, the author discussed about processing web access logs to determine the user's visits by using some simple sessionization techniques.

In the context of web usage mining, there are some relevant researches. Pabarskaite et al. [17] proposed two new approaches improving data cleaning and filtering that remove unimportant attributes from raw data to help significantly increase the quality and usefulness of web log mining results. Later studies have shown that WUM can help extract users' behavior, interaction and the usage of the information access system by using several existing techniques. In a previous study [14], the author obtains users' interests by building a graph on the basis of web logs. However, web log data called "web audience measurement data" are collected from random users' personal computers using a modified web browser that records web watching behaviors including visited URL, time of the visit and elapsed time at the URL.

Another work [11] showed that there are many logs recorded with timestamps and placed along a timeline. Analyzing the timeline archived in search logs allows us to understand not only how the search topic changes over time, but also how relevant the results are. The authors also showed that properties of the query result along the timeline are likely useful in predicting the mean average precision of the query.

Wahab et al. [24] stated that web log file analysis can help the administrators manage the server bandwidth and capability effectively. In addition, the authors also discussed several common attributes of the log file in detail and then proposed some algorithms to handle the log file, eliminate irrelevant fields before mining web log.

In a prior study [23], the authors mainly focused on analyzing the web log file to determine system's errors using WUM techniques. However, they also discussed in detail about the structure of log files and the processes for extracting information from raw data, such as data cleaning, user identification...

According to Agosti et al. [6] and Hussain et al. [10], Web Log Mining is a vast field with thousands of results. Log file of HCMUT web server is still an untapped resource. In this paper, we mainly focus on retrieving useful information about the distribution of user access patterns by geographical regions which may be applied to improve content caching algorithms. Thus, we not only combine and manipulate existing techniques into our log file but also propose a novel lookup model to extend information about the geographic location of users based on IP addresses. Moreover, we have standardized, normalized data retrieved from the two public APIs and updated them on our offline IP database, which will also create a reliable data source and can be reused in the future.

6 Discussion

In data pre-processing process, we use some techniques, such as data cleaning to remove unnecessary log entries, data reduction to get rid of unneeded attributes, pageview identification to determine real click-throughs from users and data transformation to add new attributes including country, province, ISP. Besides, there are other techniques able to be applied to the log, such as user identification to determine end-users based on IP and user agent, session identification (sessionization) to discover a set of sessions for each user based on date/time, URL and referrer after finishing user identification, path completion to recover missing access to pages or objects which may be cached at clients or web proxies based on historical access patterns, and so on. The techniques are not utilized yet because they are really not necessary for applications which we mentioned in Sect. 1.

In the pattern discovery process, we apply only basic statistics to the pre-processed data. Attained patterns after executing the process is sufficient for our analysis. Hence, other techniques are not deployed. However, with the current log, it is possible that we can definitely apply the techniques to it. The techniques comprise clustering (unsupervised learning) to classify data into distinct unlabeled groups, classification to categorize data into labels groups, association rule to find groups of items or pages that are commonly accessed together. Such other techniques are beyond the scope of the study.

7 Conclusions and Future Work

In this article, we apply the web usage mining process to log file collected from HCMUT in Vietnam. The process consists of three stages: data pre-processing, pattern discovery, and pattern analysis. In the pre-processing stage, we use techniques like data clean, data reduction, pageview identification, and data transformation. In terms of data transformation, we propose a novel model related to attribute construction strategy to construct and add new attributes encompassing country, province, ISP from the existing attribute IP. The information can be leveraged to understand user access patterns from different geographical regions which can help decide reasonable caching policies at web proxies as well as improve system performance at web servers. In the pattern discovery, we use basic statistics to identify interesting patterns. Specifically, we are interested in countries, provinces and ISPs which access most frequently the university website. Besides, the trend of accessing monthly contents during one year are paid attention.

In general, techniques in pattern discovery can be categorized into two groups: descriptive and predictive. In this study, we use only the descriptive functionality to extract knowledge for optimizing system performance as well as improve caching policies. However, if we are capable of predicting future access pattern, the system performance as well as the caching policies will be enhanced more significantly. Therefore, we intend to exploit the predictive functionality of the pattern discovery stage to extract better knowledge in the next studies.

Acknowledgements. This research was conducted within the project of Studying collaborative caching algorithms in content delivery network sponsored by TIS (IT Holding Group).

References

1. Apache Spark. https://spark.apache.org. Accessed 10 July 2018
2. DomainTools. http://research.domaintools.com/statistics/tld-counts. Accessed 10 July 2018
3. Ho Chi Minh City University of Technology, Vietnam. http://hcmut.edu.vn. Accessed 10 July 2018
4. Internet World Stats. https://www.internetworldstats.com/stats.htm. Accessed 10 July 2018
5. Tableau. https://www.tableau.com. Accessed 10 July 2018
6. Agosti, M., Crivellari, F., Di Nunzio, G.M.: Web log analysis: a review of a decade of studies about information acquisition, inspection and interpretation of user interaction. Data Min. Knowl. Discov. **24**(3), 663–696 (2012)
7. Barsagade, N.: Web usage mining and pattern discovery: a survey paper. Computer Science and Engineering Department, CSE Technical report 8331 (2003)
8. Gündüz, Ş., Özsu, M.T.: A web page prediction model based on click-stream tree representation of user behavior. In: Proceedings of the Ninth ACM SIGKDD International Conference on Knowledge Discovery and Data Mining, pp. 535–540. ACM (2003)
9. Han, J., Pei, J., Kamber, M.: Data Mining: Concepts and Techniques. Elsevier, Amsterdam (2011)
10. Hussain, T., Asghar, S., Masood, N.: Web usage mining: a survey on preprocessing of web log file. In: 2010 International Conference on Information and Emerging Technologies (ICIET), pp. 1–6. IEEE (2010)
11. Jones, R., Diaz, F.: Temporal profiles of queries. ACM Trans. Inf. Syst. (TOIS) **25**(3), 14 (2007)
12. Kosala, R., Blockeel, H.: Web mining research: a survey. ACM SIGKDD Explor. Newsl. **2**(1), 1–15 (2000)
13. Liu, B.: Web Data Mining: Exploring Hyperlinks, Contents, and Usage Data. Springer, Heidelberg (2007). https://doi.org/10.1007/978-3-540-37882-2
14. Murata, T., Saito, K.: Extracting users' interests from web log data. In: Proceedings of the 2006 IEEE/WIC/ACM International Conference on Web Intelligence, pp. 343–346. IEEE Computer Society (2006)

15. Nakajima, T., Yoshimi, M., Wu, C., Yoshinaga, T.: A light-weight content distribution scheme for cooperative caching in telco-CDNs. In: 2016 Fourth International Symposium on Computing and Networking (CANDAR), pp. 126–132. IEEE (2016)
16. Nakajima, T., Yoshimi, M., Wu, C., Yoshinaga, T.: Color-based cooperative cache and its routing scheme for Telco-CDNs. IEICE Trans. Inf. Syst. **100**(12), 2847–2856 (2017)
17. Pabarskaite, Z.: Implementing advanced cleaning and end-user interpretability technologies in web log mining. In: Proceedings of the 24th International Conference on Information Technology Interfaces, ITI 2002, pp. 109–113. IEEE (2002)
18. Pani, S.K., Panigrahy, L., Sankar, V., Ratha, B.K., Mandal, A., Padhi, S.: Web usage mining: a survey on pattern extraction from web logs. Int. J. Instrum. Control Autom. **1**(1), 15–23 (2011)
19. Perkowitz, M., Etzioni, O.: Adaptive web sites: automatically synthesizing web pages. In: AAAI/IAAI, pp. 727–732 (1998)
20. Piatetsky-Shapiro, G., Fayyad, U., Smith, P.: From data mining to knowledge discovery: an overview. Adv. Knowl. Discov. Data Min. **1**, 35 (1996)
21. Spiliopoulou, M., Faulstich, L.C.: WUM: a web utilization miner. In: International Workshop on the Web and Databases, Valencia, Spain. Citeseer (1998)
22. Spiliopoulou, M., Faulstich, L.C., Winkler, K.: A data miner analyzing the navigational behaviour of web users. In: Proceedings of the Workshop on Machine Learning in User Modelling of the ACAI 1999, Greece, July 1999
23. Suneetha, K., Krishnamoorthi, R.: Identifying user behavior by analyzing web server access log file. IJCSNS Int. J. Comput. Sci. Netw. Secur. **9**(4), 327–332 (2009)
24. Wahab, M.H.A., Mohd, M.N.H., Hanafi, H.F., Mohsin, M.F.M.: Data preprocessing on web server logs for generalized association rules mining algorithm. World Acad. Sci. Eng. Technol. **48**, 2008 (2008)
25. Yun, L., Xun, W., Huamao, G.: A hybrid information filtering algorithm based on distributed web log mining. In: Third International Conference on Convergence and Hybrid Information Technology, ICCIT 2008, vol. 1, pp. 1086–1091. IEEE .(2008)

Internet of Things and Applications

Lower Bound for Function Computation
in Distributed Networks

H. K. Dai[1(✉)] and M. Toulouse[2]

[1] Computer Science Department, Oklahoma State University,
Stillwater, OK 74078, USA
dai@cs.okstate.edu
[2] Computer Science Department, Vietnamese-German University,
Binh Duong New City, Vietnam
michel.toulouse@vgu.edu.vn

Abstract. Distributed computing network systems are modeled as graphs with which vertices represent compute elements and adjacency-edges capture their uni- or bi-directional communication. Distributed computation over a network system proceeds in a sequence of time-steps in which vertices update and/or exchange their values based on the underlying algorithm constrained by the time-(in)variant network topology. For finite convergence of distributed information dissemination and function computation in the model, we present a lower bound on the number of time-steps for vertices to receive (initial) vertex-values of all vertices regardless of underlying protocol or algorithmics in time-invariant networks via the notion of vertex-eccentricity.

Keywords: Distributed function computation
Linear iterative schemes · Information dissemination
Finite convergence · Vertex-eccentricity

1 Preliminaries

Distributed computation algorithms, decentralized data-fusion architectures, and multi-agent systems are modeled with a network of interconnected vertices that compute common value(s) based on initial values or observations at the vertices. Key computation and communication requirements for these network/system paradigms include that their vertices perform local/internal computations and regularly communicate with each other via an underlying protocol. Fundamental limitations and capabilities of these algorithms and systems are studied in the literature with viable applications in computer science, communication, and control and optimization (see, for examples, [1,3,4,7,8]). We give brief and informal descriptions of some example studies below:

1. Quantized consensus [5]: Consider an order-n network with an initial network-state in which each vertex assumes an initial (integer) value $x_i[0]$ for $i = 1, 2, \ldots, n$. The network achieves a quantized consensus when, at some later

© Springer Nature Switzerland AG 2018
T. K. Dang et al. (Eds.): FDSE 2018, LNCS 11251, pp. 371–384, 2018.
https://doi.org/10.1007/978-3-030-03192-3_28

time, all the n vertices simultaneously arrive with almost equal values y_i for $i = 1, 2, \ldots, n$ (that is, $|y_i - y_j| \leq 1$ for all $i, j \in \{1, 2, \ldots, n\}$) while preserving the sum of all initial values (that is, $\sum_{i=1}^{n} x_i[0] = \sum_{i=1}^{n} y_i$).

2. Collaborative distributed hypothesis testing [6]: Consider a network-system of n vertices (sensors/agents) that collaboratively determine the probability measure of a random variable based on a number of available observations/measurements. For the binary setting in deciding two hypotheses, each vertex collects measurement(s) and makes a preliminary (local) decision $d_i \in \{0, 1\}$ in favor of the two hypotheses for $i = 1, 2, \ldots, n$. The n vertices are allowed to communicate, and the network-system resolves with a final decision by, for example, the majority rule (that is, computes the indicator function of the event $\sum_{i=1}^{n} d_i > \frac{n}{2}$) in distributed fashion.

3. Solitude verification [3]: Consider an unlabeled network of n vertices (processes) in which each vertex is in one of a finite number of states: s_i for $i = 1, 2, \ldots, n$. Solitude verification on the network checks if a unique vertex with a given state s exists in the network, that is, computes the Boolean function for the equality $|\{i \in \{1, 2, \ldots, n\} \mid s_i = s\}| = 1$.

While there is a wide spectrum of algorithms in the literature that solve distributed computation problems such as the above, there are also studies that deal with algorithmic and complexity issues constrained by underlying time-(in)variant network topology, resource-limitations associated with vertices, time/space and communication tradeoffs, convergence criteria and requirements, etc. We present below a model of distributed computing systems and address the motivation of our study.

1.1   Model of Distributed Computing Systems

Most graph-theoretic definitions in this article are given in [2]. We will abbreviate "directed graph" and "directed path" to digraph and dipath, respectively.

We consider the topological model and algorithmics detailed in [7] for distributed function computation, and provide its abstraction components as follows:

1. Network topology: A distributed computing system is modeled as a digraph G with $V(G)$ and $E(G)$ denoting its sets of vertices and directed edges, respectively. Uni-directional communication on $V(G)$ is captured by the adjacency relation represented by $E(G)$: for all distinct vertices, $u, v \in V(G)$, $(u, v) \in E(G)$ if and only if vertex u can send information to vertex v (and v can receive information from u). Note that bi-directional communication between u and v is viewed as the co-existence of the two directed edge (u, v) and (v, u) in $E(G)$.

    Distributed computation over the network proceeds in a sequence of time-steps. At each time-step, all vertices update and/or exchange their values based on the underlying algorithm constrained by the network topology, which is assumed to be time-invariant.

2. Resource capabilities in vertices: The digraph G of the network topology is vertex-labeled such that messages are identified with senders and receivers. The vertices of $V(G)$ are assumed to have sufficient computational capabilities and local storage. Generally we assume that: (1) all communications/transmissions between vertices are reliable and in correct sequence, and (2) each vertex may, in the current time-step, receive the prior-step transmission(s) from its in-neighbor(s), update, and send transmission(s) to its out-neighbor(s) in accordance to the underlying algorithm.

   The domain of all initial/input and observed/output values of the vertices of G is assumed to be an algebraic field $\mathbb{F}$.

3. Linear iterative scheme (for algorithmic lower- and upper-bound results): For a vertex $v \in V(G)$, denote by $x_v[k] \in \mathbb{F}$ the vertex-value of v at time-step $k = 0, 1, \ldots$. A function with domain $\mathbb{F}^{|V(G)|}$ and codomain $\mathbb{F}$ is computed in accordance to a linear iterative scheme. Given initial vertex-values $x_v[0] \in \mathbb{F}$ for all vertices $v \in V(G)$ as arguments to the function, at each time-step $k = 0, 1, \ldots$, each vertex $v \in V(G)$ updates (and transmits) its vertex-value via a weighted linear combination of the prior-step vertex-values constrained by neighbor-structures: for all $v \in V(G)$ and $k = 0, 1, \ldots$,

$$x_v[k + 1] = \sum_{u \in V(G)} w_{vu} x_u[k],$$

where the prescribed weights $w_{vu} \in \mathbb{F}$ for all $v, u \in V(G)$ that are subject to the adjacency-constraints $w_{vu} = 0$ (the zero-element of $\mathbb{F}$) if u is not adjacent to v (that is, $(u, v) \notin E(G)$); equivalently,

transpose of $(x_v[k + 1] \mid v \in V(G)) = W \cdot$ transpose of $(x_v[k] \mid v \in V(G))$

where the two vectors of vertex-values and W are indexed by a common discrete ordering of $V(G)$ with $W = [w_{vu}]_{(v,u) \in V(G) \times V(G)}$.

1.2 Motivation of Our Study

Based on the framework and its variants for distributed function computation, researches and studies are focused on mathematical interplays among:

- time-(in)variance of network-topology
- granularity of time-step: discrete versus continuous
- choice of base field: special (real or complexes) versus arbitrary (finite or infinite)
- characterization of calculable functions
- convergence criteria and rates (finite, asymptotic, and/or probabilistic)
- adoption and algebraic properties of weight-matrix for linear interactive schemes: random weight-matrix, spectrum of eigenvalues, base field, etc.
- resilience and robustness of computation algorithmics for network-topology in the presence/absence of malicious vertices

– lower and upper bounds on (linear) iteration required for the convergence of calculable functions.

Summarized results, research studies, and references are available in, for examples, [7–9,11].

Sundaram and Hadjicostis [7,8] present their research findings in the finite convergence of distributed information dissemination and function computation in the model with linear iterative algorithmics stated above, among other contributions in distributed function computation and data-stream transmission in the presences of noise and malicious vertices. More specifically, (1) they employ structural theories in observability and invertibility of linear systems over arbitrary finite fields to obtain lower and upper bounds on the number of linear iterations for achieving network consensus for finite convergence of arbitrary functions, and (2) the bounds are valid for all initial vertex-values of arbitrary finite fields as arguments to the functions in connected time-invariant topologies with almost all random weight-matrices.

For a time-invariant topology with underlying digraph G and a vertex $u \in V(G)$, denote by $\deg_{G,\text{in}}(u)$ the in-degree of u in G, and by $\Gamma_{G,\text{in}}(u)$ the in-neighbor of u in G; hence $\Gamma^*_{G,\text{in}}(u)$ denotes the in-closure of u in G, that is,

$$\Gamma^*_{G,\text{in}}(u) = \cup_{\eta \geq 0} \Gamma^{\eta}_{G,\text{in}}(u)$$
$$= \{v \in V(G) \mid \text{ there exists a dipath in } G \text{ from } v \text{ to } u\}.$$

Consider all possible families of directed trees that are: (1) a vertex-decomposition of $\Gamma^*_{G,\text{in}}(u) - \{u\}$, and (2) rooted in (as subset of) $\Gamma_{G,\text{in}}(u)$. Denote by:

$$\alpha_{G,u} = \min\{\max\{\text{order}(T_i) \mid 1 \leq i \leq n\} \mid$$
$$\{T_i\}_{i=1}^n \text{ is a family of directed trees that are: (1) a vertex-}$$
$$\text{decomposition of } \Gamma^*_{G,\text{in}}(u) - \{u\}, \text{ and (2) rooted in (as sub-}$$
$$\text{set of) } \Gamma_{G,\text{in}}(u)\}.$$

Their upper-bound result for a vertex $u \in V(G)$ is stated as follows: for every linear iterative scheme with random weight-matrix over a finite base field $\mathbb{F}$ of cardinality $|\mathbb{F}| \geq (\alpha_{G,u} - 1)(|\Gamma^*_{G,\text{in}}(u)| - \deg_{G,\text{in}}(u) - \frac{1}{2}\alpha_{G,u})$, then, with probability at least $1 - \frac{1}{|\mathbb{F}|}(\alpha_{G,u} - 1)(|\Gamma^*_{G,\text{in}}(u)| - \deg_{G,\text{in}}(u) - \frac{1}{2}\alpha_{G,u})$, the vertex u can calculate arbitrary functions of arbitrary initial vertex-values $x_v[0] \in \mathbb{F}$ for all $v \in \Gamma^*_{G,\text{in}}(u)$ via the linear iterative scheme within a most $\alpha_{G,u}$ time-steps.

Sundaram conjectures in [7] that $\alpha_{G,u}$ may also serve as a lower bound on the number of time-steps for a vertex $u \in V(G)$ to receive the initial vertex-values of all $v \in \Gamma^*_{G,\text{in}}(u)$ regardless of underlying protocol or algorithmics. Hence, linear iterative schemes are time-optimal in disseminating information over arbitrary time-invariant connected networks.

Toulouse and Minh [10] refute the conjecture via the notion of rank-step sequences for linear iterative schemes over connected network with an explicit counter-example in Fig. 1.

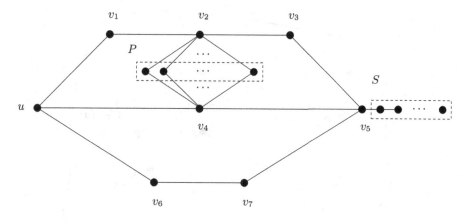

Fig. 1. A counter-example graph, in which the embedded parallel component P and serial component S satisfying $\text{order}(S) = \lfloor \frac{\text{order}(P)}{2} \rfloor + 1$, to the lower-bound conjecture in terms of $\alpha_{G,u}$ in [7].

In order to complement the explicitly constructed counter-example to the lower-bound conjecture on the number of time-steps for distributed function computation and information dissemination with respect to a given vertex, we present in this article a lower bound on the number of time-steps for a vertex $u \in V(G)$ to receive the initial vertex-values of all $v \in \Gamma^*_{G,\text{in}}(u)$ regardless of underlying protocol or algorithmics in a time-invariant network via the notion of vertex-eccentricity.

2 Revised Lower Bound for Distributed Function Computation and Information Dissemination

Consider an arbitrary vertex $u \in V(G)$, and assume a non-trivial $\Gamma^*_{G,\text{in}}(u)$ ($|\Gamma^*_{G,\text{in}}(u)| > 1$) hereinafter. We develop a lower bound on the number of time-steps required for the vertex u to receive the (initial) vertex-values of all vertices of $\Gamma^*_{G,\text{in}}(u)$ (regardless of underlying protocol, including linear iterative schemes). See Fig. 2 for an example of $\Gamma^*_{G,\text{in}}(u)$.

For two vertices u and v of G, $\overrightarrow{d}_G(u,v)$ denotes the directed distance from u to v in G, that is,

$$\overrightarrow{d}_G(u,v) = \begin{cases} \text{length of a shortest dipath from } u \text{ to } v \text{ in } G & \text{if exists,} \\ \infty & \text{otherwise.} \end{cases}$$

For a vertex u of G, $e_{G,\text{in}}(u)$ denotes the in-eccentricity of u in G, which is the maximum directed distance from a vertex to u in G, that is,

$$e_{G,\text{in}}(u) = \max\{ \underbrace{\overrightarrow{d}_G(v,u)}_{\text{minimum length of a dipath from } v \text{ to } u \text{ in } G} \mid v \in V(G)\}.$$

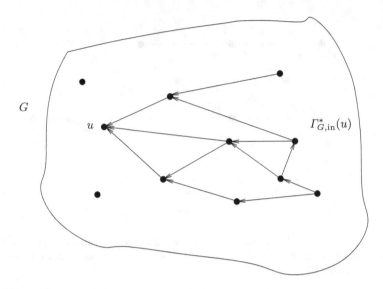

Fig. 2. For a vertex u in a digraph G: an example organization of the in-closure $\Gamma_{G,\text{in}}^*(u)$ of u in G.

Following the above-stated distributed-computation framework as in [8] and for their conjecture, we develop a lower-bound result based on the notion of eccentricity (instead of "order" or "size" as in the conjecture):

1. For every (linear or non-linear) iteration scheme, in which a vertex's value or information is transmitted to its out-neighbors via their incidence directed edges in unit time-step, requires at least $e_{G,\text{in}}(u)$ time-steps for vertex u to access values/information of all the vertices in $\Gamma_{G,\text{in}}^*(u)$. Thus, $e_{G,\text{in}}(u)$ serves as a lower bound on the number of time-steps required for function-computation by vertex u via such iteration scheme.
2. In accordance with the distributed framework for our function-computation, we show below that:

$$e_{G,\text{in}}(u) = 1 + \min\{\max\{\underbrace{e_{T_i,\text{in}}(\text{root}(T_i))}_{= \text{ depth}(T_i)} \mid 1 \leq i \leq n\} \mid$$

$$\{T_i\}_{i=1}^n \text{ is a family of directed trees that are:}$$
(1) a vertex-decomposition of $\Gamma_{G,\text{in}}^*(u) - \{u\}$, and
(2) rooted in (as subset of) $\Gamma_{G,\text{in}}(u)\}$.

We illustrate an example organization of $\Gamma_{G,\text{in}}^*(u) - \{u\}$ in a family of vertex-disjoint directed trees in Fig. 3.

To show the above equality for $e_{G,\text{in}}(u)$, we prove the two embedded inequalities in the following sections.

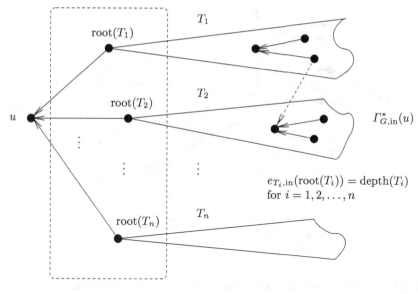

$\{\text{root}(T_i) \mid 1 \leq i \leq n\}$ is a subset (not necessarily proper) of $\Gamma^*_{G,\text{in}}(u)$

Fig. 3. For a vertex u in a digraph G: an example organization of $\Gamma^*_{G,\text{in}}(u) - \{u\}$ in a family $\{T_i\}^n_{i=1}$ of directed trees that are: (1) a vertex-decomposition of $\Gamma^*_{G,\text{in}}(u) - \{u\}$, and (2) rooted in (as subset of) $\Gamma_{G,\text{in}}(u)$.

2.1 Upper Bound for Vertex-Eccentricity

We first prove that:

$$e_{G,\text{in}}(u) \leq 1 + \min\{\max\{\underbrace{e_{T_i,\text{in}}(\text{root}(T_i))}_{= \text{ depth}(T_i)} \mid 1 \leq i \leq n\} \mid$$

$\{T_i\}^n_{i=1}$ is a family of directed trees that are: (1) a vertex-decomposition of $\Gamma^*_{G,\text{in}}(u) - \{u\}$, and (2) rooted in (as subset of) $\Gamma_{G,\text{in}}(u)\}$;

equivalently,

$$e_{G,\text{in}}(u) \leq 1 + \max\{\underbrace{e_{T_i,\text{in}}(\text{root}(T_i))}_{= \text{ depth}(T_i)} \mid 1 \leq i \leq n\}$$

for arbitrary family of directed trees, $\{T_i\}^n_{i=1}$, which are a vertex-decomposition of $\Gamma^*_{G,\text{in}}(u) - \{u\}$ and are rooted in (as subset of) $\Gamma_{G,\text{in}}(u)$.

Consider an arbitrary family of directed trees, $\{T_i\}^n_{i=1}$, which are a vertex-decomposition of $\Gamma^*_{G,\text{in}}(u) - \{u\}$ and are rooted in (as subset of) $\Gamma_{G,\text{in}}(u)$. The in-eccentricity $e_{G,\text{in}}(u)$ of u in G is realized by a dipath P from a vertex $v \in \Gamma^*_{G,\text{in}}(u) - \{u\}$ to u in G. Since $\{T_i\}^n_{i=1}$ is a vertex-decomposition of $\Gamma^*_{G,\text{in}}(u) - \{u\}$, we have $v \in V(T_i)$ for some $i \in \{1, 2, \ldots, n\}$. We depict the scenario in Fig. 4.

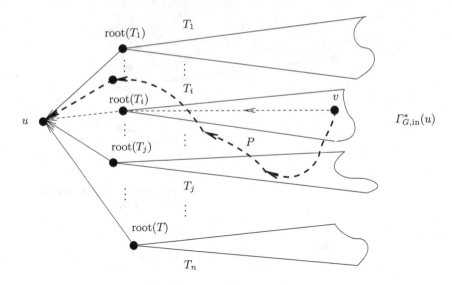

Fig. 4. For a vertex u in a digraph G: the in-eccentricity $e_{G,\text{in}}(u)$ of u in G is realized by a dipath P from a vertex $v \in \Gamma^*_{G,\text{in}}(u) - \{u\}$ to u in G.

Now,

$$\underbrace{e_{G,\text{in}}(u)}_{\text{in } G} = \underbrace{\text{length}(P)}_{\text{in } G} = \underbrace{\vec{d}_G(v, u)}_{\text{in } G}$$

$\leq$ length((unique) dipath from v to root(T_i) in T_i concatenated
  with directed edge (root$(T_i), u)$) since T_i is a sub-digraph of
  the digraph vertex-spanned by $\Gamma^*_{G,\text{in}}(u)$

$\leq$ depth$(T_i) + 1$

$\leq 1 + \max\{\text{depth}(T_i) \mid 1 \leq i \leq n\}$

as desired.

2.2   Lower Bound for Vertex-Eccentricity

To show the reverse inequality:

$$e_{G,\text{in}}(u) \geq 1 + \min\{\max\{\underbrace{e_{T_i,\text{in}}(\text{root}(T_i))}_{= \text{ depth}(T_i)} \mid 1 \leq i \leq n\} \mid$$

$\{T_i\}_{i=1}^n$ is a family of directed trees that are:
(1) a vertex-decomposition of $\Gamma^*_{G,\text{in}}(u) - \{u\}$, and
(2) rooted in (as subset of) $\Gamma_{G,\text{in}}(u)\}$,

it suffices to construct a family $\{T_i\}_{i=1}^n$ of directed trees that are a vertex-decomposition of $\Gamma^*_{G,\text{in}}(u) - \{u\}$, and are rooted in (as subset of) $\Gamma_{G,\text{in}}(u)$, such that:

$$e_{G,\text{in}}(u) \geq 1 + \max\{\text{depth}(T_i) \mid 1 \leq i \leq n\}.$$

We proceed with an inductive construction of a sequence $(P_1, P_2, \dots)$ of dipaths with common end-vertices u such that the sequence $(P_1 - \{u\}, P_2 - \{u\}, \dots)$ is organized as a family $\{T_1, T_2, \dots, T_i\}$, where $i \geq 1$, of directed trees such that:

1. The family $\{T_1, T_2, \dots, T_i\}$ consists of mutually vertex-disjoint directed trees with their roots in $\Gamma_{G,\text{in}}(u)$,
2. Each directed tree in the family provides a shortest dipath (in G) for each of its vertices to u, that is, for every vertex $v \in V(T_j)$ where $j \in \{1, 2, \dots, i\}$, the (unique) dipath from v to $\text{root}(T_j)$ in T_j yields $\vec{d}_G(v, u)$:

   length((unique) dipath from v to $\text{root}(T_j)$ concatenated with directed edge $(\text{root}(T_j), u)) = \vec{d}_G(v, u)$,

   and
3. The in-eccentricity of u in G is bounded below as:

$$e_{G,\text{in}}(u) \geq 1 + \max\{\text{depth}(T_1), \text{depth}(T_2), \dots, \text{depth}(T_i)\}.$$

See an example configuration in Fig. 5.

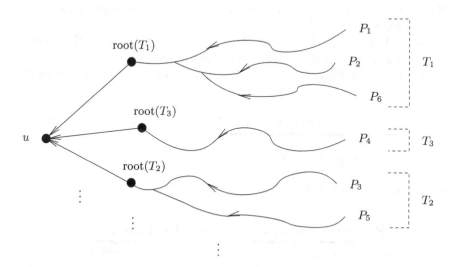

Fig. 5. For a vertex u in a digraph G: an inductive construction of a sequence $(P_1, P_2, \dots)$ of dipaths with common end-vertices u such that the sequence $(P_1 - \{u\}, P_2 - \{u\}, \dots)$ is organized as a family $\{T_1, T_2, \dots, T_i\}$, where $i \geq 1$, of directed trees that satisfies the stated conditions in items 1, 2, and 3.

Basis step: For P_1, we may employ a dipath from a vertex, say v, in $\Gamma^*_{G,\text{in}}(u) - \{u\}$ to u in G that realizes $\overrightarrow{d}_G(v, u)$. Then, designate P_1 as such a path, and $T_1 = \{P_1 - \{u\}\}$.

For the family $\{T_1\}$, we can verify the above-stated three items 1, 2 (via "shortest dipath in G" enjoys "optimal substructure property in G" by typical cut-and-paste argument), and 3.

Induction step: Assume that we have constructed a sequence $(P_1, P_2, \ldots, P_j)$ of dipaths with common end-vertices u such that the sequence $(P_1 - \{u\}, P_2 - \{u\}, \ldots, P_j - \{u\})$ is organized as a family $\{T_1, T_2, \ldots, T_i\}$, where $j \geq i \geq 1$, of directed trees that satisfies the above-stated items 1, 2, and 3.

If the family $\{T_1, T_2, \ldots, T_i\}$ yields a vertex-decomposition of $\Gamma^*_{G,\text{in}}(u) - \{u\}$, then the inductive construction is complete. Thus, we may assume that there exists a vertex $v \in (\Gamma^*_{G,\text{in}}(u) - \{u\}) - \cup_{\eta=1}^i V(T_\eta)$. We construct a desired dipath P_{j+1} from v to u in G as follows.

First, consider a dipath P from v to u in G that realizes $\overrightarrow{d}_G(v, u)$ (that is, $\text{length}(P) = \overrightarrow{d}_G(v, u)$). Observe that,

$$\text{length}(P) = \overrightarrow{d}_G(v, u) \leq \underbrace{\max\{\overrightarrow{d}_G(v, u) \mid v \in V(G)\}}_{e_{G,\text{in}}(u)}.$$

Consider the two cases of P based on its possible intersection with the constructed directed forest/family $\{T_1, T_2, \ldots, T_i\}$—which are shown in Fig. 6.

Case 1: $V(P) \cap \cup_{\eta=1}^i V(T_\eta) = \emptyset$. From the above observation that $\text{length}(P) \leq e_{G,\text{in}}(u)$, hence for P_{j+1}, we may employ P by designating $P_{j+1} = P$ and $T_{i+1} = \{P_{j+1} - \{u\}\}$ as in the basis step. We can verify the above-stated items 1, 2, and 3 for the augmented family $\{T_1, T_2, \ldots, T_{i+1}\}$.

Case 2: $V(P) \cap \cup_{\eta=1}^i V(T_\eta) \neq \emptyset$. Denote the first entrance of the dipath P into $\cup_{\eta=1}^i V(T_\eta)$ by w, say $w \in V(P) \cap V(T_k)$ for some $k \in \{1, 2, \ldots, i\}$.

With the denotations/labelings in Fig. 7, we have two possible dipaths from w to u: (1) the dipath:

$$Q = \underbrace{\text{(unique) dipath from } w \text{ to root}(T_k)}_{\text{contained in } T_k} \quad \text{concatenated} \quad \text{with} \quad \text{the}$$

directed edge $(\text{root}(T_k), u)$,

and (2) the dipath P_2 such that:

$$P = \underbrace{\text{dipath from } v \text{ to } w}_{\text{via vertices in } (\Gamma^*_{G,\text{in}}(u) - \{u\}) - \cup_{\eta=1}^i V(T_\eta)} \text{concatenated with} \underbrace{\text{dipath } P_2}_{\text{from } w \text{ to } u \text{ in } G}.$$

What can we say about $\text{length}(Q)$ versus $\text{length}(P_2)$? They must be equal—via a proof by contradiction as follows:

1. Suppose that $\text{length}(Q) < \text{length}(P_2)$: The dipath from v, via w, to u formed by the concatenation of P_1 (v to w) and Q (w, via $\text{root}(T_k)$, to u)

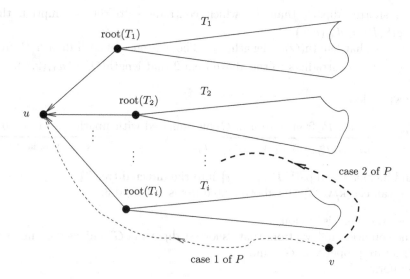

Fig. 6. For a vertex u in a digraph G: assume the inductive construction of a sequence $(P_1, P_2, \ldots, P_j)$ of dipaths that results in a family $\{T_1, T_2, \ldots, T_i\}$, where $j \geq i \geq 1$, of mutually vertex-disjoint directed trees with their roots in $\Gamma^*_{G,\text{in}}(u) - \{u\}$ that satisfies the stated conditions in items 1, 2, and 3, then, for a vertex $v \in (\Gamma^*_{G,\text{in}}(u) - \{u\}) - \cup^i_{\eta=1} V(T_\eta)$, construct a desired dipath P_{j+1} from v to u in G by considering a dipath P from v to u in G with $\text{length}(P) = \overrightarrow{d}_G(v, u)$ in two cases.

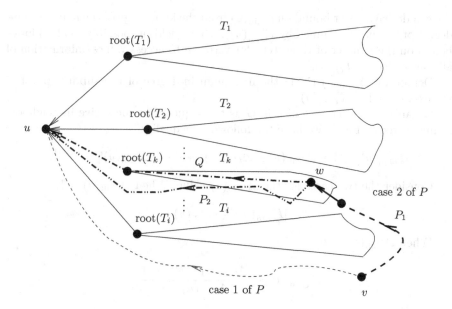

Fig. 7. For a vertex u in a digraph G: case 2 of P with $V(P) \cap \cup^i_{\eta=1} V(T_\eta) \neq \emptyset$ is considered.

is a shorter dipath than P—which contradicts to the assumption that length$(P) = \vec{d}_G(v, u)$.

2. Suppose that length$(Q) >$ length(P_2): The existence of such dipath P_2 from v to u in G contradicts to the above item 2 that length$(Q) = \vec{d}_G(v, u)$.

Now, we let:

$$P_{j+1} = \underbrace{\text{dipath } P_1 \text{ from } v \text{ to } u \text{ in } G}_{\text{via vertices in } (\Gamma^*_{G,\text{in}}(u) - \{u\}) - \cup^i_{\eta=1} V(T_\eta)} \text{concatenated with} \underbrace{\text{dipath } Q \text{ from } w \text{ to } u,}_{\text{via vertices in } T_k}$$

and include the dipath $P_{j+1} - \{u\}$ into the directed tree T_k.

We can check/verify the above-stated items 1, 2, and 3:

1. The statement is obvious,
2. The condition follows from that "shortest dipath in G" enjoys "optimal substructure property in G", and
3. By noting that:
$$e_{G,\text{in}}(u) \geq \vec{d}_G(v, u) = \text{length}(P).$$

This completes the inductive construction, and we have shown the reverse inequality.

3   Concluding Remarks

We can derive a lower bound on $e_{G,\text{in}}(u)$ from the knowledge of the maximum in-degree of G (vertex-spanned by $\Gamma^*_{G,\text{in}}(u)$), which yields a (possibly weaker) lower bound on the number of time-steps for vertex u to access values/information of all the vertices in $\Gamma^*_{G,\text{in}}(u)$.

Denote by $\Delta_{G,\text{in}}(u)$ (≥ 1) the maximum in-degree of the subdigraph of G vertex-spanned by $\Gamma^*_{G,\text{in}}(u)$.

Organize the in-closure of u in G as the sequence of successive in-neighbors as illustrated in Fig. 8, we have the following inequality:

$$|\Gamma^*_{G,\text{in}}(u)| \leq 1 + \Delta_{G,\text{in}}(u) + \Delta_{G,\text{in}}(u)^2 + \cdots + \Delta_{G,\text{in}}(u)^{e_{G,\text{in}}(u)-1}.$$

Consider the two cases for the $\Delta_{G,\text{in}}(u)$-value. When $\Delta_{G,\text{in}}(u) = 1$:

$$|\Gamma^*_{G,\text{in}}(u)| \leq e_{G,\text{in}}(u).$$

When $\Delta_{G,\text{in}}(u) \geq 2$:

$$|\Gamma^*_{G,\text{in}}(u)| \leq \frac{\Delta_{G,\text{in}}(u)^{e_{G,\text{in}}(u)} - 1}{\Delta_{G,\text{in}}(u) - 1},$$

which gives a lower bound on $e_{G,\text{in}}(u)$:

$$\log_{\Delta_{G,\text{in}}(u)}((\Delta_{G,\text{in}}(u) - 1)|\Gamma^*_{G,\text{in}}(u)| + 1) \leq e_{G,\text{in}}(u).$$

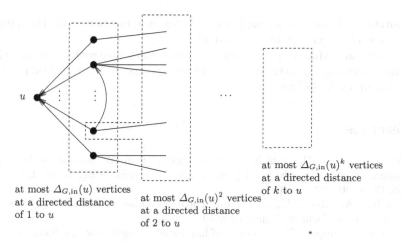

at most $\Delta_{G,\text{in}}(u)$ vertices
at a directed distance
of 1 to u

at most $\Delta_{G,\text{in}}(u)^2$ vertices
at a directed distance
of 2 to u

at most $\Delta_{G,\text{in}}(u)^k$ vertices
at a directed distance
of k to u

Fig. 8. For a vertex u in a digraph G: organize vertices of the in-closure $\Gamma^*_{G,\text{in}}(u)$ of u in G according to their directed distances to u.

We can obtain desired lower bounds in analogous fashion with similar graph-parameters such a regularity in-degree, and maximum and regularity degrees.

The lower bound of $e_{G,\text{in}}(u)$ time-steps for vertex u to collect values/information from all the vertices in $\Gamma^*_{G,\text{in}}(u)$ is stated in accordance with the min-max distributed framework as:

$$e_{G,\text{in}}(u) = 1 + \min\{\max\{\underbrace{e_{T_i,\text{in}}(\text{root}(T_i))}_{= \text{ depth}(T_i)} \mid 1 \le i \le n\} \mid$$

$\{T_i\}_{i=1}^n$ is a family of directed trees that are:
(1) a vertex-decomposition of $\Gamma^*_{G,\text{in}}(u) - \{u\}$, and
(2) rooted in (as subset of) $\Gamma_{G,\text{in}}(u)\}$.

However, it is not necessary to compute $e_{G,\text{in}}(u)$, directly or indirectly, through an underlying optimal directed forest whose maximum depth yielding $e_{G,\text{in}}(u)$. Instead, with a given/fixed topology of directed forest (described in above fashion) underlying a distributed iteration scheme, we can obtain a stronger lower bound with a distributed computation of the maximum depth D_{max} among all the directed trees in the forest, and note that:

number of time-steps for vertex u to access values/information from all vertices in the given directed forest $\ge 1 + D_{\text{max}} \ge e_{G,\text{in}}(u)$.

In addition to the probabilistic upper-bound result on the number of time-steps for (general) distributed function computation via linear iterative schemes with random weight-matrix, Sundaram and Hadjicostis [7,8] employ observability theory of linear systems to study the linear-functional case for distributed computation (of linear functions), and achieve an upper bound via the minimal polynomial of the underlying weight-matrix.

Toulouse and Minh [10] study the linear functional case with prescribed time-invariant network-topology over random weight-matrices, and obtain various empirical upper-bound results.

References

1. Ayaso, O., Shah, D., Dahleh, M.A.: Information theoretic bounds for distributed computation over networks of point-to-point channels. IEEE Trans. Inf. Theor. **56**(12), 6020–6039 (2010)
2. Bondy, J.A., Murty, U.S.R.: Graph Theory, Volume 244 of Graduate Texts in Mathematics. Springer, London (2008)
3. Fich, F.E., Ruppert, E.: Hundreds of impossibility results for distributed computing. Distrib. Comput. **16**(2–3), 121–163 (2003)
4. Hendrickx, J.M., Olshevsky, A., Tsitsiklis, J.N.: Distributed anonymous discrete function computation. IEEE Trans. Autom. Control **56**(10), 2276–2289 (2011)
5. Kashyap, A., Basar, T., Srikant, R.: Quantized consensus. Automatica **43**(7), 1192–1203 (2007)
6. Katz, G., Piantanida, P., Debbah, M.: Collaborative distributed hypothesis testing. Computing Research Repository, abs/1604.01292 (2016)
7. Sundaram, S.: Linear iterative strategies for information dissemination and processing in distributed systems. Ph.D. thesis, University of Illinois at Urbana-Champaign (2009)
8. Sundaram, S., Hadjicostis, C.N.: Distributed function calculation and consensus using linear iterative strategies. IEEE J. Sel. Areas Commun. **26**(4), 650–660 (2008)
9. Sundaram, S., Hadjicostis, C.N.: Distributed function calculation via linear iterative strategies in the presence of malicious agents. IEEE Trans. Autom. Control **56**(7), 1495–1508 (2011)
10. Toulouse, M., Minh, B.Q.: Private communication (2018)
11. Toulouse, M., Minh, B.Q.: Applicability and resilience of a linear encoding scheme for computing consensus. In: Muñoz, V.M., Wills, G., Walters, R.J., Firouzi, F., Chang, V. (eds.) Proceedings of the Third International Conference on Internet of Things, Big Data and Security, IoTBDS 2018, Funchal, Madeira, Portugal, 19–21 March 2018, pp. 173–184. SciTePress (2018)

Teleoperation System for a Four-Dof Robot: Commands with Data Glove and Web Page

Juan Guillermo Palacio Cano[1], Octavio José Salcedo Parra[1,2(✉)], and Miguel J. Espitia R.[2]

[1] Department of Systems and Industrial Engineering, Faculty of Engineering, Universidad Nacional de Colombia, Bogotá D.C., Colombia
{jgpalacioc,ojsalcedop}@unal.edu.co
[2] Faculty of Engineering, Intelligent Internet Research Group, Universidad Distrital "Francisco José de Caldas", Bogotá D.C., Colombia
{osalcedo,mespitiar}@udistrital.edu.co

Abstract. In this paper is consigned the process of implementation and validation of a teleoperation system for a robot, through a sensorial data collector glove of the hands movements. Nowadays, there are various glove type dispositives with the task of register faithfully the movements of the hand articulations, but they are very high price. The proposed glove-web page is designed on a very low budget, and although is not thought to match the performance and functioning of the commercial gloves, we intend being able of easily replacing it. On one side, the glove is based on Flex resistive sensors of low cost, located on key anatomy points of the hand, in a way that it takes register of the DOF (Degrees of freedom) of it. On the other side, the web page is a simple HTML interface host by the Arduino controller board. In this paper it is shown, finally, an application of this kind over a PhantomX Pincher robot, where it is easy verified the utility of a data glove, by implementing an Ethernet communication between the controlling board and the PC commander unit. Finally, we obtained a satisfactory teleoperation system and more important, a very low-cost one compare to its pairs. From which it is remarkable the utilization of easy access dispositives, open-source software and the adequation of a preexisting glove system.

Keywords: Command and data glove · Serial communication
PhantomX Pincher · Degrees of freedom · Matlab 2016a · Arduino
Internet of things · Teleoperation of robots

1 Introduction

The field of machines teleoperation, especially robots, is particularly important in the industry. In certain complex occasions, the robot is not capable of making decisions, thus it needs an operator's assistance. This operator, usually due to security issues, must handle the robot remotely, that is what teleoperation is all about. In this order of ideas, the challenges for the teleoperation are, overall, about two main aspects: the performance of the communication and mapping dispositives, and the correct sensing of the operator's actions. Naturally, the comfort of the operator, in the moment of handling the robot, is

© Springer Nature Switzerland AG 2018
T. K. Dang et al. (Eds.): FDSE 2018, LNCS 11251, pp. 385–404, 2018.
https://doi.org/10.1007/978-3-030-03192-3_29

a key characteristic in the functioning of the teleoperation system. One alternative already used, and that has been proved as a transparent mean to transmit the orders of the operator are the data gloves. Given the innate facility of the hands movements and the possibilities of commands and sensing variables that are possible, make of the hand an excellent choice for teleoperation systems.

Given the current state of the teleoperation systems of robots, particularly those based on sensorial data acquisition gloves, it is proposed, in this work, an obvious solution for the teleoperation field: create a practical system that achieves a similar performance, even without equating, of the systems mentioned.

2   Background

There are already several works and investigations in this subject. In [1] it is implemented a data collector glove with magnetic and inertial sensors to handle a seven DOF robot via bluetooth. There are also applications for more simple robots, such as [2] where it is established a communication based in pre-fixed commands. Alternatively [8] where the similarities of the fingers to the Gripper systems are taking advantage of. The applications are not only about handling robots, but also about virtual reality environments and communication with more abstract entities [3], such as it is done in [8], creating a graphic interface and a basic signal language between the operator and the robot.

A great part of this type of gloves work with flex resistor sensors, which allows knowing the position of an articulation by relating it with a variable resistance. However, there are different alternatives, such as the flex cable sensors, the case of [4] where a set of cables bound to specific parts of the hand sense the changes in it. In [5] is implemented a vital concept of the security of teleoperation systems: the software restrictions and optimal mapping solutions. Generally, these alternatives are implemented on a big cost, 10000 USD. For example, the higher-level applications of teleoperation data gloves are analyzed by [6], emphasizing over the concept of teaching by demonstration, and comparing these systems to traditional joystick ones. The most famous and commercial data glove is the CyberGlove, an optimal solution, but a very high cost one (30000 USD), it is used in [7] for establishing a RT-Linux communication network with a robot. A couple of works focus in optimizing even more the data gloves, in [9] the vibrotactile sensors in the fingertips complement the feedback loop, and improve the performance. Systems further more advanced are added to the glove systems, like the suit of [10], which seeks controlling two robotic arms.

In a more detailed review of the consulted sources, we can give the follow information:

In [1] Glove YoBu, implements magnetic and inertial low cost sensors. Communication via bluetooth. Thought for applications in the space, ocean and environments with visual restrictions. Calibration problems just above the articulations. Applied over a seven DOF and a four one hand. Teleoperation performed with the fingers.

In [2] the control of a board robot via bluetooth communication, using five special commands, related to five different actions. Use of CdS optic pair sensors and one LED located in the thumbs, by joining the thumb to the other fingers different combinations are possible (commands). It does not measure the hands degrees of freedom neither the movement of the fingers, only the union of them.

In [3] glove designed for virtual reality applications. A computer receives the commands via an Arduino Uno (control board). Nine degrees of freedom of the hand are registered. With CAD software, the hand is modeled in 3D, and it is related to a virtual movement depending on the sensing actions. Accelerometers play the role of position sensors.

In [4] light glove that integrates stretching cables and change the value of a potentiometer. Each finger is attached to a cable. Two applications are implemented: an interactive app with virtual reality and handling a robotic arm. It is proved that the glove is the best option for teleoperation, because it can perform the most delicate sensing actions.

In [5] two gloves and magnetic trackers to control two industrial robots. More than a teleoperation system, it is implemented a complete ROS (Robot Operating System). The movements of the hand must be map to the robot, implementing a Joint-to-Joint mapping in each sensor, in this way, each pose of the operator is related to a robot pose. Alternatively, a Point-to-Point mapping, where the robot replicates specific points. Magnetic sensors of 10000 USD are used.

In [6] use of a glove as a tool for teaching by demonstration, a very useful concept in applications where the production line changes constantly. The sensors of the industrial robots record the trajectories performed by the operator with the glove. Finally, a comparison between the performance of a joystick and the glove. The code is realized with the software RobWork. The final glove system is joined to a security supervisor system for the given industrial robots.

In [7] using the famous CyberGlove, some tactile and force sensors to improve the teleoperation action. A correction action is performed over the angle value for each articulation. RT-Linux send the data through the network and a PD control acts on the robot. They give special emphasis in the relation between the remote control and the motion control system. The location of the additional sensors is given through an anatomical analysis of the DIP, PIP and MP articulations.

In [8] control of a three finger Gripper attached to an industrial robot. The main operations are graving and holding. Furthermore, a master-slave relationship is generated for the teleoperation, based on the gestures of the hand. Implementation of a graphic interface and a first approach to a sign language between the robot and the operator. One of the main conclusions is the importance of choosing the right configuration for a given application. In this case, four operations are defined, depending of the size and the form of the object.

In [9] a low-cost sensor glove with vibrotactile feedback and multiple joint and hand motion sensing for human-robot interaction. The common data gloves have various mobility problems and they do not adjust properly to the hands kinetic. It is proposed a light, wire-less and easy to use system. The main objective is to create a low cost glove of more than one DOF, especially oriented to applications of virtual reality and teleoperation. The microcontroller board is an Arduino Mega and the flex commercial sensor are SpectraSymbol. Communication via bluetooth.

In [10] flex resistor sensors to control wirelessly a robotic arm. There are five principal parts: signal detection of the flex sensors, signal conditioning, signal processing with Arduino, transmissions and reception with Xbee series and the Cartia robotic arm. It is proved in two robotic arms, controlling a robotic suit, simulating the actions of an operator.

In [11] flex stretchable sensors, implemented by bounding cables to potentiometers, similar to joysticks, allowing to evaluate the position of the fingers. A comparison of the performance of both (flex resistor sensors and flex stretchable sensors) take place. The experiments are performed in hinge mechanisms and some fingers. Labview is used for data visualization in real time. Finally, it is cleared the performance of each, and the possible ideal applications for each of them.

3   Proposed Work – Design of the Teleoperation System

In first instance, it is proposed a limit Budget of 300000 COP for the realization of this project.

Being thus, the used elements in it must be accessible and easy to get. For this, it is decided to implement classic flex resistor sensors to measure only three articulations of the hand (given that the goal is to control a four DOF robot). The data will be registered and the control will be perform by a Arduino microcontroller board. On the other hand, Matlab will be the tool to configure a PC as a command entity and a step between the hands data and the commands sending to the Pincher robot. At last, the communication between entities will be serial, USB and via web page.

3.1   Electronic Design

Two alternatives are proposed for the resistor sensors, one cheaper than the other is, but less efficient.

3.1.1   Homemade Flex Sensors
For its making a couple of videos available in the web (shown in the references) are took as a base for the physical construction of the low cost flex resistor sensors. The main materials for its construction are metallic paper, carbon mine, paperboard, insulation tape and wires.

3.1.2   Commercial SpectraSymbol Flex Sensors
These sensors are commercially distributed with a selling value of 30000COP for unit, following the low cost guideline of this project. They present a very high performance versus the homemade sensor, besides; their datasheet is of very easy access. These kinds of sensors are finished dispositive, to those it must only be attach a signal conditioning stage for a correct functioning.

3.2 Mechanical Design

Since the glove is not intended to be subjected to shock, stress or manipulation of other objects, it is not necessary to mechanically design the glove. This means that here it is not important to create a robust glove. Thus, knowing the ease with which flex sensors can be incorporated into a glove-type configuration, any glovetype device already built can be used with the sensors.

3.3 Kinematics Considerations of the Robot

For this application, it is decided to characterize the robot solving the problem of its inverse kinematics. That is to say, by giving a goal position, thanks to the geometry of the joints, one or more possible configurations of arrival at this point are generated. By solving the inverse kinematics, we will work together with the direct kinematics to simulate and concretely realize the movements of the robot. Matlab is used with a tool set developed by Peter Corke and the libraries of Dynamixel (developer of the Pincher Phantom robots) (Fig. 1).

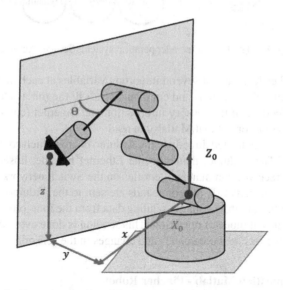

Fig. 1. Input parameters of the inverse kinematics. Source: Authors

3.4 Communications Design

The main connection in the data glove system is between the Arduino card and the PC. Largely because the other connections are between relatively close and simple entities. The only communication to configure is between the PC and the card, so it is proposed a possible solution for this communication, through an Ethernet module for Arduino. With this, we seek to simulate a connection over a wired network. In other words, the

Arduino will be one more entity of the network in question. This Ethernet module implements the TCP/IP protocol (Fig. 2).

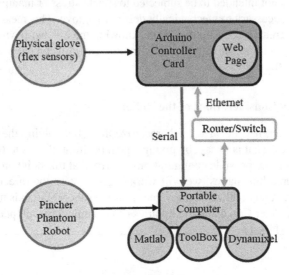

Fig. 2. Data flow in the teleoperation system. Source: Authors

The robot makes feedback of several important variables at each moment, resulting in quite simple control algorithms and command sends to the robot. The only problem to solve is the encoding of the sensory information and Ethernet (commands) to valid information in a serial format that Matlab can read.

The data glove is connected directly to the Arduino board, which constantly registers finger deflection. The Arduino, thanks to the Ethernet module, hosts and shares an HTML page that receives commands of entities on the Switch network. On the laptop, connected to the same network, the commands are sent to the Arduino, which communicates serially with Matlab, constantly sending data from the four joints, opening of the Gripper and the current mode of operation. This sending is done every 0.5 s, being able to act in front of the received commands and changes in the sensors.

3.4.1   Communication Matlab - Pincher Robot

This communication stage is done through libraries of Dynamixel for Matlab. It should be clarified that this link is serial and therefore not simultaneous on all engines of the robot. For example, if a command is sent to the motors, it will be multiplexed one by one. Figure 3 shows the connection diagram for this communication stage.

It is enough to create a script in Matlab that initiates the connection with the Pincher, followed by certain aspects of initial configuration of work (speed, torque, etc.). The fundamental aspect of this communication is that it receives five values from 0 to 1023, each representing a position value for each motor.

```
#include <SPI.h>
#include <Ethernet.h>

// MAC address (since the connection is local the MAC is suposed)
byte mac[] = { 0xDE, 0xAD, 0xBE, 0xEF, 0xFE, 0xED };
IPAddress ip(192, 168, 0, 30); // IP address.
IPAddress gateway(192, 168, 0, 1);
IPAddress subnet(255, 255, 255, 0);


EthernetServer server(80);   // create a server at port 80
```

Fig. 3. Configuration of Arduino network entity. Source: Authors

3.4.2 Ethernet Communication Arduino - Matlab

To configure the Ethernet Shield module, which will communicate the controller card and the computer, an initialization process must be followed in the Arduino programming environment.

First a LAN is created housing the computer and other entities (in this case the Arduino card). The Arduino is assigned an IP address, available within the network created, and a MAC address, composed of 48 bits.

We proceed to explain each part of this initialization:

Using Arduino's SPI and Ethernet libraries: To complete the communication process, some libraries already worked and developed by Arduino are used. These implement the Ethernet protocols, and create methods of sending, configuration, opening, client status, etc. The application of the Arduino in this case will be to act as a web server that will host a page and handle the orders of a client that arrives at it. Port 80 is generally responsible for such actions.

MAC Address: The MAC address can be arbitrary in an application like this. Generally the first 24 bits (OUI Organizationally Unique Identifier) correspond to the associated manufacturer and the general product, the remaining 24 (NIC Network Interface Controler Specific) are associated with the specific devices of that product. A communication is implemented with a single receiving NIC, or unicast. A locally managed address is given, which is indicated by setting the second least significant bit of the first bit octet (1 local address). This causes the remaining bits in the OUI to be 0.

IP Address: corroborating pool of available addresses by the Router, or simply analyzing the network created we give a valid IP address to the Arduino (Fig. 4).

HTML Website: The controller card houses a web page in basic HTML language. It generates a user-friendly interface, composed of text labels, a checkbox to choose the mode of operation and submit buttons to enter values. The appearance of this page is denoted in Fig. 5. When Arduino does not meet the request of a client from its web server, it will be constantly sending a set of 16 numbers. They represent the values of the five robot actuators (three digits each) and a boolean representing the mode of operation.

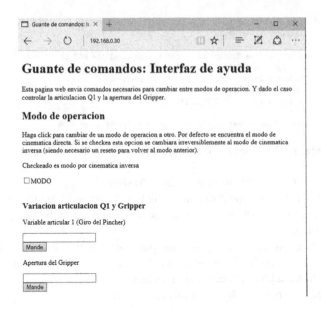

Fig. 4. Web page of the teleoperation system for the Pincher Robot. Control of the operating mode, articulation one and opening of the Gripper. Source: Authors

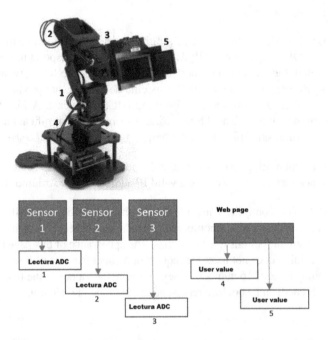

Fig. 5. Operation diagram of the free operating mode. Source: Authors

3.5 Design of the Manipulation Modes

Thinking about the future use of this device to teach some basic principles of robotics, two modes of use are generated, analogous to the two basic ways of moving a robot.

Free mode (direct kinematic movement): each sensor of the glove will represent a value in degrees of the robot's main rotational joints. This means that, given certain ranges of motion, each finger will move the motor two, three and four of the robot. Setting a position by varying the values of the joint variables. Stop and security parameters must be implemented by software, in a way that they do not allow dangerous movements in the workspace, nor actions that damage the robot. The value of joint one and of the opening of the final effector is configured by the web page.

This working mode interacts in real time with the user of the glove. The three resistive sensors will represent the three most representative joints of the robot, while the parameters of the other two engines are varied on the web page with a submit button (represent the joint one and the Gripper). In this mode, the serial message decrypted in Matlab is sent directly.

Target positions mode (reverse kinematic movement): several target positions previously set in a database are proposed. The user, through the glove, will give activation commands for the routines that fix trajectories to these points. Finally, when the robot reaches the position, the user can control the closing or opening of the end effector with the web page (Fig. 6).

When this mode is activated, the reading of the sensors and the opening of the end effector will be the only ones taken into account. The sensors in the glove determine which

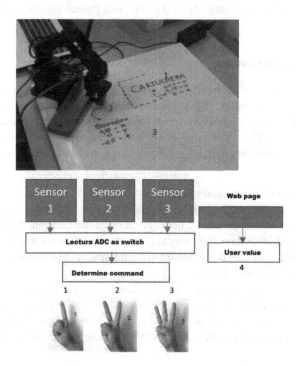

Fig. 6. Operating mode diagram of the operation mode by commands. Source: Authors.

command is being made at that moment; this operation is very simple since the flex sensors arrive at the end in deflection, acting like switches. For this reason, identifying one command from the other is relatively easy. Once it is clear which command was performed, the system executes a unique path related to that command, stored inside the Matlab code. When the robot reaches the desired point, the user will have a waiting time to handle the opening of the Gripper and grab the object desired (Fig. 7).

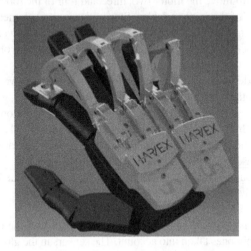

Fig. 7. Harvex Exoskeleton. Source: Applied Engineering Project Reports, National University of Colombia.

3.6 Glove Characterization

In this case, and following the indications given in the design section, it will be used an already built glove configuration. It is the Harvex Exoskeleton (Haptic Augmented Reality Visualization Exoskeleton), currently on development at the Universidad Nacional de Colombia (See Fig. 8). This glove incorporates three resistive flex sensors and several force sensors, and will be very easy to accommodate to our teleoperation application.

3.6.1 Validation of Homemade Sensors

A test glove is implemented with the homemade sensors, and the relationship defined above is evaluated. These sensors are not very robust, and a constant deflection worsens their behavior, in addition there is no way to guarantee that they present similar behaviors to each other, spite of having been created in a similar way. The characterization of one of the sensors is shown in Table 1.

As can be seen Table 1, the range of use of these sensors is extremely small (12 ADC value) allowing many errors in a measurement from 0 to 90°. Clearly the sensor does not fit the desired teleoperation application. Mainly because its operation is similar to that of a switch, only when completely doubled it presents a significant variation in its value.

Sensors	
Commercial sensor 1	
Angle	*ADC*
0	*642*
0,9	*664*
17,7	*670*
27,11	*685*
41,2	*696*
56,6	*730*
72	*737*
Commercial sensor 2	
Angle	*ADC*
0	*630*
10,8	*650*
25,6	*679*
42,4	*707*
58,6	*725*
74	*742*

Fig. 8. Characterization of commercial sensors in the Harvex Exoskelexton

Table 1. Homemade sensor characterization

Angle	ADC2
0	1010
30	1012
45	1009
90	1022

3.6.2   Validation of SpectraSymbol Flex Sensors

As for commercial sensors, a similar experiment is performed with the Harvex exoskeleton. This test is done in conjunction with the authors of this project. The value of ADC is measured as in the homemade sensors test, the angle on the other hand is measured by means of photographic analysis performed during the test. Next, the results are presented for two of the sensors, for continuous variations of the position angle.

It is clear that these sensors are much more reliable and give greater advantages to the teleoperation application than the homemade sensors. Its behavior has a much greater range of operation (approximately 100 ADC), and its behavior is almost completely linear (correlation coefficient of 0.98). Figures 9 and 10 show two configurations that exemplify the experiment.

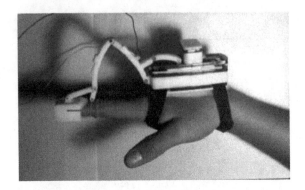

Fig. 9. Harvex exoskeleton experimental unit, in a 0° position. Source: [3].

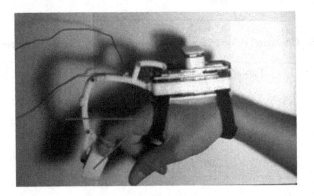

Fig. 10. Harvex exoskeleton experimental unit, in a 42.4° position. Source: [3].

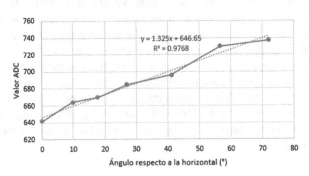

Fig. 11. Experimental graph of commercial sensor 1 in the Harvex exoskeleton. (Deflection angle vs. ADC value). Source: Authors

Therefore, it is decided to continue this project only with SpectraSymbol Flex resistor sensors. In addition, work with the Harvex exoskeleton (Figs. 11 and 12).

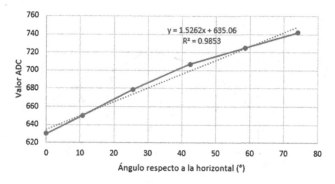

Fig. 12. Experimental graph of commercial sensor 2 in the Harvex exoskeleton. (Deflection angle vs. ADC value). Source: Authors

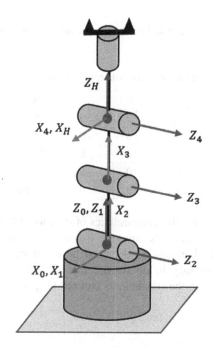

Fig. 13. Assignment of link frames in the Pincher. Source: Authors.

3.7 Implementation of the Kinematic Model

As described above, a classical procedure is followed to obtain the forward and reverse kinematic model, which will subsequently be related to the sensor glove commands. Craig [12] describes this procedure in the book Robotics. Following the regular robotic work convention, frames are assigned to each link in the serial chain, and the Denavit-Hartenberg link parameters are obtained. Figure 14 shows the location of the frames and then the table DH that represents the Pincher robot (Table 2).

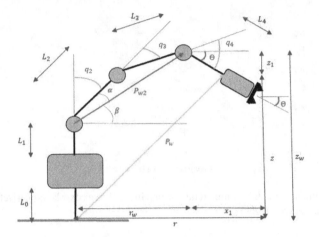

Fig. 14. Geometric model for the inverse solution. Source: Authors

Table 2. Modified DH parameters that represents the Pincher Phantom Robot

i	θ_i	d_i	$a_i - 1$	$a_i - 1$
1	q_1	0	0	0
2	q_2	0	0	$\pi/2$
3	q_3	0	L_3	0
4	q_4	0	L_4	0
Gripper	q_5	0	0	$-\pi/2$

With this development the direct kinematics of the robot has been described, that is to say, by varying the joint variables, any point in the working space can be reached. As a next step, the inverse kinematic model must be developed. Figure 14 shows the geometric modeling necessary to obtain the inverse relationships (the process of obtaining them is detailed in the attachments section) (Fig. 13).

$$\theta_3 = a \, \tan 2\left(\sin \theta_3, \cos \theta_3\right) \tag{1}$$

$$\theta_2 = \frac{\pi}{2} - (\beta + \alpha) \tag{2}$$

$$\theta_1 = a \, \tan 2 \, (y, x) \tag{3}$$

$$\theta_4 = \frac{\pi}{2} - \theta_1 - \theta_2 - \theta_3 \tag{4}$$

It should be clarified that to implement this model successfully, points of the desired trajectory must be interpolated.

4 Simulation and Analysis

4.1 Simulation Work in the Communication System

Before trying the proposed work in the real robot, it must be performed some simulation tasks. The first one is corroborating that the IP address assigned to the Arduino card is the IP address of the device. For this, a simple ping action is realized between the Arduino card and the PC. In the Fig. 15 is shown this corroboration.

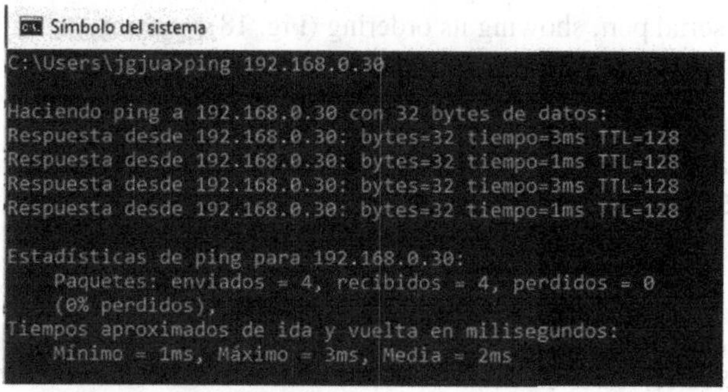

Fig. 15. Verification of the connection to the Arduino entity by ping. Source: Authors.

```
GET /?MODO=2 HTTP/1.1
Accept: text/html, application/xhtml+xml, image/jxr, */*
Referer: http://192.168.0.30/
Accept-Language: es-MX,es-ES;q=0.8,es;q=0.6,fr-FR;q=0.4,fr;q=0.2
User-Agent: Mozilla/5.0 (Windows NT 10.0; Win64; x64) AppleWebKit/537.36
Accept-Encoding: gzip, deflate
Host: 192.168.0.30
Connection: Keep-Alive

GET /Q1?art1=0555 HTTP/1.1
Accept: text/html, application/xhtml+xml, image/jxr, */*
Referer: http://192.168.0.30/
Accept-Language: es-MX,es-ES;q=0.8,es;q=0.6,fr-FR;q=0.4,fr;q=0.2
User-Agent: Mozilla/5.0 (Windows NT 10.0; Win64; x64) AppleWebKit/537.36
Accept-Encoding: gzip, deflate
Host: 192.168.0.30
Connection: Keep-Alive

GET /GR?gril=0989 HTTP/1.1
Accept: text/html, application/xhtml+xml, image/jxr, */*
Referer: http://192.168.0.30/
Accept-Language: es-MX,es-ES;q=0.8,es;q=0.6,fr-FR;q=0.4,fr;q=0.2
User-Agent: Mozilla/5.0 (Windows NT 10.0; Win64; x64) AppleWebKit/537.36
Accept-Encoding: gzip, deflate
Host: 192.168.0.30
Connection: Keep-Alive
```

Fig. 16. Possible types of requests by the web server client. Source: Authors

The handling of commands is done through an analysis of the characteristic 'request' message of HTTP. Every time there is an action of the client, different requests and therefore different actions in the serial communication are handled. Figure 16 shows the possible HTTP client requests in the created application. The program will constantly check in the 'loop' section if there is a client with a request. We can see that the HTTP actions work ideally for this relatively simple system. The change in the request from one work mode to the other is done correctly and the values of each articulation are sent in a correct way.

Knowing that the articulation values and the work mode are given by a serial communication, and that these represents the most important variables of the system, it must be done a simulation that test this exchange. In Fig. 17, and with the help of a serial visualizer, the type of command and the articulation values can be obtained from the request message. In the following figures are shown portions of the data sent (every half second) by the serial port, showing its ordering (Fig. 18).

```
50072072472750005007207277275000!
```

Fig. 17. Half second serial port capture. Value of q1 = 500, q2 = 720, q3 = 724, q4 = 727, q5 = 500, Modo = 0. Source: Authors.

```
55572272972998915557227277299891!
```

Fig. 18. Half second serial port capture. Value of q1 = 555, q2 = 722, q3 = 729, q4 = 729, q5 = 989, Modo = 1. Source: Authors.

Finally, to connect the two communications previously made, we make a simple Matlab script that reads the information (from the Ethernet communication with Arduino) and do some action on it by invoking the methods of the Dynamixel library. Thus, the data received from the Arduino influences the commands given to the Pincher robot. When the port is open and the information is read, the system decides both the mode of operation and the values that will be sent to the Pincher (Fig. 19).

```
Q =

   500   724   724   731   500     1

Q =

   555   720   718   727   500     1

Q =

   555   722   722   729   500     1
```

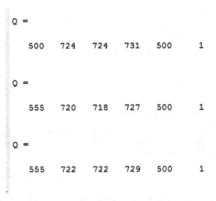

Fig. 19. Vector in constant change with the information coming from the serial port. Source: Authors.

4.2   Results Obtained in Each Work Mode

The work area consists of an anchor board, the data glove, sensor circuitry, Arduino with Ethernet module, PC and network connections to the Switch. It should be noted that each time the serial port is disconnected or shut down, the Arduino is restarted.

4.2.1   Free Mode (Direct Kinematic Movement)

With the half-second delay between sending and given the asynchronous communication of the robot, the sensation of movement or teleoperation in real time is limited, but quite good (Fig. 20).

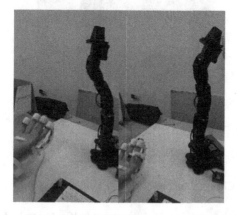

Fig. 20. Sequence of change of the joint two (related to the index finger) in the Pincher robot. Source: Authors.

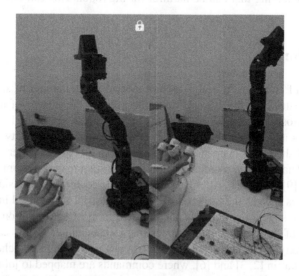

Fig. 21. Sequence of change of the joint three in the Pincher robot (related to the middle finger). Source: Authors.

Fast movements of the fingers in the glove can generate confusion in the decision and sending system to the robot. For this reason at the time of handling the glove, the movements must be made slowly and stably, in this way the change of position value in each engine will be much more evident (Figs. 21 and 22).

Fig. 22. Change joint one (web page command). Source: Authors.

This type of operation receives values every half a second. If an error is made in sending the command of the web page, this will cause a total desynchronization of the serial buffer, something that can be harmful for the robot. This can only be fix with a system reset.

5   Conclusions

The first point of comparison in this type of system is the number of samples; this is because it indicates how the systems take into account the actions of the user in the system itself. Most applications, such as [5] and [8], have a very high sampling rate, respectively 144 and 60 samples per second. Higher than the rate of two samples per shipment established. This happens mainly because, in these works, there are additional control loops. That when implemented require a constant response from the sensor, as in the case of [9] and [11]. Alternatively, also due to the high sensitivity of more advanced devices such as the CyberGlove or the industrial robots used in [5] and [6].

Although the size of the buffer is not specified in most teleoperation jobs, this is a good indicator of the type of data transfer that is happening. In this case, given the sending synchrony, the data controls the state of each union in the Pincher. Something similar must occur in [2, 3] and [6], where commands are mapped to joint values.

The sending speed of 9600 baud is lower than that registered in the references, whose value is usually 115200 baud. In this project, it would be a computational loss to impose such a high delivery rate, when only a 120-bit buffer is sent sporadically every 0.5 s. In fact, the 115200 baud communication was tested, obtaining the same results.

Several developers prefer to use Bluetooth wireless communication, such as [1] and [2], due to its ease of operation and connection advantages. No work was contrasted, except that [8] implements complementary web interfaces to the teleoperation system such as the present one in this work. This means handling additional protocols such as HTTP (TCP IP).

Deepening the suitability of the command glove used. A satisfactory mapping of the movement of the fingers to the movement of the two and three articulations of the robot is determined. The work [8] presents a ratio of 0.37° per finger change to achieve a change in the opening of a clamp. In this case, the 1.6° ratio is more than sufficient for the average human perception and the average movement of the fine finger.

Finally, the experiments on the teleoperation system itself, derived from those carried out in [5] and [6], were evaluated, but on a very small scale (only two test subjects). The success rate of the system is very high, probably due to this factor, but it also shows the relative simplicity of the robot's handling through this medium. In [5] a success rate lower than 77.08% was recorded, considering that it was performed in 20 test subjects. On the other hand, in [6] the same test is performed, but with subjects from different contexts, that is, alien to the robot's mapping area.

References

1. Fang, B., Guo, D., Sun, F., Liu, H., Wu, Y.: A robotic hand-arm teleoperation system using human arm/hand with a novel data glove, Zhuhai, China, December 2015
2. Jaemyung, R., Yeongyun, K., Wang, H., Hun, K.: Wireless control of a board robot using a sensing glove, Kuala Lumpur, Malaysia, November 2014
3. Pereira, L.C., Aroca, R., Dantas, R.: FlexDGlove: a low cost dataglove with virtual hand simulator for virtual applications, Natal, Brasil (2013)
4. Yeongyu, P., Jeongsoo, L., Joonbum, B.: Development of a wearable sensing glove for measuring the motion of fingers using linear potentiometers and flexible wires, February 2015
5. Rosell, J., Suárez, R., Pérez, A.: Safe teleoperation of a dual hand-arm robotic system. In: Armada, M.A., Sanfeliu, A., Ferre, M. (eds.) ROBOT2013: First Iberian Robotics Conference. AISC, vol. 253, pp. 615–630. Springer, Cham (2014). https://doi.org/10.1007/978-3-319-03653-3_44
6. Kuklinski, K., Fischer, K., Marhenke, I., Kirstein, F., aus der Wieschen, M.: Teleoperation for learning by demonstration: data glove versus object manipulation for intuitive robot control, Denmark (2014)
7. Kobayashi, F., et al.: Multiple Joints reference for robot finger control in robot teleoperation, Fukuoka, Japan, December 2012
8. Tabassum, M., Ray, D.D.: Intuitive control of three fingers robotic Gripper with a Data hand glove. In: 2013 International Conference on Control, Automation, Robotics and Embedded Systems (CARE), Jabalpur, pp. 1–6 (2013)
9. Weber, P., Rueckert, E., Calandra. R., Peters, J., Beckerle, P.: A low-cost sensor glove with vibrotactile feedback and multiple joint and hand motion sensing for human-robot interaction, Columbia, United States of America, August 2016
10. Pititeeraphab, Y., Sangworasil, M.: Design and construction of system to control the movement of the robot arm, Pathumthani, Thailand (2015)

11. Sbernini, L., Pallotti, A., Saggio, G.: Evaluation of a Stretch Sensor for its inedited application in tracking hand finger movements. In: 2016 IEEE International Symposium on Medical Measurements and Applications (MeMeA), Benevento, pp. 1–6 (2016)
12. Craig, J.: Robótica, México (2006)

Design of PHD Solution Based on HL7 and IoT

Sabrina Suárez Arrieta[1], Octavio José Salcedo Parra[1,2(✉)],
and Roberto Manuel Poveda Chaves[1]

[1] Faculty of Engineering, Intelligent Internet Research Group,
Universidad Distrital "Francisco José de Caldas", Bogotá D.C., Colombia
ssuareza@correo.udistrital.edu.co,
{osalcedo,rpoveda}@udistrital.edu.co
[2] Department of Systems and Industrial Engineering, Faculty of Engineering,
Universidad Nacional de Colombia, Bogotá D.C., Colombia
ojsalcedop@unal.edu.co

Abstract. The communication architecture for different data management platforms between healthcare organizations through HL7 standards with the introduction of an application for manipulating information obtained from a device for recording vital signs in different programs.

Keywords: HL7 · PHD · IoT · Solution

1 Introduction

Different health entities in the country (especially EPS) currently provide a PHD home hospitalization service which is provided to different types of patients, which usually belong to one of the following categories; patients with sequelae of cerebrovascular accident, sequelae of spinal cord injuries or senile degenerative diseases; Chronic diseases such as COPD, hypertension, degenerative arthritis, diabetes, AIDS or cancer; Pre-established pathologies that require the administration of medications intravenously or administration of parenteral nutrition; Complicated wounds, open abdomens in the process of healing; Articular replacements that require intravenous antibiotic management and pain management; Scheduled post-surgical patients that do not require permanent hospitalization [1].

Although the services are quite complete, the patients do not have a record of their vital signs between the visits of the specialist staff, and it could be through a tool that allows the constant monitoring of their general health to expand the types of people to which can be provided with the PHD service; improving general attention through early detection of possible attacks and/or symptoms.

HL7 (Health Level-7) is a set of protocols that allows to establish, through flexible standards, lines of work and methodologies, an effective and safe communication between different health entities, in addition, guarantees that the information that each health system it is transparent to the other [2]. In the work presented in this paper, the interoperability capabilities of the standard are used, proposing the design of an HL7 interface, which allows the management systems of the application developed during this paper to interoperate, for the management of the information collected in the

© Springer Nature Switzerland AG 2018
T. K. Dang et al. (Eds.): FDSE 2018, LNCS 11251, pp. 405–409, 2018.
https://doi.org/10.1007/978-3-030-03192-3_30

provision of PHD services, with the current data management systems used in IPS (Providers of Health Services), EPS (Health Promoting Entities) and ARS (Subsidized Regime Administrators).

Another technology involved in the development of the project is the "IoT"; According to the Internet Business Solutions Group (IBSG) of Cisco, it is estimated that the term IoT "was born" sometime between 2008 and 2009, the concept was simple but powerful: if all the objects of our daily life were equipped with identifiers and wireless connectivity, these objects could communicate with others and be managed by computers, this way you could follow and tell everything, and greatly reduce waste, losses and costs. Provide computers to perceive the world and get all the information we need to make decisions [3]. Through this concept, the collection of the information provided by the constant readings of vital signs is achieved by means of a device that allows the measurement of these, for the patients who require it due to their corresponding condition, allowing a system of notifications both for the patient, as for the health professional in charge of the follow-up and attention of the case.

2 Background

The Internet of Things (IoT) comes from the Internet industry, conceptually is to connect one device to another through a network of data. M2M, for his part, arrives from a universe more technical. In M2M there is no human intervention, that is its principle. The difference between IoT and M2M is that the M2M technology is what brings to the Internet of Things the connectivity you need and without which it would not be possible. M2M technologies can convert almost any thing in a component of the IoT that can be monitored at a distance and allowing interaction via an interface M2M.

2.1 Internet of Things

One of the most well known of the term referring to the Internet of things comes from the first report of the ITU in 2005 [4] where is described as:

"A promise of a interconnected devices world that provide relevant content to users".

That relevant content can be information of a product in a store, the contents of a medicine or the location of a particular device.

2.2 IoT Architectures

Ideally the concept of Internet of Things must be able to offer, from the technical point of view a common framework on which define any solution required in a given application domain, taking into account the need to interoperate with other solutions.

The search for an architecture IoT of reference that support different environments and contexts of the world of things has been a challenge from the same appearance of the concept of IoT.

Below is a review architectures that have been proposed and defined in a specific context and that are a solution to a part of the world of the things. It is taken as reference for the work the architectural project for the Internet of Things (IoT-A) [5] and describes the components and the scope of the reference architecture proposed by IoT-A.

The IoT-A Project made a series of works related to the IoT environments, among which is an assessment of platforms and solutions that have been developed in the world of IoT. These solutions are classified into: (1) solutions provided by public projects such as those of the FP7 of the European Commission and solutions arising from existing commercial products and (2) attempts to normalization of the various standardization organisms [6].

Below are the features that are taken into account for the analysis of the IoT Architectures:

- Description of the Architecture: addressed as the vision of the architecture is provided in relation to the project/product/standard.
- Style of architecture: addresses on the style that governs the structure of the architecture.
- Model and distribution of the information: addresses the issue of how the information is processed by the project/product/standard and how it is distributed in the system and the way in which becomes accessible.
- Horizontality: refers to the ability of the system to the reuse of the same building blocks to provide different functionalities of the top layer. For example, the horizontality applies to a framework for the provision of virtual services for the construction of applications.
- Knowledge of the context and semantic capacities: Refers to the possibility of improving the information exchanged through systems with descriptors, which allow the data to be categorized and perform complex queries to be answered.
- Technological Specification and Interoperability: the projects/Products/standards depend on a particular technology and as focus on interoperability.

3   Proposed Architecture

For the construction of the application, an SOA architecture is proposed, based on the construction of a web service that allows the communication between devices through the World Wide Web, through the transfer of files in XML and JSON format and based on the use of the HTTP, this service goes in favor of machine-to-machine interoperability over a network, looking for any application to be scalable.

The basic actors of the system are represented through 3 main roles: the doctor (can be any other type of health specialist), the patient and an agency or control entity of the health secretary, each of these actors has permission to different services, the control agency of the secretary of health has access to a service of notification of risks which will allow diagnoses of problematic conditions that weigh on a whole population, on the other hand the services to which the patient accesses are the of the consultation of his own clinical history and the registration of his vital signs for the subsequent consignment in his clinical history. The consultation service is exclusively for the role of the doctor, he is the only one with the permissions to carry out a consultation in the web service.

To provide each of these services, business processes are performed, for the registration of a query (Fig. 1), starting from the event that is the consultation itself, verifying the existence of a clinical history (business object) and in if necessary, creation; later there is an aggregation to that clinical history of the symptomatic data, and then it is passed to another process that is the registration of the vital signs.

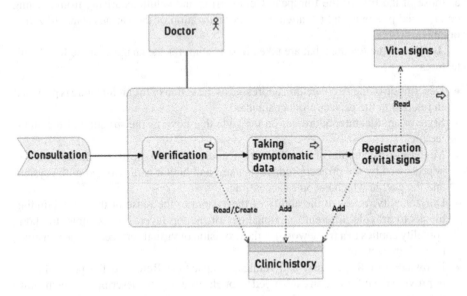

Fig. 1. Business process query record. Source: Authors

The business process of registering vital signs (Fig. 2) begins with the event of electing the option for taking vital signs.

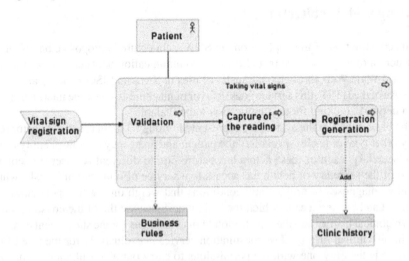

Fig. 2. Business process vital signs registry. Source: Authors

The patient participates in this business process and the first thing that is done is the validation that the business rules are being complied with (patient's condition, number of records taken per day, etc.), once the compliance of the patient is guaranteed. The business rules are made taking the vital signs, and immediately the record is drawn up that will be added to the clinical history.

For the realization of the processes, it is necessary a deployment by logical modules that will be responsible for implementation, in the diagram of structure of the application, we see how the different modules are integrated.

The structure of the application is based on the construction of a web service which is the one that provides the services through the consumption of the information provided by other modules, the vital module Scanner is in charge of the execution of the taking of the vital signs, this module is the one that validates the business rules and later builds the CDA so that it can be consumed by the web service and later stored, the analysis module is the one that has the algorithms for the elaboration of statistics based on logic models diffuse [], which will be the only service to which the control entity of the ministry of health will have access. The last module is the one of the connection with the Mirth Connection database, it is the one that allows us through hl7 protocols to store and later consult the information.

4   Conclusions

Through the study and implementation of multiple architectures, we achieved the design of a sufficiently robust to ensure security and interoperability. Due to its modular pattern, the proposal allows the addition of new functionalities (different from those presented in this document), for the use of information for the generation of new studies and reports, starting from the data provided.

This architecture is scalable to any other type of sensors that wish to be used, as sensor nanos, since it is based on architectures that support the use of them.

The proposal manages to unite concepts and focus on the improvement of medical care currently received by this type of patients, for the prevention and timely treatment of the most fateful symptoms of their disease.

References

1. Hospitalización en casa – FAMICARE, Disponible en: http://famicareclinicadia.com/hospitalizacion-en-casa.html
2. HL7 About us - HL7, Disponible en: http://www.hl7.org/
3. Jara, A.J., Zamora-Izquierdo, M.A., Skarmeta, A.F.: Interconnection for Health and Remote Monitoring Based on the Internet of Things – IEEE
4. ITU: ITU Internet Reports 2005: The Internet of Things (2005)
5. IoT-A: Internet of Things – Architecture (2013). http://www.iot-a.eu/
6. IoT-A (Internet of Things Architecture): Project Deliverable D1.1 - SOTA report on existing integration frameworks/architectures for WSN, RFID and other emerging IoT Related Technologies (2011)

Smart City: Data Analytics and Security

Analysis of Diverse Tourist Information Distributed Across the Internet

Takeshi Tsuchiya[1]([⊠]), Hiroo Hirose[1], Tadashi Miyosawa[1], Tetsuyasu Yamada[1], Hiroaki Sawano[2], and Keiichi Koyanagi[3]

[1] Suwa University of Science, Chino, Nagano, Japan
{tsuchiya,hirose,miyosawa,yamada}@rs.sus.ac.jp
[2] Aichi Institute of Technology, Toyota, Aichi, Japan
sawano@aitech.ac.jp
[3] Waseda University, Kitakyushu, Fukuoka, Japan
keiichi.koyanagi@waseda.jp

Abstract. Herein, we propose and discuss a new method for analyzing various types of tourist information about the Suwa area of Nagano Prefecture, Japan, available on the Internet. This information includes not only long sentences that can be found on web pages and in blogs, but also short sentences comprising a few words posted on social media. In this paper, we propose a novel method based on a neural network, called paragraph vector, for expressing relationships between words included in sentences. Our method achieves high retrieval accuracy even across social media posts comprising just a few words. Based on our evaluation results, the proposed method outperforms the conventional information retrieval technique wherein sufficient accuracy cannot be achieved as it is based on the occurrence probability of words in sentences. This improvement is achieved by using the word order as an input feature to the neural network model.

Keywords: Tourist information · SNS · Paragraph vector

1 Introduction

We have built an online portal that contains regional tourist information about the Suwa area of Nagano Prefecture, Japan, where our organization is located. In this portal, tourist information on sightseeing around the Suwa area is spread via blogs and social network service (SNS). The visitors of the portal can search for information using keywords. This information targets both the tourists who are already aware of and use the portal as well as those who are merely browsing. Thus, we expect to revitalize sightseeing in the Suwa area by increasing the number of tourists and by promoting relevant information over social media through tourist feedback.

The portal provides information regarding various tourist spots, events, and general information on mountains and seasons, as shared by actual tourists.

© Springer Nature Switzerland AG 2018
T. K. Dang et al. (Eds.): FDSE 2018, LNCS 11251, pp. 413–422, 2018.
https://doi.org/10.1007/978-3-030-03192-3_31

Local tourist associations and tour operators help to generate the latest tourist trends, relevant information, and list of keywords based on the interests of their past and potential customers. To help the portal visitors find the required information, developing a mechanism to retrieve relevant content based on the input keywords, then analyzing and expressing that content is necessary.

The currently implemented information retrieval mechanism is based on vectorizing the target content and calculating term frequency-inverse document frequency (TF-IDF) for each search word. TF-IDF is derived from the frequency of a word in each article and its total weight across the entire content. Latent Dirichlet allocation (LDA) [1] is employed to reduce the dimensions based on the appearance probability distribution of the vocabulary and topic estimation of the content. While conventional information such as that posted on web pages and in blogs can be retrieved using TF-IDF, more information is present in SNS posts such as photos and related comments, which contain just a few words. This information in SNS posts is difficult to acquire using the conventional TF-IDF approach owing to the rare appearance of keywords in short comments as compared with entire pages or blogs.

In this article, we propose a method for retrieving relevant tourist information about the Suwa area not only from conventional web content but also from short SNS posts. We discuss two versions of the proposed method: one is based on the word context information for distributed representation of the content using paragraph vector, while the other is based on the similarity of a group of words (i.e., bag of words (BoW)) included in the search query and the content. We then conduct a comparative evaluation of the proposed method against the conventional TF-IDF method using information regarding the national holidays in Japan for May 2017, called the Golden Week (GW), as the target content.

2   Tourist Information About the Suwa Area

In this section, we discuss the characteristics and preprocessing of the regional tourist information collected from the Internet.

2.1   Regional Tourist Information About Suwa

The Suwa area in Nagano Prefecture, Japan, which is the target area of this research, is famous for sightseeing with scenic nature and historical sites such as Suwa-Taisha (the Suwa Shrine), Mt. Yatsuga-take, and hot springs. According to the investigation of the Tourism Bureau of Nagano Prefecture, 70% visitors of the Suwa area are over 40 years old. Therefore, majority online tourist information concerning the Suwa area features in blogs written by experienced tourists and sightseeing operators promoting the area for this older generation. However, the information thus distributed is not very verbose and may not appeal to younger generations. This situation has completely changed after a popular animation film was shot here last year using images photographed around Lake Suwa. The number of posts on SNS by youngsters about the different spots associated with the animated story has suddenly increased. However, retrieving the desired information from these posts, collectively called "pilgrimage to scared places," is difficult owing to the above mentioned problem of post brevity.

Based on the results of a survey conducted by us in March 2015 among 30 tour operators representing hotels of different business scales, the reviews posted on web pages and in blogs do impact the customer attraction rate. Therefore, visualizing tourist experiences and the interests of potential tourists by capturing their behavior patterns and keywords is important to find new customers.

2.2 Collecting the Suwa Area Tourist Information

The target content concerning sightseeing in the Suwa area of the Nagano Prefecture was crawled and collected from the Internet on April 28 and May 8, 2017. These dates represent the day before and the day after the national holiday in Japan, i.e., the GW. We chose some public websites provided by the local tourist association and business operators such as hotels, the Trip Advisor website [2], and representative SNS Twitter and Facebook accounts. The information was collected from these sites within 3 hops from the first targeted page.

To extract information related only to the Suwa area, a dictionary managing the entries of "place names" and "tourist spot names" in the Suwa area was generated in advance to distinguish among the entire online content; the relevance of the target content is decided on the basis of this dictionary. We expect that this approach to improve retrieval accuracy by reducing the noise and calculation cost associated with irrelevant information. In particular, we expect that it would improve the retrieval performance of the proposed method in terms of data retrieval from SNS content such as Twitter posts.

2.3 Preprocessing of Regional Tourist Information

The acquired target content is processed in advance as follows. All words within the target content are extracted and classified using morphological analysis. All words that are likely to become noise in the retrieval service are removed; specifically, the extracted words are classified as verbs, adjectives, and nouns, while numerals are excluded. Each piece of content is quantified (vectorized) using extracted features based on the proposed method. To identify these features, user retrieval queries are quantified (vectorized) in the same manner, and then, the similarities between each query and a piece of content is calculated.

3 Method Using Paragraph Vector

In this section, we propose and discuss our approach for analysis and retrieval of tourist information about the Suwa area.

3.1   Manner Using Paragraph Vector

First, we provide a brief summary of the paragraph vector algorithm and propose our method for analyzing regional tourist information.

Paragraph vector [3] has two models: the PV-DM (paragraph vector with distributed memory) and PV-DBoW (paragraph vector with distributed BoW) models.

The PV-DM model facilitates the prediction of the next word $w(t)$ using a neural network based on a sequence of words indicating the context and topics of the paragraph. The range of this sequence is called the window size parameter. Sequences of words are learned as characteristics of paragraphs and words for their vectorization. Specifically, all target paragraphs are quantified (vectorized) by analyzing the included words and the order of their appearance using a neural network. Consequently, a learning model of these paragraphs is generated, which predicts the next most probable word $w(t)$ that maximizes the inner product with the context window of the paragraph. Then, the word to be predicted is modified by sequentially moving the window, and the same derivation process is repeated.

In this study, we adopt paragraph vector for content analysis through the PV-DM model that enables us to express the content characteristics based on the sequences of words, regardless of the length of the content. The context (sequences of words) and keywords indicating the topic of interest are expected to match even for content containing just a few words such as SNS posts. When the user inputs a query, it is quantified (vectorized) using the above mentioned learning model, and the similarity between the query and the posts' content is calculated.

The window size, which determines the range of the word sequence from a small number of words in each SNS post, can be regulated with a parameter. The value of this parameter will be examined and clarified in the Evaluation section (Sect. 4).

3.2   Method Using Similarity of BoW

Here, we explain the method that uses the similarity of BoW.

In the conventional approach based on word frequencies, the term frequency (tf) indicates the frequency of a word in each piece of content, and the inverse document frequency (idf) indicates the frequency of the word across all target documents. The products of these frequency values represent the features of each word across all target documents. LDA [1] is applied to reduce the number of feature dimensions, which improves the accuracy of the retrieval rate. Using this approach the system semantically combines similar words that are treated as different dimensions. In the case of deriving these values from the content posted on SNS, the value of tf is only 1 for nearly every word within the SNS post. Therefore, the feature values of words are determined by the value of idf in most cases, which makes it difficult to achieve accurate information retrieval. In contrast, our proposed method uses groups of words included in the content and executes the following two steps. First, content similar to that of the query is determined. Second, word frequencies from content with high similarities are calculated. The user query typically comprises a few words indicating the most important features of the content. The similarity calculated between this query and each piece of content determines the target to be retrieved. The similarity being discussed in this article refers to the Jaccard similarity coefficient expressed by Eq. 1 and is derived for the sets X and Y; the value thus derived judges the similarity. The Jaccard coefficient value indicates the ratio of common elements between the two sets (in this case, the query and the words included in each content). If these sets are completely identical, the Jaccard coefficient value is 1; when the sets do not match at all the value is 0. The similarity of these two sets is quantified in the range of 0–1. Herein, the words in each piece of content that correspond to a set indicate each feature; the similarity is judged based on this value.

$$Jaccard = \frac{|X \cap Y|}{|X \cup Y|} \tag{1}$$

The value of the Jaccard coefficient is significantly affected by the number of words in the content owing to the denominator in Eq. 1. Therefore, the denominator value for a SNS post comprising a few words is smaller as compared with that of web pages and blogs that contain numerous words. Even if a SNS post and a blog are both queried using the same words, the content of the SNS post will exhibit higher similarity based on Eq. 1. As both types of sets (contents) contain the same words included in the query, the current implementation of the proposed method adopts loose judgment of the similarity between the sets. This problem is not limited to this paper; it occurs in information retrieval of any content with different lengths. In this case, this problem can be dealt with, in general, by normalizing and judging it by the similarity rate per word. If the content has a larger similarity with the query according to Eq. 1 than the threshold value of α ($\alpha = 0.75$ in the current implementation), then it is classified as a similar candidate content. Frequencies of the words in the query are measured from all similar candidate contents. Then, the cosine similarity is calculated to judge the priority of each piece of content.

4   Evaluation

The proposed method is evaluated using actual data acquired from web pages and SNS posts.

4.1   Content for Evaluation

The content used for this evaluation was acquired on April 28 and May 8, 2017 using the crawling procedure described in Sect. 2.2. The content accumulated on servers is not considered within the scope of this study, and only the data acquired on both mentioned days is used for this evaluation.

Table 1 shows the statistics of the acquired data. The row "Web and blog" contains the number of documents acquired on April 28, 2017 and the number of documents appended thereafter, until May 8, 2017. "SNS" contains the estimated number of unique posts (i.e., multiple appearances such as re-tweets and shares are deleted) acquired on both days. Thus, total content comprising about 10,000 documents was used for this evaluation. "Word Average" denotes the average number of words in the content after preprocessing. By comparing the word averages, it is obvious that the word average of "Web and blog" is larger than that of "SNS".

Table 1. Statistics of the acquired data

	April 28th	May 8th	Word Average
Web & blog	2312	120	137.1
SNS (twitter)	約 3600	約 4600	13.0
Total	5912	4720	114.4

Table 2. Parameters

Environment	Specification
Paragraph vector model	PV-DM, Dim:400, Window size:5 words, learning rate:0.01, iter:200
Word similarity model	Jaccard:0.75
LDA model	tf·idf, topics:70

4.2   Scenario of Evaluation

The performance of the proposed information retrieval method was evaluated using 50 queries that have the same ratios as that of Table 1. Each variation of the proposed method is evaluated from the perspective of retrieval performance using the similarity between the query and the content. We compare our method to the conventional one based on LDA and TF-IDF.

The environment of the evaluation is described in Table 3, which was developed using Python. In particular, the implementation of paragraph vector is based on Doc2Vec developed in gensim [4] for brief topic analysis.

4.3   Evaluation Results

Figure 1 shows the evaluation results of the considered methods' retrieval performance. The x axis indicates the precision rate, and the y axis indicates the recall rate. These lines indicate the average retrieval results over five runs. "PV" is the proposed paragraph vector method, "WS" is the proposed method using the similarity of BoW, and "LDA" refers to the conventional method. Both versions of

Table 3. Evaluation environment

Environment	Specification
Paragraph vector model	gensim [4] Doc2Vec, Python 3.6.2
Word similarity model	Python 3.6.2
LDA model	gensim [4]tfidf, LDA, Python 3.6.2
OS	ubuntu 16.10
System	Docker 17.09

our method (PV and WS) achieve higher accuracy and recall rates as compared to that achieved by the conventional method (LDA). Thus, PV demonstrates better performance as compared to WS.

Table 4 shows the use of the top-five words within three user queries ("Taisha" (shrine), "Onsen"(hotspring), "Seichi" (Pilgrim) + "Eiga" (movie)), which are words related to sightseeing in Suwa. All queries are represented in Japanese characters, and words show the meaning of each query in English.

The table shows the difference of similar words before and after the GW. The results for the word "Taisha" do not indicate the general name of "Suwa-Taisha" (Suwa Shrine), rather the formal name "sha" and the detailed names of the shrine "kami-sha," "Shimo-sha," "Haru-miya," and "Aki-miya" are extracted as semantically similar words.

On the other hand, in the query of "Holy Land" + "movies," the name of the animation and the characters that have become a topic of discussion in recent years, appeared after GW.

5 Discussion

This section discusses and clarifies the characteristics of our proposed method based on the evaluation results.

Table 4. Similar words in queries

date	2017/04/28			2017/05/07		
words	大社 ('tai-sya')	温泉 ('Onsen')	聖地'+ 映画 ('Seichi'+'Eiga')	大社	温泉	聖地'+ 映画
Paragraph Vector	春宮	朝風呂	咲	春宮	湯	アニメ
	下社	足湯	恋人	オーラス	源泉	咲
	上社	活火山	幕開け	下社	旅館	アート
	神米	円野	巡礼	上社	足湯	記事
	祭神	すべすべ	アニメ	男神	風呂	ダンス
Similarity BoW	諏訪	諏訪湖	諏訪湖	諏訪	諏訪湖	諏訪湖
	諏訪湖	蓼科	映画	諏訪湖	勧め	聖地
	観光	宿	聖地	観光	上諏訪	映画
	柱	諏訪	諏訪	宮	ホテル	名
	上社	信州	時間	上社	蓼科	君
LDA	大社	上諏訪	聖地	宮	勧め	君
	宮	そば	兄弟	柱	車山	名
	秋	地	浅間	春宮	車山	聖地
	春宮	食べ	玲	上社	白樺湖	富士見
	下社	酒	君	下社	峰	巡礼

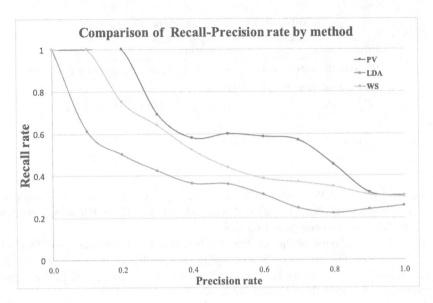

Fig. 1. Comparison of search performance

5.1 Discussion of Results

The evaluation results show that both variations of our proposed method enable the expression of the content's characteristics and features.

The proposed method using paragraph vector calculates the similarity of contents by considering both words and their sequence in the content as features. This variation of our method demonstrated better retrieval performance as compared with the other proposed variation and the conventional method. However, paragraph vector showed unstable results as compared with the other methods (see Fig. 1). We believe that such unstable performance is caused by concentrating on some specific topics in the content. In particular, increasing the number of SNS posts containing just a few words is necessary for feature extraction. Although we included posts from only one day, the performance of paragraph vector is expected to improve with an increase in the considered period. The proposed variation that uses the similarity of BoW calculates the similarity of a group of words included in the content using word co-occurrences. This method exhibited better retrieval performance as compared to the conventional method that uses frequencies of words in the content. This could be attributed to the target content being limited to tourism in the Suwa area. In particular, the words included in each preprocessed content are limited to nouns—such as spots and events—as well as nouns and verbs referring to sightseeing, nature, and seasons. Therefore, there exists a high possibility for matched word co-occurrences in the content. Thus, the second variation of our proposed method is more effective in the case of limiting the target words included in the content to a specific topic as compared to the conventional method. However, it is necessary to execute the two previously mentioned steps (see Sect. 3.2) for similarity calculation in our proposed method.

The similarity is calculated for every query and content, which increases the processing load. Simultaneously, such an approach enables instantaneous content retrieval, especially for SNS posts because previously set models are not required. It would be useful for real systems to filter re-tweets and shares that have the same content as the original post. However, content containing a large number of words such as blogs or web pages has higher similarity and co-occurrence values as compared to content with a low number of words such as SNS; this is due to the large number of candidate words.

In this evaluation, we only use the fixed parameters shown in Table 2. While this paper shows only the best results, we have tried and evaluated the proposed method several times with many parameters. In the case of applying our method to a real system, parameter tuning is necessary based on the situation. In particular, window size, which represents the range of the word sequences, and iteration, which is the number of iterations required to train the neural network, significantly affect system performance.

5.2 Future Works

To facilitate the application of the proposed method in practical services and systems, it is necessary to solve several problems as part of future work. One such problem is parameter optimization for paragraph vector based on the content being considered. Another problem is reducing the amount of calculations in the case of content retrieval using the similarity BoW.

For parameter optimization, the discovery method in [5] uses about 5000 parameters with low calculation and efficiency using a combination of dichotomy and random search. This method enables the adaptive calculation of the optimum values of the parameters in the context of this paper. With respect to the reduction of the calculation amount, vectorizing the words contained in the target content in advance is expected to reduce the load of calculating the similarity and number of word occurrences. The proposed method is deemed feasible even now because the remaining issues only appear at implementation level.

6 Conclusion

In this article, we proposed and discussed a method for analysis and retrieval of online content with tourist information about the Suwa area in the Nagano Prefecture, Japan.

The conventional method, which uses frequencies of word occurrences in online content, often fails to express the features for model training from frequencies of words occurrences in SNS-based content. Our proposed method uses paragraph vector that extracts features by learning the sequence of word ordering in the content. To compare the effectiveness of the method using word ordering as a feature, a variation that uses the similarity of groups of words (BoW) in the content is also discussed. According to our evaluation results, the proposed method achieves better performance as compared with conventional methods. For future work, considering the problem of parameter optimization for the proposed method is necessary.

Acknowledgements. This work was partly supported by MEXT KAKENHI Grant Number 17K01149

References

1. Blei, D.M., Ng, A.Y., Jordan, M.I.: Latent Dirichlet allocation. J. Mach. Learn. Res. **3**, 1107–1135 (2003)
2. https://www.tripadvisor.jp/
3. Le, Q., Mikolov, T.: Distributed representations of sentences and documents. In: Proceedings of International Conference on Machine Learning, Beijing, China, vol. 32 (2014)
4. Topic modelling for humans. https://radimrehurek.com/gensim/
5. Hara, Niizuma, Ota: Research on Effective Parameter Search Method for Paragraph Vector. DEIM Forum (2017)

Improving the Information in Medical Image by Adaptive Fusion Technique

Nguyen Mong Hien[1,2](✉), Nguyen Thanh Binh[2], Ngo Quoc Viet[3], and Pham Bao Quoc[2]

[1] Faculty of Engineering and Technology, Tra Vinh University, Tra Vinh, Viet Nam
hientvu@tvu.edu.vn, 8141212@hcmut.edu.vn
[2] Faculty of Computer Science and Engineering, Ho Chi Minh City University of Technology,
VNU-HCM, Ho Chi Minh City, Viet Nam
{ntbinh,1585008}@hcmut.edu.vn
[3] Faculty of Information Technology, Ho Chi Minh City University of Pedagogy,
Ho Chi Minh City, Viet Nam
vietnq@hcmup.edu.vn

Abstract. Image fusion plays a huge role in many fields, especially in medical image processing because the visual interpretation of the image can enhance by using the fusion technique. The result shows the important detail which is very useful for doctor to diagnose health problems. In the paper, we proposed a method for image fusion. The guided filter is used to enhance the detail of the input image and then the cross bilateral filter is applied to extract detail image from the enhanced image. The image result is made by weighted average using the weights calculated from the detailed images. The experimental results showed that the proposed method can work well with medical image as well as other kinds of image. In addition, our result is better than the other recent methods based on compared objective performance measures.

Keywords: Fusion technique · Medical image · Weighted average · Guided filter
Bilateral filter

1 Introduction

In the modern life, the diagnosing result is improved significantly thanks to the support of image processing technique. One of the useful techniques is image fusion which is very helpful for determining medical problems. Because without using the technique the doctor usually uses the images with lack of information to detect the problems. As a result, the doctor has to spend a lot of time to find out the important clues, created by small details in medical images. If the image is not full information, the tiny details will not be showed in the image. The solution based on image fusion is a good way to solve the limitation. In this technique, many images with different focus points are combined to create a new image with all objects fully focused. As a result, all the important visual information which belongs to the input images will be presented in the fused image.

© Springer Nature Switzerland AG 2018
T. K. Dang et al. (Eds.): FDSE 2018, LNCS 11251, pp. 423–432, 2018.
https://doi.org/10.1007/978-3-030-03192-3_32

Recently, many researchers combined the algorithms to make novel methods. Yin [1] proposed the medical image fusion using a new weighted average rule, which computed from the sensor sparse coefficient. Its result can preserve the 3-D structure of medical volume, and decrease the low contrast and artifacts. Guo [2] built a conceptual architecture for the image fusion schemes in supervised biomedical image analysis at many levels and then these schemes are combined into a single scheme based on the convolutional neural network. Liu [3] proposed the image fusion method based on multiscale transform and sparse representation. But the result of method is not good because most energy of the input image is distributed in low pass. Kumar [4] proposed the image fusion method at pixel level. However, the method is limited with low intensity images. Li [5] proposed a new image fusion method using guided filter. But, the gray level of input image may cause some limitations for the method. Bai [6] used a new quadtree-based algorithm to make a novel image fusion method with support of the strategy of effective quadtree decomposition. In which, some blocks with optimal sizes in a quadtree structure are made by decomposing the input images. Zhang [7] combined two independent process such as registration and fusion to enhance fusion performance. Because the registration errors ignored in the fusion process influence on the fusion quality. Sale [8] used hybrid multi resolution image fusion to improve image quality. The discrete wavelet transforms (DWT) is used to images with frequency distortion to supply better frequency resolution. Different fusion rules are used to low and high frequency coefficients to combine wavelet coefficients of source images. Javed [9] proposed a fusion scheme which combined the useful information in the source images. The scheme used guided filter to extract and combine the information showing at different frequencies. Kavitha [10] combined wavelet with genetic algorithm in order to optimize the weight estimation and fusion process. Himanshi [11] combined curvelet transform, principal component analysis (PCA) and maximum selection rule together to improve fusion method for medical image. Shuaiqi [12] proposed a novel medical image fusion using the rolling guidance filter (RGF) and spiking cortical model (SCM) each other. RGF is used to capture saliency of medical images. Geng [13] fused multi-modal medical images by using the modified local contrast information. The pixel with large modified local contrast used to fuse images. However, majority methods are complex.

In this paper, we have proposed an approach to fuse image using guided filter (GF) to enhance the detail of input images and cross bilateral filter (CBF) to fuse input images. The rest of the paper is organized as follows: we described the background of fusion performance measure, guided filter and bilateral filter in Sect. 2; the proposed method is shown in Sect. 3; the result experiments and conclusion of the paper are presented in Sects. 4 and 5 respectively.

2 Related Knowledge

2.1 Fusion Performance Measure

The fusion performance [14–16] is analyzed deeply by quantifying: total fusion perform-ance, fusion artifacts and fusion loss. The parameter computed by the procedure in [15] and their symbols shown below:

$Q^{AB/F}$: total information in output image is transferred by input images
$N^{AB/F}$: artifacts are added in output image by fusion process
$L^{AB/F}$: total information lost by fusion process.

The parameters are complimentary indicating that the sum of them should result in unity [15]. It means:

$$Q^{AB/F} + N^{AB/F} + L^{AB/F} = 1 \qquad (1)$$

These parameters reviewed in the cases this value may not reach to unity. Based on reviewing, the fusion artifact parameter has modified. During the fusion process, a fusion artifact can make into the result image. It could become a benign object considered as a valid target or a threat. Thus, a fusion method is efficient if its result is minimum artifacts. A fusion artifact measure proposed in [15] given as:

$$N_m^{AB/F} = \frac{\sum_{\forall i} \sum_{\forall j} AM_{i,j}\left[\left(1 - Q_{i,j}^{AF}\right)w_{i,j}^A + \left(1 - Q_{i,j}^{BF}\right)w_{i,j}^B\right]}{\sum_{\forall i} \sum_{\forall j}\left(w_{i,j}^A + w_{i,j}^B\right)} \qquad (2)$$

where, $AM_{i,j} = \begin{cases} 1, & g_{i,j}^F > g_{i,j}^A \text{ and } g_{i,j}^F > g_{i,j}^B \\ 0, & \text{otherwise} \end{cases}$
indicating locations of fusion artifacts where fused gradient is stronger than input.

$g_{i,j}^A, g_{i,j}^B$ and $g_{i,j}^F$ are the edge strength of A, B and result image respectively,

$Q_{i,j}^{AF}$ and $Q_{i,j}^{BF}$ are the gradient information preservation estimates of input images A and B respectively,

$w_{i,j}^A$ and $w_{i,j}^B$ are the perceptual weights of input images A and B respectively.

The parameters $g_{i,j}^A, g_{i,j}^B, g_{i,j}^F, Q_{i,j}^{AF}, Q_{i,j}^{BF}, w_{i,j}^A$ and $w_{i,j}^B$ calculated by the procedure in [15]. Further, the fusion performance evaluated by various parameters that presented in [16].

With the newly modified fusion artifact measures $N_m^{AB/F}$, Eq. (1) is able to rewritten as:

$$Q^{AB/F} + N^{AB/F} + N_m^{AB/F} = 1 \qquad (3)$$

2.2 Guided Filter

A local linear model between the guided image I and the filter result q considered as a main assumption of guided filter [5]. The guided filter assumed that the filtering result q is a linear transformation of the guided image I in local window w_k centered at pixel k.

$$q_i = a_k I_i + b_k, \forall i \in w_k, \tag{4}$$

where, w_k is a square window with size $(2r + 1 \times (2r + 1)$. In w_k, the values of linear coefficient a_k and b_k are unchange. The square difference between the result image q and the input image p minimized to estimate these values.

$$E(a_k, b_k) = \sum_{i \forall w_k} \left((a_k I_i + b_k - p_i)^2 + \varepsilon a_k^2 \right) \tag{5}$$

where ε is supplied by user and is a regularization parameter. The linear regression [20] can use to solve the coefficients a_k and b_k directly as follows:

$$a_k = \frac{\frac{1}{|w|} \sum_{i \in w_k} I_i p_i - \mu_k \bar{p}_k}{\sigma_k^2 + \varepsilon} \tag{6}$$

$$b_k = \bar{p}_k - a_k \mu_k \tag{7}$$

where the mean and variance of I in w_k are named μ_k and σ_k^2 respectively, the number of pixels in w_k is $|w|$ and the mean of p in w_k is $\bar{p}_k$. After that based on (4) the result image can be computed. The pixel i will locate in all local windows centered at pixel k in the window w_k. So, the p_i value in (4) will vary when it is calculated in different windows w_k. To handle the problem, firstly, we average all the possible values of coefficients a_k and b_k. Secondly, estimating the filter result as follows:

$$qi = \bar{a}_i I_i + \bar{b}_i \tag{8}$$

where $\bar{a}_i = \frac{1}{|w|} \sum_{i \in w_k} a_k, \bar{b}_i = \frac{1}{|w|} \sum_{i \in w_k} b_k$.

2.3 Bilateral Filter

Bilateral filter (BF) is a local and non-iterative as well as nonlinear technique that combines a classical low-pass filter with an edge-stopping function which attenuates the filter kernel when the intensity difference between pixels is huge [4]. When both gray level similarities and geometric closeness of the neighboring pixels are considered. Both Euclidian distance and the distance in gray/color space influence on the weights of the filter. The benefit of the filter is that, it not only smoothes the image but also preserve edges thank to neighboring pixels. Mathematically, for an image A, the BF output at a pixel location p is computed as follow [18]:

$$A_F(p) = \frac{1}{w} \sum_{q \in S} G_{\sigma_s}(\|p - q\|) G_{\sigma_r}\left(\left|A_{(p)} - A_{(q)}\right|\right) A_{(q)} \tag{9}$$

where, $G_{\sigma_s}(\|p - q\|) = e^{-\frac{\|p - q\|^2}{2\sigma_r^2}}$ is a geometic closeness function,

$G_{\sigma_r}(|A(p) - A(q)|) = e^{-\frac{|A(p) - A(q)|^2}{2\sigma_r^2}}$ is a gray level similarity/edge-stopping function

$W = \sum_{q \in S} G_{\sigma s}(\|p - q\|) G_{\sigma s}(|A(p) - A(q)|)$ is a normalization constant,

$\|p - q\|$ is the Euclidean distance between p and q. And S is a spatial neighborhood of p.

When the behavior of BF is controlled by σ_s and σ_r, the dependence of σ_s/σ_r values and derivative of the input signal on the behaviors of the BF are analyzed in [19]. The desired amount of low-pass filtering is used to optimize the σ_s value and blurs more for lager σ_s, as the values from more distant image locations are combined. Based on the scale of image the σ_s must change to have equivalent results. It shows that the ideal range for the σ_s value is from 1.5 to 2.1. However, the amount of preserved edges influences on the optimal σ_s value. The σ_s must be changed by amplified or attenuated image to retain the same result.

3 Improving the Information in Medical Image

In this section, we propose an efficient method for magnetic resonance image fusion, which is based on guided filter combined with cross bilateral filter. The generalized block diagram of the proposed method given in Fig. 1. The proposed method includes two stages: enhancing the detail in input image and fusing image. The stages will explain in the following subsections.

3.1 Enhancing the Detail in Image

In the first stage, GF [5] applied to enhance the detail in input image. This task should do to improve the quality of source images because the real medical image usually has low quality. It described as follows:

(i). To compute the coefficients a_k and b_k by Eqs. (6) and (7).
(ii). To average all the possible values of coefficients a_k and b_k. They are $\bar{a}$ and $\bar{b}$.
(iii). To compute the output image by Eq. (8)

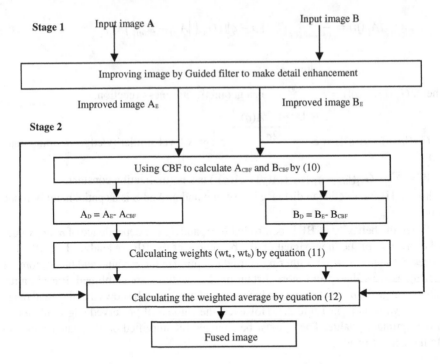

Fig. 1. Block diagram of proposed method.

3.2 Fusing Image

In the fusion process, the result image of the first stage is used as the input image for the second stage. The result is very useful for CBF. Because both gray level similarities and geometric closeness of neighboring pixels are considered, the weights of the filter depend on both Euclidian distance and the distance in gray space. The CBF will fuse input images based on weighted average. The stage described as follows:

Not only gray level similarities but also geometric closeness of neighboring pixels in image A are considered in order to shape the filter kernel and filter the image B. CBF output of image B at a pixel location p is computed as [19]:

$$B_{CBF}(p) = \frac{1}{W} \sum_{q \in s} G_{\sigma_s}(\|p - q\|) G_{\sigma_r}(|A(p) - A(q)|) B(q) \tag{10}$$

where, $G_{\sigma_r}(|A(p) - A(q)|) = e^{-\frac{|A(p) - A(q)|^2}{2\sigma_r^2}}$: a gray level similarity/edge-stopping function and $W = \sum_{q \in S} G_{\sigma s}(\|p - q\|) G_{\sigma s}(|A(p) - A(q)|)$: a normalization constant.

These detail images are obtained by subtracting CBF result from the respective input image, for enhanced image A_E and B_E is given by $A_D = A_E - A_{CBF}$ and $B_D = B_E - B_{CBF}$ respectively.

The weight of particular detail coefficient computed by adding horizontal detail strength (HdetailStrength) and vertical detail strength (VdetailStrength).

$$wt(i,j) = \text{HdetailStrength}(i,j) + \text{VdetailStrength}(i,j) \tag{11}$$

When wt_a and wt_b are the weights for the detail coefficient A_D and B_D belong to the respective input enhanced images A_E and B_E, then the weighted average of both calculated as the fused image by:

$$F(i,j) = \frac{A(i,j)wt_a(i,j) + B(i,j)wt_b(i,j)}{wt_a(i,j) + wt_b(i,j)} \tag{12}$$

4   Experiments and Evaluation

To evaluate the performance of the methods, we applied these methods to data obtained from many sources such as [21, 22]. Fused image by our proposed method is compared to other methods discussed in [3, 4] with the same simulation parameters. The proposed method uses the parameters such as $\sigma_s = 1.8$, $\sigma_r = 25$, neighborhood window $= 11 \times 11$ (for CBF) and neighborhood window $= 5 \times 5$ (finding detail strength).

The result of CBF subtracted from the respective input image to capture the detail image. Two detail images captured by using two input images. The results used to compute the weights wt_a and wt_b by measuring the detail strengths. The weights applied to find fused image by weighted average. We test many cases with various images. In here, we present some cases. The fused images of pairs are shown in Figs. 2 and 3.

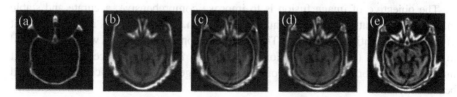

Fig. 2.   Improving the information for first pair source medical images. (a) and (b) first pair source medical images, (c) Improving the information by Liu method [3], (d) Improving the information by Kumar method [4], (e) Improving the information by the proposed method.

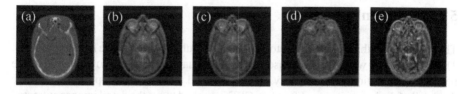

Fig. 3.   Improving the information for second pair source medical images. (a) and (b) second pair source medical images, (c) Improving the information by Liu method [3], (d) Improving the information by Kumar method [4], (e) Improving the information by the proposed method.

According to the Figs. 2 and 3, we can say that the quality of the result images by our proposed method in (e) is better than that of two mentioned methods. Because, our results consist of detailed images from both of the input images with less information loss and artifacts.

The objective performance measures: $Q^{AB/F}$, $L^{AB/F}$, $N^{AB/F}$ and $N_m^{AB/F}$ and the respective sum are shown in Table 1. The quality of the result image is better when $Q^{AB/F}$ has higher value. It means that the total information in the output image transferred by input images is high. And $L^{AB/F}$, $N^{AB/F}$ and $N_m^{AB/F}$ should be lower. In other word, the total information lost by fusion process and artifacts are low. In the Table 1, the higher values were presented by using *italics* except for $L^{AB/F}$, $N^{AB/F}$ and $N_m^{AB/F}$ *where italics shows their lower values*.

Table 1. Objective image fusion performance measures

Measure	$Q^{AB/F}$	$L^{AB/F}$	$N^{AB/F}$	Sum	$N_m^{AB/F}$	Sum
Source images	The first image pair					
Liu method [3]	0.824	0.159	0.179	1.161	0.017	1
Kumar method [4]	0.883	0.106	0.103	1.091	0.012	1
Proposed method	*0.906*	*0.088*	*0.081*	1.075	*0.006*	1
Source images	The second image pair					
Liu method [3]	0.775	0.201	0.178	1.163	0.024	1
Kumar method [4]	0.802	0.184	0.069	1.055	0.014	1
Proposed method	*0.829*	*0.161*	*0.052*	1.042	*0.010*	1

The objective of image fusion is to improve comprehensive, accurate and stable information so that it is easier for human perception. Visual analysis is also important in addition to quantitative analysis. There are three popular conditions, which are applied to visual analysis (i) the information is transferred from each source image to result image, (ii) the information is lost from the input images and (iii) artifacts are introduced because of fusion.

In Table 1, for the all image pairs it observed that, the values of all parameters in our proposed method are the best. In other words, our result image consists of most information from the source images.

5 Conclusion

The proposed method supplied doctors with the useful tool to diagnose health problems. The method will work well because the doctor usually faces with low quality image because of limitations of machinery. In the proposed method, GF is applied to enhance the detail of the input images. After that, CBF is applied to get the detail images extracted from the input images. Based on the detailed images the fused image has been created by applying weighted average. The method performs better because it supplies good detail images thanks to the GF technique. As a result, the proposed method gives a good result, which presents objective performance measures.

References

1. Yin, H.: Tensor sparse representation for 3-D medical image fusion using weighted average rule. IEEE Eng. Med. Bio. Soc. 1–12 (2018)
2. Guo, Z., Li, X., Huang, H., Guo, N., Li, Q.: Medical image segmentation based on multimodal convolutional neural network: study on image fusion schemes. In: The 15th International Symposium on Biomedical Imaging, Washington, D.C., pp. 903–907 (2018)
3. Liu, Y., Liu, S., Wang, Z.: A general framework for image fusion based on multi-scale transform and sparse representation. Inf. Fus. **24**, 147–164 (2015)
4. Shreyamsha Kumar, B.K.: Image fusion based on pixel significance using cross bilateral filter. Signal, Image Video Process. **9**(5), 1193–1204 (2015)
5. Li, S., Kang, X., Hu, J.: Image fusion with guided filtering. IEEE Trans. Image Process. **22**(7), 2864–2875 (2013)
6. Bai, X., Zhang, Y., Zhou, F., Xue, B.: Quadtree-based multi-focus image fusion using a weighted focus-measure. Inf. Fus. **22**, 105–118 (2013)
7. Zhang, Q., Cao, Z.-G., Hu, Z., Wu, X.: Joint image registration and fusion for panchromatic and multispectral images. IEEE Geosci. Remote Sens. Lett. **12**(3), 467–473 (2015)
8. Sale, D., Patil, V., Joshi M.A.: Effective image enhancement using hybrid multi resolution image fusion. In: IEEE Global Conference on Wireless Computing Networking (GCWCN), pp. 116–120 (2014)
9. Javed, U., Riaz, M.M., Ghafoor, A., Cheema, T.A.: Weighted fusion of MRI and PET images based on fractal dimension. Multidimensional Systems and Signal Processing **28**, 679–690 (2015)
10. Kavitha, S., Thyagharajan, K.K.: Efficient DWT-based fufion techniques using genetic algorithm for optimal parameter estimation. Soft Comput. Fusion Found., Methodol. Appl. **21**, 3307–3316 (2016)
11. Himanshi, Bhateja, V., Krishn, A., Sahu, A.: Medical image fusion in curvelet domain employing PCA and maximum selection rule. In: Satapathy, S., Raju, K., Mandal, J., Bhateja, V. (eds.) Proceedings of the Second International Conference on Computer and Communication Technologies. Advances in Intelligent Systems and Computing. AISC, vol. 379, pp. 1–9. Springer, New Delhi (2015). https://doi.org/10.1007/978-81-322-2517-1_1
12. Shuaiqi, L., Jie, Z., Mingzhu, S.: Medical image fusion based on rolling guidance filter and spiking cortical model. In: Computational and Mathematical Methods in Medicine, Hindawi (2015)
13. Geng, P., Liu, S., Zhuang, S.: Multimodal medical image fusion by adaptive manifold filter. In: Computational and Mathematical Methods in Medicine, Hindawi (2015)
14. Nguyen, Mong Hien, Nguyen, Thanh Binh: Efficient framework for magnetic resonance image fusion using histogram equalization combined with cross bilateral filter. 6th International Conference on the Development of Biomedical Engineering in Vietnam (BME6). IP, vol. 63, pp. 351–356. Springer, Singapore (2018). https://doi.org/10.1007/978-981-10-4361-1_59
15. Petrovic, V., Xydeas, C.: Objective image fusion performance characterization. In: IEEE International Conference on Computer Vision, pp. 1866–1871 (2005)
16. Shreyamsha Kumar, B.K.: Multifocus and multispectral image fusion based on pixel significance using discrete cosine harmonic wavelet transform. Signal Image Video Process. **7**, 1125–1143 (2012)
17. Gonzalez, R.C., Woods, R.E.: Digital Image Processing, 3rd edn. Prentice Hall, Upper Saddle River (2002)

18. Tomasi, C., Manduchi, R.: Bilateral filtering for gray and color images. In: Proceedings of International Conference on Computer Vision, pp. 839–846 (1998)
19. Zhang, M., Gunturk, B.K.: Multi resolution bilateral filtering for image denoising. IEEE Trans. Image Process. **17**, 2324–2333 (2008)
20. Draper, N., Smith, H.: Applied Regression Analysis, 3rd edn. Wiley, Hoboken (2014)
21. www.metapix.de. Accessed 15th June 2018
22. www.medinfo.cs.ucy.ac.cy. Accessed 15th June 2018

Resident Identification in Smart Home
by Voice Biometrics

Minh-Son Nguyen[1(✉)] and Tu-Lanh Vo[2(✉)]

[1] Faculty of Computer Engineering, University of Information Technology in VNUHCM,
Ho Chi Minh City, Vietnam
sonnm@uit.edu.vn
[2] School of Computer Science and Engineering, International University in VNUHCM,
Ho Chi Minh City, Vietnam
votulanh@gmail.com

Abstract. In smart home environments, it is highly useful to know who is performing what actions. This knowledge allows the system to make intelligent decisions and control the end devices based on the current resident. However, this is extremely challenging to take the personalized action in the multi-resident situation without individuals identification. This research work introduces the use of voice biometrics as a means to identify individuals. Which especially suitable for the system that uses speech as the command to control smart devices. The results of this research will provide the knowledge base for residents behavior learning and prediction.

Keywords: Resident identification · Smart home · Voice biometrics · Open-set Text-independent speaker identification · Adaptive threshold

1 Introduction

Home automation with voice control can achieve a high level of performance in real-world environment. However, such performance drops significantly in the multi-resident situation. In this circumstance, the smart home needs to address the problems of identifying the residents in some manner. Speaker recognition or voice recognition is a biometric modality that use characteristics of voice (voice biometrics) to identify a person. Voice recognition is a popular choice to identify individuals due to the availability of tools for gathering voice samples and its ease of integration. The challenge of modern smart homes is how to create an intelligent environment for multiple residents, which is extremely difficult due to the complexity of individuals identify [1]. One of the possible solutions is using voice biometrics which is unique among people to identify residents inside smart home environment. Therefore, this research introduces the use of voice biometrics as a means to identify smart home residents. In this work, the open-set and text-independent speaker identification with the adaptive threshold will be proposed and evaluated.

The process of identifying residents in the smart home by using voice biometrics is expressed as follows. In the training phase, the resident utterances are given as an input

© Springer Nature Switzerland AG 2018
T. K. Dang et al. (Eds.): FDSE 2018, LNCS 11251, pp. 433–448, 2018.
https://doi.org/10.1007/978-3-030-03192-3_33

for the system. Next, those speech signal is extracted to produce feature vectors that convey speaker information. After that, the training features have been used to create the speaker model and the corresponding imposter model. Furthermore, speaker models are created by using probability density function [2]. During speaker modeling, probability density function parameters are estimated from the training features. Finally, those models have been stored in the database as the reference models.

In the testing phase, the speech signal is first processed to extract the testing feature. In feature matching, the process of open-set, text-independent speaker identification with the adaptive threshold will be applied to decide whether the test utterances has been produced by the registered speaker or not by comparing the likelihood [3] between the testing feature and reference models. Finally, the decision will be taken based on the best match (Fig. 1).

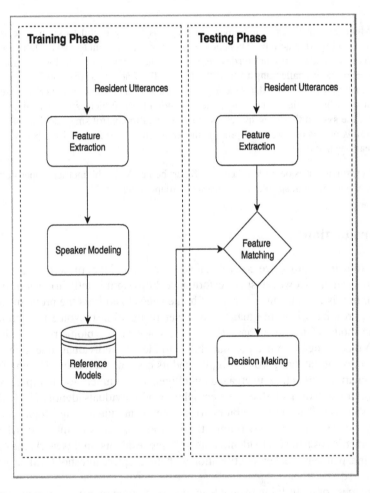

Fig. 1. The process of identifying residents in the smart home by using voice biometrics.

The rest of the paper is organized as follows. Section 2 describes the approaches for speaker recognition. The methodology of the study is provided in Sect. 3, which is followed by result and discussion in Sect. 4, and finally concluding remarks are given in Sect. 5.

2 Approaches for Speaker Recognition

Speaker Recognition can be divided into verification and identification [4]. Speaker verification is the process of accepting or rejecting the identity claim of a speaker. Speaker identification is the process of selecting a registered speaker who provides a given utterance. The process of speaker recognition includes feature extraction and feature matching. In feature extraction, the speaker utterances are extracted to produce voice features. In feature matching, the unknown speaker is identified by comparing the extracted features from a given utterance with the reference models.

2.1 Speaker Identification

Speaker identification is the process of determining the identity of an unknown speaker by comparing a given voice with voices of registered speakers [5]. It is a one to many comparisons or 1: N match where the voice is compared against N templates (Fig. 2).

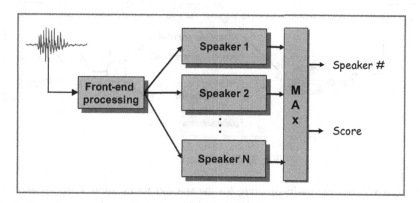

Fig. 2. Speaker identification.

The speaker identification problem may further be subdivided into closed-set and open-set. The closed-set refers to a case where the speaker belongs to a set of registered speakers. In the open set case, the speaker may be out of the set of registered speakers. Another distinctive aspect of speaker identification systems is that they can either be text-independent or text-dependent. In the text-independent case, there is no restriction on the sentence or phrase to be spoken. In the text-dependent case, the input sentence or phrase is fixed for each speaker.

The speaker identification process is shown in Fig. 2. The speech signal is first processed to extract features conveying speaker identity. After that, these features are compared to a bank of models, obtained from previous enrollment. For the closed-set identification, the highest scoring model is selected as the identified speaker. For the open-set identification, the score of the highest scoring model is then compared with the pre-defined threshold to decide whether the identified speaker belongs to a set of registered speakers or not.

In this project, the open-set text-independent speaker identification will be implemented to identify residents in the smart home environment.

2.2 Speaker Verification

Speaker verification is used to determine whether a person claims to be according to his/her voice sample [6]. This task is also known as voice verification and speaker detection. Speaker verification is a one to one match or 1:1 match where one speaker's voice is matched to one template (also called a "voice print" or "voice model") or in other sense Pattern Matching between the claimed speaker model registered in the database and the imposter model (Fig. 3). If the match is above a certain threshold, the identity claim is verified. Using a high threshold, the system gets high safety and prevents impostors to be accepted, on the other hand, it also takes the risk of rejecting the correct person, and vice versa. Therefore, selecting a suitable threshold is extremely important for the accuracy of recognition.

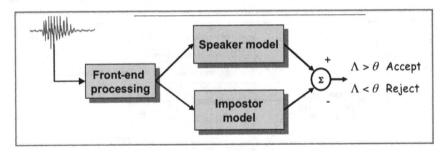

Fig. 3. Speaker verification

Figure 3 shows the basic structure for the speaker verification. Features extracted from a given utterance are compared to a model of claimed speaker, obtained from the training phase, and to a model representing potential imposter speakers. The ratio of speaker and imposter match scores is the likelihood ratio statistic (Λ), which is then compared to a threshold (θ) to decide whether to accept or reject the speaker.

2.3 Feature Extraction

The purpose of this module is to convert speaker utterances into voice features for further analysis, which is also known as the front-end processing.

There are several methods for parametrically representing the speech signal which used for the speaker recognition purpose. These include Linear Predictive Coefficients (LPC), Mel Frequency Cepstrum Coefficients (MFCCs) and several others.

Linear Predictive Coefficients (LPCs). LPCs is defined as a digital method for encoding an analog signal in which a particular value is predicted by a linear function of the past values of the signal. Human speech is produced in the vocal tract which can be approximated as a variable diameter tube. The linear predictive coding model is based on a mathematical approximation of the vocal tract represented by this tube of a varying diameter. At any particular time (t) the speech sample s(t) is represented as a linear sum of the previous samples (p). The most important aspect of LPCs is the linear predictive filter which allows the value of the next sample to be determined by a linear combination of previous samples. Linear predictive coding may reduce bit rate significantly and at this reduced rate the speech has a distinctive synthetic sound and there is a noticeable loss of quality [7]. However, the speech is still audible and it can still be easily understood. Since there is information loss in linear predictive coding, it is a lossy form of compression. LPCs does not represent the vocal tract characteristics from the glottal dynamics and also it takes more time and computational cost to create the model of each speaker.

The fundamental steps of LPCs [8] are expressed as follows (Fig. 4).

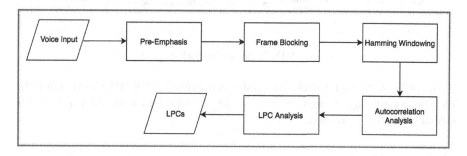

Fig. 4. LPCs extraction block diagram

The figure above describes the process of LPCs extraction. In the pre-emphasis step, the signal passes a filter to emphasize the magnitude in the high frequencies. In frame blocking, the signal will be split into some small frames. In order to reduce the signal discontinuities at the start and end of each segment, windowing is performed. Autocorrelation analysis is achieved to auto-correlate each frame of the windowed signal. Finally, each frame will be converted to LPCs.

Mel Frequency Cepstral Coefficients (MFCCs). The most popular spectral based parameter used in recognition approach is the Mel Frequency Cepstral Coefficients (MFCCs). MFCCs is based on human hearing perceptions which cannot perceive frequencies over 1 kHz. In other words, in MFCCs is based on the known variation of the human ear's critical bandwidth with frequency. MFCCs has two types of filters which are spaced linearly at the low frequency below 1000 Hz and logarithmic spacing above 1000 Hz.

The process of MFCCs extraction [9] is described as follows (Fig. 5). The digitized speech waveform has a high dynamic range and suffers from additive noise. In order to reduce this range and spectrally flatten the speech signal, Pre-Emphasis is applied. In Frame Blocking, the speech signal is split into several frames such that each frame can be analyzed in the short time instead of analyzing the entire signal at once. In order to avoid spectral leakage, the next step that is Windowing is undertaken. Fast Fourier Transform is applied to convert each frame of N samples from the time domain into the frequency domain. In Mel Filter Bank, the obtained spectrum is later wrapped and converting the frequency spectrum to Mel spectrum. In the Discrete Cosine Transform, the Mel spectrum is converted back to time domain. The result is called the Mel Frequency Cepstrum Coefficients (MFCCs).

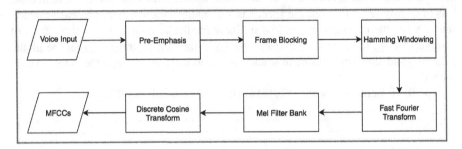

Fig. 5. MFCCs extraction block diagram

According to recent research, the significant improvement of MFCCs extraction [10] has a huge advantage of high accuracy in the real-world environment which will be applied for feature extraction.

2.4   Feature Matching

There are several popular classification techniques (pattern matching): Dynamic Time Warping (DTW), Neural Networks, Hidden Markov Models (HMMs),... The selection of techniques is largely dependent on the type of speech to be used, the expected performance, the ease of training and updating, and storage and computation considerations.

Template Matching. In the template matching technique, the model consists of a template that is a sequence of feature vectors from a fixed phrase. During verification, a match score is produced by using dynamic time warping (DTW) to align and measure the similarity between the test phrase and the speaker template [11]. This approach is used almost exclusively for text-dependent applications.

Neural Networks. The particular model used in this technique can have many forms, such as multi-layer perceptions or radial basis functions. The main difference with the other approaches described is that these models are explicitly trained to discriminate between the speaker being modeled and some alternative speakers [12]. The main advantages of Neural Networks include their discriminant-training power, a flexible architecture that permits easy use of contextual information. However, the training process can be computationally expensive and models are sometimes not generalizable due to the temporal structure of speech signals [13].

Hidden Markov Models. This technique uses HMMs, which encode the temporal evolution of the features and efficiently model statistical variation of the features, to provide a statistical representation of how a speaker produces sounds. During enrollment HMM parameters are estimated from the speech using established automatic algorithms. During verification, the likelihood of the test feature sequence is computed against the speaker's HMMs. For text-dependent applications, whole phrases or phonemes may be modeled using multi-state left-to-right HMMs [14]. For text-independent applications, single state HMMs, also known as Gaussian Mixture Models [15], are used.

Gaussian Mixture Models is a classic parametric method best used to model speaker identities due to the fact that Gaussian components have the capability of representing some general speaker dependent spectral shapes. Gaussian classifier has been successfully employed in several text-independent speaker identification applications since the approach used by this classifier is similar to that used by the long-term average of spectral features for representing a speaker's average vocal tract shape [16].

The purpose of this project is to identify the speaker by using the text-independent method. Therefore, the Gaussian Mixture Models (GMMs) will be applied for feature matching.

3 Methodology

In this section, the open-set, text-independent speaker identification with the adaptive threshold will be proposed and evaluated.

Given a set of registered speakers and a sample test utterance, the open-set speaker identification process can be divided into two successive stages of identification and verification. This is because firstly, it is required to identify the speaker model in the set, which best matches the test utterance. Secondly, it must be determined whether the test utterance has actually been produced by the speaker associated with the best-matched model, or by some unknown speaker outside the registered set.

Open-Set, Text-Independent Speaker Identification with the Adaptive Threshold.
Speaker identification involves representing a set of registered speakers using their corresponding statistical model descriptions, i.e. $\lambda_1, \lambda_2, \ldots, \lambda_N$, where N is the number of speakers in the set. Each model description is developed using the short-term spectral features extracted from the utterances produced by the registering speaker. Based on such speaker modeling, the process of speaker identification in the open-set mode [17] can be stated as

$$\max_{1\leq n\leq N}\{p(\mathbf{O}|\lambda_n)\} \gtrless \theta \;\rightarrow\; \mathbf{O} \in \left\{ \begin{array}{l} \lambda_i,\, i = \arg\max_{1\leq n\leq N}\{p(\mathbf{O}|\lambda_n)\} \\ \text{unknown speaker model} \end{array} \right. \tag{1}$$

Where O denotes the feature vector sequence extracted from the test utterance, and θ is a pre-determined threshold. In other words, O is assigned to the speaker model that yields the maximum likelihood over all other speaker models in the set, if this maximum likelihood score is greater than the threshold θ. Otherwise, it is declared as originated from an unknown speaker (Fig. 6).

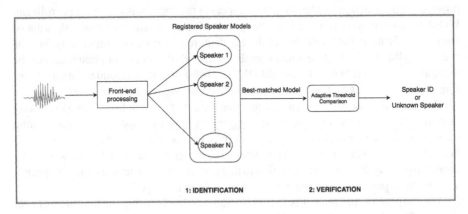

Fig. 6. The open-set, text-independent speaker identification with the adaptive threshold process

This figure above summarizes the process of open-set, text-independent speaker identification (OSTI-SI) with the adaptive threshold. The process can be divided into two successive stages of identification and verification. The test utterance is first processed to extract features conveying speaker information.

In the identification stage, these features are compared to the registered speaker models. The highest scoring model is selected as the best-matched model. In the verification stage, features extracted from the test utterance are compared to the best-match model, and to the imposter model to produce the likelihood or match score. The match score of the best-matched model is then compared with the match score of the imposter model - adaptive threshold (2) to decide whether the test utterance has actually been produced by the speaker associated with the best-matched model, or by some unknown speaker outside the registered set.

$$p(\mathbf{O}|\lambda^{BM}) - p(\mathbf{O}|\lambda^{IM}) > 0 \tag{2}$$

where λ^{BM} is the best-matched model in identification stage, and λ^{IM} is the imposter model. In order to deploy Eq. (2), $p(\mathbf{O}|\lambda^{IM})$ has to be determined accurately. However, in practice λ^{IM} is unavailable. Therefore, the best option is to determine an appropriate replacement for $p(\mathbf{O}|\lambda^{IM})$. This technique involves approximating $p(\mathbf{O}|\lambda^{IM})$ with $p(\mathbf{O}|\lambda^{Adaptive})$, where $\lambda^{Adaptive}$ is a model generated using utterances from the registered speakers except the utterances of the best-matched model.

Adaptive Threshold Comparison. In the verification stage, the match score (likelihood) of the best-matched model is then compared with the pre-defined threshold to decide whether to accept or reject the speaker. In this stage, selecting a suitable threshold is extremely important due to using a high threshold, the system gets high safety and prevents impostors to be accepted, on the other hand, it also takes the risk of rejecting the correct person, and vice versa. Therefore, the adaptive threshold method will be applied in order to improve the accuracy of recognition. In this methodology, the pre-defined threshold will be replaced by an adaptive threshold which is different among speakers. The adaptive threshold is the match score of the test utterance with the imposter model which is defined by the training utterances of all registered speaker models except the utterances of the best-matched model.

4   Result and Discussion

4.1   Datasets

In this project, the CSTR VCTK [18] and VoxCeleb [19] datasets have been used in the training phase (Tables 1, 2 and 3) and testing phase (Tables 2, 3 and 4). Both datasets contain the native English speaker with various accents, ages.

Table 1.   Training data - CSTR VCTK dataset

Process	Description
Speaker	40 speakers: 20 males + 20 females
Utterance	10 utterances/speaker
Duration	3–10 s/utterance
Environment	Clean Speech - Laboratory
Format	.wav
Sample frequency	48 kHz

Table 2.   Testing data - CSTR VCTK dataset

Process	Description
Speaker	80 speakers: 40 males, 40 females
Utterance	10 utterances/speaker
Duration	3–10 s/utterance
Environment	Clean Speech - Laboratory
Format	.wav
Sample frequency	48 kHz

In the datasets above, each speaker includes 10 utterances. The duration of each utterance varies from 3 to 10 s. All utterances have been recorded in the clean speech environment with 48 kHz for sample frequency and in wav format. The number of speakers varies from 20 to 80 depends on each dataset.

Table 3. Training data - VoxCeleb dataset

Process	Description
Speaker	20 speakers: 10 males + 10 females
Utterance	10 utterances/speaker
Duration	3–10 s/utterance
Environment	Clean speech - Laboratory
Format	.wav
Sample frequency	48 kHz

Table 4. Testing data - VoxCeleb dataset

Process	Description
Speaker	40 speakers: 20 males + 20 females
Utterance	10 utterances/speaker
Duration	3–10 s/utterance
Environment	Clean Speech - Laboratory
Format	.wav
Sample frequency	48 kHz

4.2 Results

In the CSTR VCTK Dataset, the training data include the utterances from 40 speakers. The testing utterances are provided by 40 registered speakers and 40 unregistered speakers.

In the VoxCeleb Dataset, the training data include the utterances from 20 speakers. The testing utterances are provided by 20 registered speakers and 20 unregistered speakers.

In both datasets, there are 20 utterances for each speaker which is divided equally for the training and testing phases. Therefore, the utterances for each speaker in the training phase are different from the testing phase which also known as text-independent. Furthermore, the testing phase includes both registered and unregistered speakers which also known as the open-set, text- independent speaker identification.

Three types of error [20] can be recorded for a given test utterance as follows:

- Mislabelling (ML): A test utterance from a registered speaker is associated with an incorrect speaker identity.
- False Rejection (FR): A test utterance from a registered speaker is declared to have been produced by an unknown speaker.
- False Acceptance (FA): A test utterance from an unknown speaker is associated with a registered speaker identity.

In order to obtain the overall performance of OSTI-SI, a measure for combining all the possible types of errors is required. The Error Rate (ER) is expressed as:

$$ER = 100 \times \frac{ML + FR + FA}{T} \qquad (3)$$

In the Eq. (3) above, T is the total number of test utterances. ML, FR, and FA are three error types identified in this methodology. The Error Rate is the ratio of the total number of decision errors and the total number of test utterances.

The following sections will be discuss about the results of two test case which include the adaptive threshold test case and the predefined threshold test case.

The Adaptive Threshold Test Case

In this test case, both two datasets (CSTR VCTK and VoxCeleb) has been serially applied for training and testing. In feature matching, the process of open-set, text-independent speaker identification with the adaptive threshold will be applied to decide whether the test utterances has been produced by the registered speaker or not. The result of the open-set, text-independent speaker identification with the adaptive threshold - CSTR VCTK Dataset is shown in the table below:

Table 5. The result of open-set, text-independent speaker identification with the adaptive threshold - CSTR VCTK dataset

Open-set, text-independent speaker identification (CSTR VCTK Dataset)	
Test utterances	800
Mislabelling	3
False acceptance	11
False rejection	4
ER	2.25%

The result in the table above has been shown that the Error Rate is around 2% which mean that the accuracy of identification is approximate 98%.

The result of the open-set, text-independent speaker identification with the adaptive threshold - VoxCeleb Dataset is shown in the table below:

Table 6. The result of open-set, text-independent speaker identification with the adaptive threshold - VoxCeleb dataset

Open-set, text-independent speaker identification (VoxCeleb dataset)	
Test utterances	400
Mislabelling	0
False acceptance	6
False rejection	4
ER	2.5%

The result in the table above has been shown that the Error Rate is around 3% which mean that the accuracy of identification is approximate 97%. In this case, the error rate is slightly higher than the previous case due to the training and testing utterances included the background noises (multiple speakers at the same time).

The Predefined Threshold Test Case

In this test case, both two datasets (CSTR VCTK and VoxCeleb) has been serially applied for training and testing. In feature matching, instead of using the adaptive threshold, the predefined threshold will be used to decide whether the test utterances has been produced by the registered speaker or not. In addition, different predefined thresholds will be applied until the minimal error rate has been found. Furthermore, the first threshold is randomly chosen, after that this threshold will be changed until the equal error rate (EER) has been reached. Based on this threshold, the next thresholds will be calculated and applied.

The variation of error rates in Open-Set, Text-Independent Speaker Identification with different thresholds - CSTR VCTK Dataset is shown in Fig. 7. In this figure, Mislabelling Rate (MLR), False Acceptance Rate (FAR) and False Rejection Rate (FRR) are the rates of ML, FA, and FR errors respectively. As observed in this figure, the FA error decrease by increasing the threshold whereas the FR error shows an increasing trend with an increase in the threshold. The ML error pretty low and slightly changes via different thresholds.

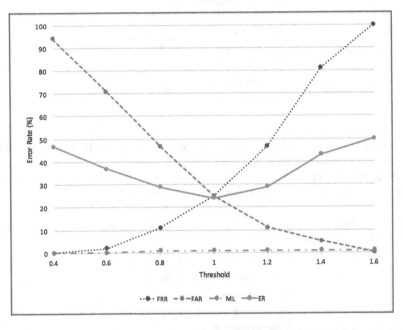

Fig. 7. Variation of error rates in open-set, text-independent speaker identification with different thresholds - CSTR VCTK dataset

The result of the open-set, text-independent speaker identification with the predefined threshold in which the minimal error rate has been achieved - CSTR VCTK Dataset is shown in the table below:

Table 7. The result of open-set, text-independent speaker identification with the predefined threshold - CSTR VCTK dataset

Open-set, text-independent speaker identification (CSTR VCTK Dataset)	
Test utterances	800
Mislabelling	3
False acceptance	100
False rejection	102
ER	26%

The result in the table above has been shown that the Error Rate is around 26% which mean that the accuracy of identification is approximate 74%. The Error Rate in this table is much higher than the Error Rate in Table 5 (with the same dataset) which is only around 2%. Therefore, we can conclude that the Open-Set, Text-Independent Speaker Identification with the adaptive threshold is more effective than the predefined threshold in case of using the CSTR VCTK Dataset.

The variation of error rates in Open-Set, Text-Independent Speaker Identification with different thresholds - VoxCeleb Dataset is shown in Fig. 8.

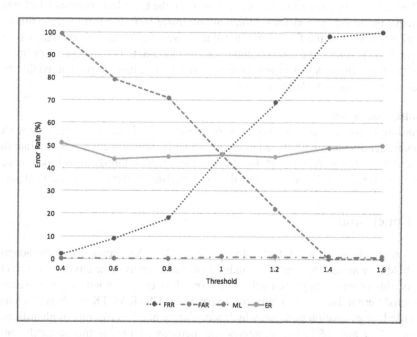

Fig. 8. Variation of error rates in open-set, text-independent speaker identification with different thresholds - VoxCeleb dataset

As observed from the figure above, the FR error increased by increasing the threshold whereas the FA error shows a decreasing trend with an increase in the threshold. The ML error notable low and insignificantly changes via distinct thresholds.

According to Figs. 7 and 8, choosing a proper threshold is greatly significant due to using a low threshold, prevents registered speakers to be rejected (low false rejection rate), on the other hand, it also takes the risk of accepting the incorrect person (high false acceptance rate), and vice versa.

The result of the open-set, text-independent speaker identification with the predefined threshold in which the minimal error rate has been achieved - VoxCeleb Dataset is shown in the table below:

Table 8. The result of open-set, text-independent speaker identification with the predefined threshold - VoxCeleb dataset

Open-set, text-independent speaker identification (VoxCeleb dataset)	
Test utterances	400
Mislabelling	2
False acceptance	157
False rejection	17
ER	44%

The result in the table above has been shown that the Error Rate is around 44% which mean that the accuracy of identification is approximate 56%. The Error Rate in this table is critical larger than the Error Rate in Table 7 (with the same dataset) which is only around 3%. Therefore, we can conclude that the Open-Set, Text-Independent Speaker Identification with the adaptive threshold is more effective than the predefined threshold in case of using the VoxCeleb Dataset.

Results Evaluation

According to the results from Tables 5, 6, 7, and 8, the Open-Set, Text-Independent Speaker Identification with the adaptive threshold is more effective and accurate than the predefined threshold method. In addition, the adaptive threshold method has been achieved a significant low error rate with high consistency through different datasets.

5 Conclusion

In this research, an open-set, text-independent speaker identification with the adaptive threshold is proposed for identifying individuals in the smart home environment. As the results, this proposed approach achieved a very low error rate with high consistency through different datasets. Furthermore, two datasets CSTR VCTK and VoxCeleb have been used for evaluation purpose, which accordingly has the error rate of identification around 2.25% and 2.5%. The strategies and methods adopted in this research can be extended to be used with multiple languages due to its advantages of high accuracy and simplicity.

The project is very useful and owns a large potential use in different industries. Although the application primary concerns more about how to identify resident in the smart home environment by using voice biometrics, the concept of voice recognition can be applied in different industries as in many situations such as authentication for banking systems, multi-speaker tracking for video conferencing.

Future work will focus on improving noise robustness which not included in this project. Speaker identification with the adaptive threshold can achieve a high level of performance. However, such performance drops significantly in noisy conditions. Furthermore, the concern of the security and privacy of smart home's resident will be addressed due to its essential needed for this project. Besides that, the proposed approach in this research needs to be verified in more datasets with higher volume in both training and testing data.

References

1. Mohamed, R., Perumal, T., Sulaiman, M., Mustapha, N.: Multi-resident activity recognition using label combination approach in smart home environment. In: 2017 IEEE International Symposium on Consumer Electronics (ISCE), Kuala Lumpur, pp. 69–71 (2017)
2. Necioglu, B.F., Ostendorf, M., Rohlicek, J.R.: A Bayesian approach to speaker adaptation for the stochastic segment model. In: 1992 IEEE International Conference on Acoustics, Speech, and Signal Processing, San Francisco, CA, pp. 437–440 (1992)
3. Sankar, A., Lee, Chin-Hui: A maximum-likelihood approach to stochastic matching for robust speech recognition. IEEE Trans. Speech Audio Process. 4(3), 190–202 (1996)
4. Reynolds, D.A.: An overview of automatic speaker recognition technology. In: IEEE International Conference on Acoustics, Speech, and Signal Processing (ICASSP) (2002)
5. Togneri, R., Pullella, D.: An overview of speaker identification: accuracy and robustness issues. IEEE Circuits Syst. Mag. 11(2), 23–61 (2011)
6. Fazel, A., Chakrabartty, S.: An overview of statistical pattern recognition techniques for speaker verification. IEEE Circuit Syst. Mag., 62–81 (2011)
7. Raja, M.N., Jangid, P.R., Gulhane, S.M.: Linear predictive coding. Int. J. Eng. Sci. Res. Technol. (2015)
8. Dave, N.: Feature extraction methods LPC, PLP and MFCC in speech recognition. Int. J. Adv. Res. Eng. Technol., 1–4 (2013)
9. Muda, L., Begam, M., Elamvazuthi, I.: Voice recognition algorithms using mel frequency cepstral coefficient (MFCC) and dynamic time warping (DTW) techniques. J. Comput. 139–140 (2010)
10. Nguyen, M.S., Vo, T.L.: Vietnamese voice recognition for home automation using MFCC and DTW techniques. In: 2015 International Conference on Advanced Computing and Applications (ACOMP), Ho Chi Minh City, pp. 150–156 (2015)
11. Xihao, S., Miyanaga, Y.: Dynamic time warping for speech recognition with training part to reduce the computation. In: International Symposium on Signals, Circuits and Systems ISSCS2013, pp. 1–4 (2013)
12. Pawar, R.V., Kajave, P.P., Mali, S.N.: Speaker Identification Using Neural Networks. In: World Academy of Science, Engineering and Technology (2005)

13. Kamble, B.C.: Speech recognition using artificial neural network—a review. Int. J. Comput. Commun. Instrum. Eng. **3**(1) (2016)
14. Shahin, I., Botros, N.: Text-dependent speaker identification using hidden Markov model with stress compensation technique. In: Proceedings IEEE Southeastcon 98 Engineering for a New Era, Orlando, pp. 61–64 (1998)
15. Maesa, A., Garzia, F., Scarpiniti, M., Cusani, R.: Text independent automatic speaker recognition system using mel-frequency cepstrum coefficient and gaussian mixture models. J. Inf. Sec., 335–340 (2012)
16. Campbell, J.P.: Speaker recognition: a tutorial. Proc. IEEE **85**(9), 1437–1462 (1997)
17. Fortuna, J., Sivakumaran, P., Ariyaeeinia, A., Malegaonkar, A.: Open-set speaker identification using adapted gaussian mixture models. In: INTERSPEECH 2005 - Eurospeech, 9th European Conference on Speech Communication and Technology, Lisbon, Portugal, 4–8 September 2005
18. Veaux, C., Yamagishi, J., MacDonald, K.: CSTR VCTK Corpus: English Multi-speaker Corpus for CSTR Voice Cloning Toolkit. University of Edinburgh. The Centre for Speech Technology Research (CSTR)
19. Nagrani, A., Chung, J.S., Zisserman, A.: VoxCeleb: A large-scale speaker identification dataset. INTERSPEECH (2017)
20. Malegaonkar, A., Ariyaeeinia, A.: Performance evaluation in open-set speaker identification. In: Vielhauer, C., Dittmann, J., Drygajlo, A., Juul, N.C., Fairhurst, M.C. (eds.) BioID 2011. LNCS, vol. 6583, pp. 106–112. Springer, Heidelberg (2011). https://doi.org/10.1007/978-3-642-19530-3_10

Modeling and Testing Power Consumption Rate of Low-Power Wi-Fi Sensor Motes for Smart Building Applications

Cao Tien Thanh[✉]

Ho Chi Minh City University of Foreign Languages and Information Technology,
Ho Chi Minh City, Vietnam
thanh.ct@huflit.edu.vn

Abstract. Power consumption is the major concern in the development of Wireless Sensor Network (WSN) applications, especially in smart building applications. Therefore, many techniques have been proposed to investigate the power consumption of these types of applications. These techniques can estimate lifetime of network, thereby providing some recommendations for developers in optimizing the energy consumption. This paper presents an approach for assess power consumption of WSN with low-power sensor Wi-Fi motes by using simulation models. Starting from propose a new power consumption model of sensor motes, due to have been implemented this model in Network Simulation 2 (NS-2) for testing and evaluation, used GNU plot and Trace Graph tool for plot the graphs for NS-2. Consequently, we can show results, compare its in detail.

Keywords: Wireless sensor network · Power consumption · Smart building
Ns2

1 Introduction

Energy management is a critical concern in wireless sensor networks. Without it, the lifetime of a wireless sensor node is limited to just a few short weeks or even days, depending on the type of application it is running and the size of batteries it is using. With efficiency energy management, the same node, running the same application, can be made to last for months or years on the same set of batteries. The goal of this paper is to study firstly an existing low-power and optimized algorithms and models to represent the power consumption rate of WSNs motes. Secondly, develop a new model for our developed low-power Wi-Fi sensor motes in smart building applications. Thirdly, the implementation, and the performance evaluation of the proposed/extended algorithms/models in NS2 are needed. The content of this paper consists of the table, figures, data, diagrams which are generated by testing a full-featured existing application, a description of methodologies that were used to manage energy consumption and the implementations of low power WSN applications.

© Springer Nature Switzerland AG 2018
T. K. Dang et al. (Eds.): FDSE 2018, LNCS 11251, pp. 449–459, 2018.
https://doi.org/10.1007/978-3-030-03192-3_34

2   Overview on Power Consumption Model

This section describes what previous researchers have proposed and demonstrated. From which we will analyze and compare the techniques that they applied and make a judgment which effectively, which is not effective to seek development consistent with reality device in [6] that we are.

They tend to find routing protocols or MAC protocols. Some of them show methodologies related to the operation system for nodes and its into radio structure. Around there we will see some of these works, and we will give a key's segment choices presented in theses portion.

In next section, the MAC protocols were proposed by several earlier researchers will be introduce. Therefore, we can see how the problem with energy saving can be resolved.

2.1   Energy Saving at MAC Layer

Generally, MAC protocols are classified into two categories: contention based and scheduled based protocols.

Contention based protocols allow many users to use the same radio channel without pre-coordination. The main idea of these protocols is to listen the channel before sending the packet, IEEE 802.11, ALOHA and CSMA (Carrier Sense Multiple Access) are the most well-known contention-based protocols. Compared to the scheduled based protocols, the contention one is simple, because they don't require global synchronization, or topology knowledge which allows some nodes to join or to leave the network few years after deployment. Message collisions, overhearing and idle listening are the main drawbacks of this approach. Thereafter we will present these basic protocols (IEEE 802.11, ALOHA and CSMA), and in the next sections other contention-based approaches more suitable for WSNs will be discussed [6].

Sensor MAC in WSNs (SMAC). Authors in [1, 2] depict SMAC can reduce energy consumption of network by using three transceiver modes and SMAC allows node transmit, receive and sleep periodically with a low duty cycle. The energy consumption of node is proportional of the activity time. A low duty cycle has a low energy consumption.

Fundamental concept of SMAC is periodic sleep listen schedules which are handled locally by the sensor network. Nodes which are adjacent form clusters virtually and they apportion prevalent schedule. So two nodes are contiguous and fall in two different clusters they wake up at listen schedule of both clusters.

This also results in more energy consumption as nodes wake up to two different schedules. The schedules are also needed to be communicated to different nodes of virtual cluster which is accomplished by SYNC packets and time in which it is sent is known as synchronization period. Figure 1 represents a sample sender-receiver communication. Carrier Sense (CS) helps in collision avoidance. In addition to it, unicast data packets transmission is done using RTS/CTS. A new and innovative feature of SMAC is message passing through which a long message is sent in burst by dividing it into small messages. This helps in energy saving by using common overhead.

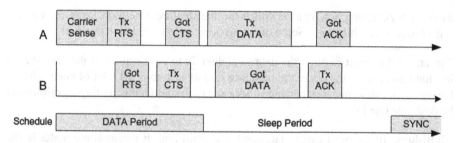

Fig. 1. A successful data transmission between two nodes

However, this concept of sleeping schedule may also result in high delay termed as latency which will be significant in case of multi-hop routing algorithms, as each node in between will have their own sleep schedules. This is known as sleep delay. This disadvantage can be overcome by using adaptive listening technique. In that technique, the overhearing node wakes up for a small duration at the end of the transmission. So, if this node is the next-hop node, it can take the data immediately from the transmitting/passing node. Figure 2 illustrates how SMAC works.

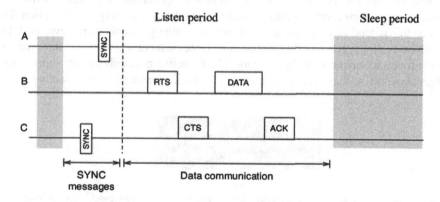

Fig. 2. Sensor MAC in wireless sensor network

Advantages: The battery utilization is increased implementing sleep schedules. This protocol is simple to implement long messages can be efficiently transferred using message passing technique.

Disadvantages: RTS/CTS are not used due to which broadcasting which may result in collision. Adaptive listening causes overhearing or idle listening resulting in inefficient battery usage. Since sleep and listen periods are fixed variable traffic load makes the algorithm efficient.

2.2 Energy Saving at Network Layer

The previous section was about sensor nodes elements, data collision in wireless channel. The selection of optimal data routing and communication protocol which costs a low energy consumption for a network is the main feature of network layer in a wireless sensor network architecture. Some of data routing have included in this part.

Direct. In the direct routing, the data send directly to base station. If the base station lies long away from the node, the traveling of data is consumed a lot of energy but in the cases with short distance between nodes and base station, this routing algorithm can be desirable one [4].

Multi-hop. In the most cases, data send in a multi-hop. It means some nodes in the network act such as intermediate nodes. Each node can collect data from environment and transmit data to neighboring nodes. The router nodes are responsible for routing and disseminating data. The existence of many router nodes increases the number of short hops in the network. Clearly, power consumed by using N short hops is N times smaller than the power consumed in a long hop [3].

Figure 3 shows some popular protocols in WSN. We focused on flat routing protocols, each node typically plays the same role and sensor nodes collaborate to perform the sensing task. Flooding is flat type of routing protocol in which each sensor node receives data and then sends them to the neighbors by broadcasting, unless a maximum number of hops for the packet are reached or the destination of packet is achieved. Disadvantage of this routing technique is data redundancy and energy consumption. To reduce this redundancy Proactive and Reactive routing protocols were developed. In Proactive routing, each node has one or more tables that contain the latest information of the routes to any node in the network. The Proactive protocols are not suitable for larger networks, as they need to maintain node entries for each node in the routing table.

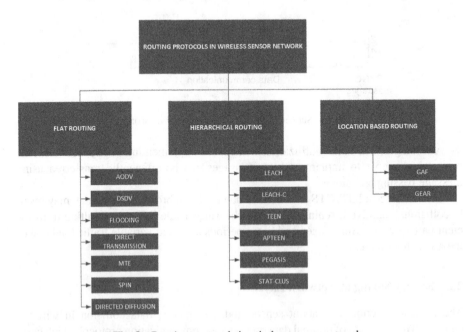

Fig. 3. Routing protocols in wireless sensor network

This causes more overhead in the routing table leading to increases energy consumption. Destination-Sequenced Distance-Vector routing (DSDV) is one of the table-driven routing schemes for sensor networks based on the Bellman-Ford algorithm. To reduce this regular update, Reactive Routing protocols were developed. Reactive routing protocols Dynamic Source Routing (DSR) and Adhoc On-demand Distance Vector (AODV) discover route only when a source node wants to communicate with destination [1, 2].

Energy efficient routing is possible by means of cluster based routing or hierarchical schemes.

3   New Power Consumption Model in WSN

Based on basic model [2]. If we assume a channel model, which only includes path loss then a multi-hop routing scheme will perform the best in a simple 1-D linear WSN topology [5]. The single-hop 1-D linear WSN consists of a source node S and a destination node D separated by a distance R, and multi-hop 1-D linear WSN has an additional n-1 intermediate identical relay nodes Ni, i = 1, ...n − 1, placed in a line from S to D. In Fig. 4(b), the relay nodes are placed an arbitrary distance apart, and in Fig. 4(c) the relay nodes are placed equidistantly. The objective of the WSN is the reliable.

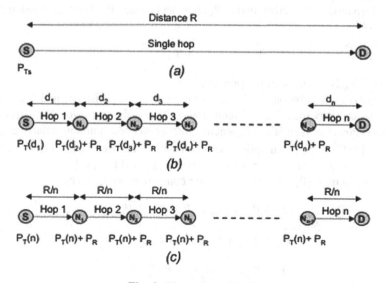

Fig. 4. Network model [4]

P_R describes the power consumption for receiving. $P_T(d_i)$ denotes the power consumption for transmitting over a distance d. i is an integer from 1 to the total number of hops, n. $P_T(R/n)$ denotes power consumption for transmitting over a distance R/n. We use $P(n)$ to denote the total power consumption for sending from S to D with n-hops. We ignore the power consumption in the destination node D, because it is assumed to be connected to an external power supply and is not resourced constrained. Based on the network model in Fig. 4(b), we can obtain the multi-hop power consumption model with arbitrary distance between nodes as follows:

$$P_{(n)} = (n-1)P_{R0} + nP_{T0} + \frac{\varepsilon}{\eta}\sum_{i=1}^{n}d_i^{\alpha} \tag{1}$$

Similarly, based on the network model in Fig. 4(c), we can obtain the multi-hop power consumption model with equal distance between nodes as follows:

$$P_{(n)} = (n-1)P_{R0} + nP_{T0} + \frac{n \times \varepsilon \times (R/n)^{\alpha}}{\eta} \tag{2}$$

In particular, this model of WSN power consumption clearly shows the dependence of the power amplifier performance, which differs from other power consumption models widely cited by the WSN research community.

Based on the model of author [4], we proposed new power consumption model with parameters described as: The total power consumption P(n) of a node is a result of all steps of the operation: sensing, data processing and radio transmission from source (S) to destination (D) with n-hops. Assume that transmission without interference. We have a new model for power consumption and a new algorithm to control power in transition modes: sleep mode - P_{sl}, active mode - P_{ac}, transient mode - P_{tr}. The transmission period T is given by:

$$T = T_{tr} + T_{ac} + T_{sl} \tag{3}$$

$T_{ac} \leq T$, T_{ac} is a parameter to optimize.

In details of transmission, we defined some parameters are used in model (equation): P_t – Transmit power; P_{cp} – Circuit power; P_{mix} – mixer power; P_{fs} – frequency synthesizer power consumption; P_{fil} – active filter power consumption at the transmitter. P_{DAC} – the DAC power consumption; $\alpha = nd/\eta$ with η - drain efficiency [4]. Pac consists Pt and Pcp in signal path. While, P_{cp} consists P_{mix}, P_{fs} and P_{LNA} [4]

Since $P_{ac} = \max \{P_{ac}, P_{tr}, P_{sl}\}$ The power constraints are given by:

$$P_{ac} = P_t + P_{cp} = (1+\alpha)P_t + P_{cp} \leq P_{max} \tag{4}$$

$$P_{cp} = P_{mix} + P_{fs} + P_{fil} + P_{DAC} \tag{5}$$

$$E = P_{ac}T_{ac} + P_{sl}T_{sl} + P_{tr}T_{tr} \tag{6}$$

E is total energy consumption.

Smart Building Applications. In this section, we present the model to real scenario and describe how to reduce power consumption in low-power Wi-Fi sensor motes in smart building applications. The main idea is in smart building application, the 802.11 technology provides larger coverage range compared to other technologies such as ZigBee, Bluetooth, etc., allowing us to decrease the number of immediate nodes in multi-hop connections for data transmitting and receiving, and thus, increasing the probability of receiving data correctly, as well as reducing the latency. This will allow the users, managers, re-act instantly and faster once the troubles and problems occur.

Figure 5 shows the main features of smart building application. Which are increasingly being completed for intelligently and flexible management.

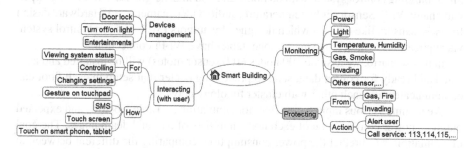

Fig. 5. Smart building application

We propose one real model smart building control system (SBCS) with many types of sensor. Such as camera, audio, light, temperature sensor, ... each with their own responsibilities is different, however, they have the same general purpose: collect data and reporting to processing center (sink). Figure 6 shows the different sensors in SBCS.

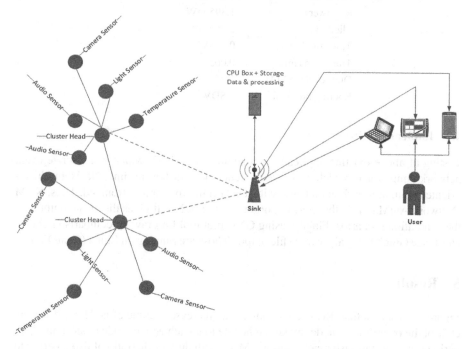

Fig. 6. Multimedia sensor nodes in smart building application

4 Experiment

4.1 Scenario

Smart building control system with 60 floors and 200 m in height. Each floor is equipped with many Wi-Fi Sensor (light, temperature, audio, video, camera, …). Hardware design for these sensors like devices which designed by authors [3]. Users can control system via applications installed on smart phone, tablet have Wi-Fi connection.

In this scenario, we assume 100 nodes (100 sensor motes) will be installed and activated and ready for collect data, send to processing center that setup in each floor can communicate with each other with cluster topology network.

We assume various input parameters for simulation as Table 1 below, and expected output is power consumption of each node, numbers of live and death nodes. The goal of simulation is to predict the power consumption, comparing the different between to MAC protocols: MAC 802.11 and SMAC in power saving.

Table 1. Input parameters for simulation

Data rates	11 MB
Initial energy	10 mW
TxPower	1.660 mW
RxPower	1.395 mW
IdlePower	0.25 mW
TransitionPower	0.2 mW
TransitionTime	0.005
Duty-cycle (SMAC)	10%
Routing protocol	DSDV

4.2 Implementation

Testing of above existing applications is simulated on NS2, which is a discrete event network simulator in which each event occurs at an instant in time. NS is an object-oriented simulator, written in C++, with an OTcl interpreter as a frontend. Thus, NAM (Network AniMator) is the most important feature of NS. It gives the visual outputs of the simulation scenarios. Finally, using GNU plot tool for drawing comparison graphs and Trace Graph for analyze trace file output. Those are running on Linux (CentOS 6.5).

5 Result

Figures 7, 8 shows that: Reduce idle power can increase lifetime of node. So if we can control the operation of node: make a schedule to switch active mode to sleep mode in period time, we can save much energy. SMAC with duty cycle (ratio of listen period to a complete sleep and listen cycle) is the good way to control.

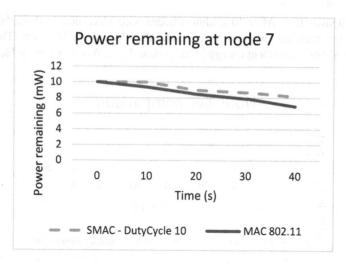

Fig. 7. Result with SMAC duty cycle 10% and idle power 0.05 mW

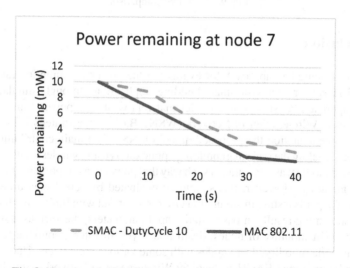

Fig. 8. Result with SMAC duty cycle 10% and idle power: 0.25 mW

For contention-based MAC protocols, such as IEEE 802.11 ad-hoc mode, in order to perform effective carrier sense against possible collisions, it puts nodes to listen to the channel all the time when nodes are idle. And radio will consume almost the same power as in receiving state. A considerable percentage of energy will be wasted on idle listening, especially when the traffic load on the network is light. Among those factors for energy waste, idle listening is a dominant one.

The results indicate that: SMAC save more energy than IEE 802.11 MAC in same environment.

Figure 9 show that: After 40 s, almost nodes with MAC 802.11 dead, but SMAC nodes still live with the low power level, until after 70 s they will be die. That can be concluded: SMAC is efficient energy MAC protocols for Wireless Sensor Networks.

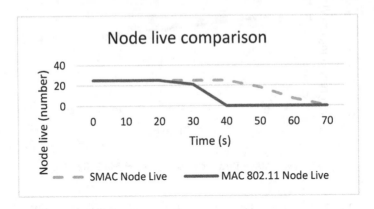

Fig. 9. Node live comparison

6 Conclusion

This paper presented an approach for evaluating the power consumption rate of low-power Wi-Fi sensor motes in smart building applications through simulation. The proposed approach consists a new model that represents the power consumption of sensor nodes in Wireless Sensor Network (WSN). Basic in the strategy is the development and implementation the proposed model on NS2. The main contribution of this paper is the modeling the real smart building applications using wireless sensor network. The proposed model was evaluated in such way that power consumption results obtained through simulating. However, this should be evaluated by actual measurement. The limitation of paper is testing in simulation environment and with limited parameters, we have just shown the results in basic, static input parameters, the various distance, not mentioned about mobility of nodes, death nodes yet. In future, continue to simulate WSNs with node mobility, death nodes with some various parameters. Understanding, running and simulating LEACH protocol for Wireless Sensor Networks. Simultaneously combined with MAC protocols to solve the problem of power consumption more efficiently.

In the future, based on the research of Thien et al. [7–11], we are also interested in integrating speech processing techniques as well as natural language processing techniques for smart building applications.

References

1. van Dam, T., Langendoen, K.: An adaptive energy-efficient MAC protocol for wireless sensor networks. In: ACM SenSys, Los Angeles, CA, November 2003
2. Vuran, M.C., Akan, B., Akyildiz, I.F.: Spatio-temporal correlation: theory and applications for wireless sensor networks. Comput. Netw. J. **45**(3), 245–259 (2004)
3. Le-Trung, Q., Engelstad, P.E., Skeie, T., Eliassen, F.: Information storage, reduction, and dissemination in wireless sensor networks: a survey: In: 6th IEEE Consumer Communications and Networking Conference, CCNC 2009, pp. 1–6. IEEE (2009)
4. A Realistic Power Consumption Model for Wireless Sensor Network Devices. http://ece.drexel.edu/mhempstead/pubs/SECON2006-power-model.pdf
5. Sikora, M., Laneman, J.N., Haenggi, M., Costello, Jr, D.J., Fuja, T.: On the optimum number of hops in linear wireless networks: In: ITW 2004 (2004)
6. Cristescu, R., Beferull-Lozano, B., Vetterli, M., Wattenhofer, R.: Network correlated data gathering with explicit communication: NP-completeness and algorithms. IEEE/ACM Trans. Netw. **14**(1), 41–54 (2006)
7. Tran, T.K., Pham, M.D., Huynh, V.B.: Towards building an intelligent call routing system. Int. J. Adv. Comput. Sci. Appl. **7**(1), 2016 (2016)
8. Tran, T.K., Phan, T.T.: An upgrading SentiVoice—a system for querying hotel service reviews via phone. In: Proceedings of the 19th International Conference on Asian Language Processing (IALP), Suzhou, China, 24–25 October 2015
9. Tran, T.K., Phan, T.T.: Mining opinion targets and opinion words from online reviews. Int. J. Inf. Technol. **9**(3), 239–249 (2017)
10. Tran, T.K., Phan, T.T.: A hybrid approach for building a Vietnamese sentiment dictionary. J. Intell. Fuzzy Syst. **35**(1), 967–978 (2018)
11. Tran, T.K., Phan, T.T.: Towards a sentiment analysis model based on semantic relation analysis. Int. J. Synth. Emot. (IJSE) **9**(2), 54–75 (2018)

Emerging Data Management Systems and Applications

Distributed Genetic Algorithm on Cluster of Intel Xeon Phi Co-processors

Nguyen Quang-Hung[✉], Anh-Tu Ngoc Tran, and Nam Thoai

Faculty of Computer Science and Engineering, Ho Chi Minh City University of Technology, Vietnam National University Ho Chi Minh City, Ho Chi Minh City, Vietnam
{nqhung,51304672,namthoai}@hcmut.edu.vn

Abstract. This paper presents a study to parallelize genetic algorithm on cluster of Intel Xeon Phi co-processors. Our study investigates using hybrid MPI and OpenMP to parallelize genetic algorithm (GA) onto high performance computing (HPC) servers included Intel Xeon Phi co-processors. We will revise sequential GA and parallelize the GA into parallel GA, which a highly portable, modern code and optimized in any MIC architectures like Knights Corner (KNC), Knights Landing (KNL), or even the up-coming processors of Intel like Knights Hill by using Intel compiler, MPI and several OpenMP directives. Our implementation is experimented on built-in 50-TFlops Supernode-XP, which is cluster of HPC servers (each server has two Intel Xeon Phi cards). Our simulated result shows that the proposed distributed GA onto cluster of Intel Xeon Phi co-processors (MICs) reduces significant execution time in comparison with sequential GA.

Keywords: Distributed Genetic Algorithm · Intel Xeon Phi · MIC

1 Introduction

Genetic Algorithm (GA) is an algorithm that can be highly parallelized. For the purpose of decreasing execution time, Intel Xeon Phi co-processor or Graphic Processing Unit (GPU) is usually used to achieve that purpose. Based on the work of my predecessors [5,6] that we try to solve task assignment problems again in a different approach which is an implementation on GPU and Intel Xeon Phi. We will revise a sequential GA and turn it into DGAXP, which a highly portable, modern code and optimized in any MIC architectures like Knights Corner (KNC), Knights Landing (KNL), or even the up-coming processors of Intel like Knights Hill by using Intel compiler, MPI and several OpenMP directives.

Our previous GAPhi [5] is implemented in offload mode. To be specific, CPU will do the initialization, and KNC will do the intensive computation. During the process of doing, we encountered some cases which cause the program slow. Therefore, we will show you some fundamental techniques you need to know to resolve them when you program with Intel Xeon Phi. In this paper, we present a

© Springer Nature Switzerland AG 2018
T. K. Dang et al. (Eds.): FDSE 2018, LNCS 11251, pp. 463–470, 2018.
https://doi.org/10.1007/978-3-030-03192-3_35

new method, which is symetric hybrid MPI and OpenMP to benefit the parallel GA from cluster of HPC servers and multiple Intel Xeon Phi co-processors.

The remainder of the paper is organized as follows. In Sect. 2, we discuss related works to our study. In Sect. 3, we will introduce you to some basic knowledge that you need to know how to program with Intel Xeon Phi and use OpenMP directives to parallelize your code. Detail of implementation GA on Intel Xeon Phi and the experimental results of execution time for parallel GA are also presented. Finally, we give some conclusions in Sect. 4.

2   Related Works

Genetic algorithm is used to find an optimal solutions for various applications in many previous works [3–5]. Genetic algorithm is also used to solve other problems such as two-dimensional panel codes [4]. The author tries to find an ideal airfoil geometry for engineering application. However, this is a nonlinear optimization problem, and traditional optimization algorithms are ineffective. Genetic algorithm is used as an alternative to find optimal solution. Based on his description, the problem itself can be well parallelized, and it can significantly benefit from accelerators such as GPU or Intel Xeon Phi. He solves his problem using multiple GPUs. The Power Aware Task Scheduling (PATS) problem in high performance computing (HPC) cloud has been described carefully in [5], they have been successfully in making the program faster by using GPU or Intel Xeon Phi [6].

Algorithm 1. Distributed DGAXP's Main

1: Initialize some populations of chromosomes with random values in the search space
2: Call MPI_Init()
3: Call MPI_Comm_rank(MPI_COMM_WORLD, &rank)
4: Call MPI_Comm_size(MPI_COMM_WORLD, &nproc)
5: **for** Some running i **do**
6:     Call *GAEvolution*(generation)
7:     Call MPI_Gather(&best_sol, 1, MPI_FLOAT, solutions, 1, MPI_FLOAT, MASTER, MPI_COMM_WORLD)
8:     **if** (rank == MASTER) **then**
9:         Choose the best solutions among nodes
10:    **end if**
11:    exe_time = omp_get_wtime() - exe_time;
12: **end for**
13: Call MPI_Finallize()

---

3   Implementation and Simulation

3.1   Algorithms

In this section, we present our proposed distributed GA onto cluster of Intel Xeon Phi co-processors (DGAXP). The Algorithms 1 and 2 are pseudo-code for the

DGAXP's main() and GA_Evolution() functions. Initialization of chromosomes, evaluation of fitness, crossover, mutation and selection will be run on KNC. Memory on KNC is only allocated one time and reused many times to reduce offload latency (data retention). Furthermore, in order to avoid time consumed by memory traffic, the result is only loaded back to CPU memory when the computation finishes; intermediate results will not be stored or loaded back to CPU memory.

On KNC, data parallelism will be done on all of genetic operations by using OpenMP directives. Random number is a key function in GA, but function like **rand()** can not be parallelized, which causes the program extremely slow, even slower than SGA. The function rand() is not reentrant or thread-safe, since it uses hidden state that is modified on each call [2]. We try to resolve this problem by using **lrand48()** [1], on the other hand, GPU version uses CUDA library **curand.h** to resolve it [5].

To optimize code, the DGAXP uses pragma OpenMP directives to take advantage of MIC architecture. When we do programming with Intel Xeon Phi, to maximize performance, particularly in KNC, we do optimization techniques such as vectorization (use **#pragma omp simd**), to take advantage of 512-bit vector register, multi-threading (use **#pragma omp parallel** or **#pragma omp for**), to use up 61 cores, 4 threads per core, and finally memory (**#pragma target offload**), to offload job and data to KNC. We present pseudocode of the DGAXP operations in the Function 1 to 5. We add some directives mentioned above as a clarification for techniques we used Algorithm 2.

Algorithm 2. GAEVOLUTION

1: Initialize some clones using pragma omp parallel for simd
2: **for** (int i =0 ; i¡ sizePop*numTask ; i++) **do**
3:    NSToffspring[i] = mangNST[i];
4: **end for**
5: Call MPI_Init()
6: Using **#pragma omp parallel for simd** to create list of events (starting, ending) of tasks
7: Call evaluation for parents
8: Cloning the fitness array fitnesses of offspring using (**#pragma omp parallel for simd**)
9: **for** some N generations **do**
10:    Call initization some chromosomes using the Function 1
11:    Call crossover for offsprings using the Function 3
12:    Call evaluation all chromosomes using the Function 2
13:    Call selection some offsprings using the Function 5
14:    Call mutation offsprings array using the Function 4
15:    Call evaluation some offsprings using the Function 2
16:    Call selection some offsprings using the Function 5
17: **end for**

Codename	CPU	Coprocessor
Model	Intel Xeon E5-2680V3	Intel Xeon Phi 7120P
Microarchitecture	Sandy Bridge EP	Intel Many Integrated Core
Clock frequency	2.50/3.30 GHz	1.24/1.33 GHz
Memory Size	128 GB	16 GB
Cache	30.0 MB SmartCache	30.5 MB L2
Max Memory Bandwidth	68 GB/s	352 GB/s
Core/Threads	12/24	61/244

Fig. 1. System environment configuration.

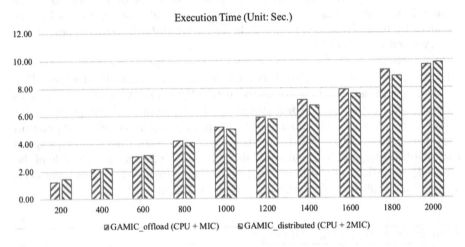

Fig. 2. Execution time (unit: seconds) with problem size of 500 machines and 500 tasks.

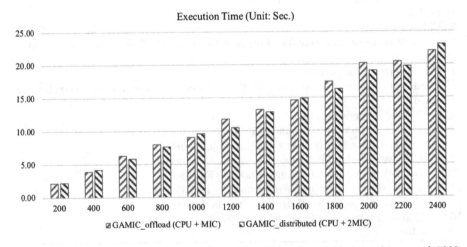

Fig. 3. Execution time (unit: seconds) with problem size of 1000 machines and 1000 tasks.

Function 1: Init chromosome

Memory allocation of array of chromosomes on KNC
#pragma offload target(mic)
{
  #pragma omp parallel for simd
  Initalize array of chromosomes with random values
}

Function 2: Evaluation

#pragma omp parallel for
FOR (still in array of chromosomes)
  powerOfDatacenter = 0
  FOR each host in a chromosome
    utilizationMips = host.getUtilizationOfCpu()
    powerOfHost = getPower (host, utilizationMips)
    powerOfDatacenter = powerOfDatacenter + powerOfHost
  ENDFOR
  Evaluation value (chromosome) = 1.0/powerOfDatacenter
ENDFOR

Function 3: Crossover

#pragma omp parallel for
FOR (still in array of chromosomes)
  Find a cross point randomly
  Find a chromosome randomly and different than current
  chromosome doing crossover
  #pragma omp simd
  Doing crossover between two chromosomes
ENDFOR

Function 4: Mutation

#pragma omp parallel for simd
FOR (still in array of chromosomes)
  IF (Mutaition happens)
    Modify chromosomes with random values
  ENDIF
ENDFOR

Function 5: Selection

Sort array of Fitness of Parents and Offspring populations from high to low
Pick up the chromosomes that have high fitness until reaching the number of populations

3.2 Experimental Results

We present experimental results of the serial GA, GAMIC with Offload mode, and DGAXP using the system environment configuration (Fig. 1) in the Tables 2 and 3. The Table 2 and the Fig. 2 show the execution times of serial GA and parallel GA with input PATS problem size of 500 machines and 500 tasks. The Table 3 and the Fig. 3 show the execution times of serial GA and parallel GA with input PATS problem size of 1000 machines and 1000 tasks.

The experimental results are quite interesting. The DGAXP program on Intel Xeon Phi is implemented and optimized by using MPI, Intel compiler and OpenMP. The execution time of the DGAXP is less than the execution time of the serial GA on CPU/MIC, and is equal to the GAPhi enabled offload mode. The DGAXP, which proposed hybrid MPI and OpenMP can use cluster of HPC servers with multiple Intel Xeon Phi co-processors, returns a better solution (results in 100% found solutions) than serial GA and our previous GAPhi [6] enables only offload mode. Table 1 shows the number of found solutions in 5 trials. The GAPhi enabled offload mode is suitable for system that has slow speed CPU, where combined with Intel Xeon Phi co-processor to speedup the offload programs. On the other hand, the DGAXP is good for cluster of HPC servers included Intel Xeon Phi co-processors to speedup parallelized HPC applications, which is programmed in hybrid MPI and OpenMP. The execution time of the DGAXP is the maximum execution time of the total execution time on slowest Intel Xeon Phi co-processor in the cluster.

Table 1. The number of found solutions in 5 trials

Population size	GA_serial (1 core CPU)	GA_serial (1 core MIC)	GAMIC_offload (1 CPU + 1 MIC)	DGAXP (1 CPU + 2 MIC)
200	0/5	0/5	2/5	3/5
400	0/5	0/5	3/5	5/5
600	0/5	1/5	1/5	5/5
800	2/5	3/5	4/5	5/5
1000	4/5	5/5	4/5	5/5
1200	4/5	2/5	4/5	5/5
1400	5/5	5/5	5/5	5/5
1600	5/5	5/5	5/5	5/5
1800	5/5	5/5	5/5	5/5
2000	5/5	5/5	4/5	5/5

Table 2. The execution times of serial GA and parallel GA with input PATS problem size of 500 machines and 500 tasks.

Population size	Serial GA (CPU) (1)	Serial GA (MIC) (2)	DGAXP Offload (OpenMP) (3)	Distributed DGAXP (MPI + OpenMP)			Speedup (7) = (2) / (6)
				Compute time (4)	Communication time (5)	Execution time(6)	
200	2.06609	31.67890	1.22781	1.47036	0.00042	1.47078	21.54
400	4.57777	63.76990	2.15777	2.22273	0.00012	2.22285	28.69
600	6.52907	95.88890	3.08261	3.16122	0.00011	3.16133	30.33
800	8.89266	127.36900	4.21093	4.06109	0.00008	4.06117	31.36
1000	11.88310	159.71800	5.17535	5.03362	0.00021	5.03383	31.73
1200	13.83870	190.69700	5.87861	5.72198	0.00015	5.72213	33.33
1400	17.12340	223.52600	7.13194	6.71412	0.00031	6.71443	33.29
1600	19.30890	255.33800	7.87513	7.57635	0.00018	7.57653	33.70
1800	21.80300	287.41000	9.29850	8.83208	0.00009	8.83217	32.54
2000	24.17800	319.34000	9.67927	9.83907	0.00026	9.83933	32.46

Table 3. The execution times of serial GA and parallel GA with input PATS problem size of 1000 machines and 1000 tasks.

Population size	Serial GA (CPU) (1)	Serial GA (MIC) (2)	DGAXP Offload (OpenMP) (3)	Distributed DGAXP (MPI + OpenMP)			Speedup (7) = (2) / (6)
				Compute time (4)	Communication time (5)	Execution time (6)	
200	2.07	31.68	2.16646	2.17587	2.23975	0.00029	2.24004
400	4.58	63.77	3.95930	3.89215	4.16522	0.00010	4.16532
600	6.53	95.89	5.66242	6.30222	5.79641	0.00033	5.79674
800	8.89	127.37	7.77659	7.98724	7.63343	0.00010	7.63353
1000	11.88	159.72	9.66490	9.06827	9.58244	0.00001	9.58245
1200	13.84	190.70	10.81020	11.77430	10.46230	0.00026	10.46256
1400	17.12	223.53	13.01400	13.20780	12.84890	0.00006	12.84896
1600	19.31	255.34	14.77830	14.60490	14.89150	0.00001	14.89151
1800	21.80	287.41	16.08040	17.41820	16.32530	0.00014	16.32544
2000	24.18	319.34	17.69780	20.19760	19.09010	0.00002	19.09012

4   Conclusions and Future Works

In a conclusion, our propose symmetric hybrid MPI and OpenMP to reduce significantly execution time in compared to sequential program. Some of the results are quite interesting. The DGAXP program on Intel Xeon Phi is implemented and optimized by using MPI, Intel compiler and OpenMP. In addition, the proposed hybrid MPI and OpenMP can use cluster of HPC servers with multiple Intel Xeon Phi co-processors to return a better solution than our previous GAPhi with only offload mode.

In the future work, we would like to evaluate the proposed solution with bigger testbed. We will run the distributed GA (DGAXP) on many different KNC cards to increase correctness, decrease time execution in order to solve bigger and more complicated problems.

Acknowledgments. This research is funded by Vietnam National University Ho Chi Minh City (VNU-HCM) under grant number C2017-20-09.

References

1. lrand48(3) - Linux man page_2017 (2017)
2. rand(3): pseudo-random number generator - Linux man page_2017 (2017)
3. Andreolli, C., Thierry, P., Borges, L., Yount, C., Skinner, G.: Genetic algorithm based auto-tuning of seismic applications on multi and manycore computers. In: EAGE Workshop on High Performance Computing for Upstream (2014)
4. Einkemmer, L.: Evaluation of the Intel Xeon Phi 7120 and NVIDIA K80 as accelerators for two-dimensional panel codes. PloS One **12**(6), e0178156 (2017)
5. Quang-Hung, N., Tan, L.T., Phat, C.T., Thoai, N.: A GPU-based enhanced genetic algorithm for power-aware task scheduling problem in HPC cloud. In: Linawati, M.M.S., Neuhold, E.J., Tjoa, A.M., You, I. (eds.) ICT-EurAsia 2014. LNCS, vol. 8407, pp. 159–169. Springer, Heidelberg (2014). https://doi.org/10.1007/978-3-642-55032-4_16
6. Quang-Hung, N., Tran, A.-T.N., Thoai, N.: Implementing genetic algorithm accelerated by Intel Xeon Phi. In Proceedings of the Eighth International Symposium on Information and Communication Technology, SoICT 2017, New York, NY, USA, pp. 249–254. ACM (2017)

Information Systems Success: Empirical Evidence on Cloud-based ERP

Thanh D. Nguyen[1,2(✉)] ⓘD and Khiem V. T. Luc[1]

[1] Banking University of Ho Chi Minh City, Ho Chi Minh City, Vietnam
thanhnd@buh.edu.vn, khiemltv@gmail.com
[2] Bach Khoa University, Ho Chi Minh City, Vietnam

Abstract. Cloud-based ERP is one of the new trends of state-of-art information systems. This paper integrates the illustrious IS success model of DeLone and McLean and the related concepts of trust and perceived risk in an extending success model of cloud-based ERP. Data is collected from 182 participants who have used the cloud-based ERP in Vietnam. A structural equation modeling is effectuated by the maximum likelihood estimation for analysis and evidence. The findings manifest that the constructs of system quality, information quality, IT service quality, perceived risk, trust, and intention to use which have the structural relationships with the net benefit. The research model accounts for 37% of the success of cloud-based ERP.

Keywords: Cloud-based ERP · IS success · Perceived risk · Trust

1 Introduction

Information Technology (IT) is renovating the relationships among the organizational stakeholders and taking an important role in the sustainable growth in GDP of the countries in the world [4]. Cloud-based ERP can support organizations achieve the ERP benefits with little regard about IT infrastructure [22]. Cloud-based ERP solutions also help the organizations reduce the pressure on IT departments, requiring only a cost for ERP software [23]. For instance, the cost of cloud-based ERP is 15% which lower than traditional ERP and implementation time has fallen around 50 to 70% [2]. Hence, the organizations only choose a standard software package or a custom software package to suit their business needs from a supplier. However, the solutions are highly dependent on suppliers, so choosing the right supplier is very important [33]. Several organizations around the world have deployed or are currently in the process of deploying the cloud-based ERP, the market share of ERP systems rose from 11 to 27% in just one year [40]. In Vietnam, there are only 1.1% of organizations using ERP solutions, of which most the ERP projects in these organizations do not achieve as the desired target [38]. Although cloud-based ERP solutions are being considered by many organizations, whether they are accepted or not depends on a variety of factors [1, 2, 22].

Regardless of how the economy, organizations need to consolidate that the investments in Information Systems (IS) are successful. The selection of successful elements depends on the feature and purpose of the IS [13]. Hence, in order to measure the success

© Springer Nature Switzerland AG 2018
T. K. Dang et al. (Eds.): FDSE 2018, LNCS 11251, pp. 471–485, 2018.
https://doi.org/10.1007/978-3-030-03192-3_36

of IS, which is necessary to base on the specific context of IS (e.g., e-commerce, ERP…). The purpose of this study that approaches the background of cloud-based ERP, the literature reviews of the IS success theory of DeLone and McLean [10–13], and other IS theories such as the technology adoption such as TAM of Davis [9], UTAUT of Venkatesh et al. [55], perceived risk and trust in e-CAM of Park et al. [41] and of Bauer [6], Pavlou [42], and the related works. This study proposes and investigates an empirical evidence of the success of IS in the context of ERP in cloud computing. Data is collected from the respondents who have used the cloud-based ERP in Vietnam. A structural model is analyzed by SEM (Structural Equation Modeling). Research results provide the information for organizations in developing the ERP system in cloud computing, and also grant the knowledge of the IS theories. There are five parts of this study: (1) introduction, including the research problem; (2) background, including introduction about cloud-based ERP, literature review about the IS success and related works; (3) research model, including the theoretical framework and hypotheses, and the research method; (4) research result, including the analyses of exploratory factor, confirmatory factor, structural equation modeling, and the discussion; (5) conclusion and future work.

2 Background

2.1 Cloud-Based ERP

Cloud Computing
Cloud Computing (cloud) is becoming more and more popular in the global and it is the growing trend in modern IT [39]. Nevertheless, cloud computing is not completely new technology, it is the combination among the constituents of existing IT services [18]. Some researchers believed that the cloud represents the future for the IT use. In particular, the power of cloud computing has a profound impact on the IT industry, organizations do not need to install software on their systems and not need to buy the hardware or software, which simply hires the IT service from the vendor [54]. Thus, the services of cloud-based can excommunicate the hardware and software cost, it countenances the organizations focus more on business than IT [5].

Enterprise Resource Planning
Enterprise Resource Planning (ERP) is also a new trend of IS, this is an IS that helps organizations manage their resources and operations, ERP integrates different modules into one system, that supports the core functions of the organization [29]. Enterprise resource planning combines single activities into the multi-function and integrates the whole of the activities of the organization into one system [45]. ERP brings many benefits, such as improving business efficiency by keeping the business process running smoothly, supporting management in providing decision-making information, enabling the operation of the organization or business more flexible [46].

ERP in Cloud Computing
ERP in Cloud Computing (cloud-based ERP) is the enterprise resource planning system has been stored in the cloud computing environment [35]. Cloud-based ERP has

deputized a model variation in the possessions of IS, it accommodates flexibility, adaptability, scalability, availability, cost efficiency, and configurable for the corporation [2], it also enraptured presto growth global [36].

Cloud-Based ERP Infrastructures. Three main kinds of cloud-based ERP services are described in detail in Fig. 1: (1) IaaS for ERP: organizations use IaaS for their ERP operations, they can typically lease servers from cloud computing providers. Accordingly, the organizations can still choose the ERP vendors and purchase ERP licenses [47]. (2) PaaS for ERP: platform services at this level are used for software development, software testing, software distribution, the PaaS for ERP is not for an integrated system and packaging software of the ERP system [47]. (3) SaaS for ERP: the role of providers of cloud-based ERP service are linked together, which allows the organizations to choose their preferred model, such as the operating ERP system on the internal or external cloud [47].

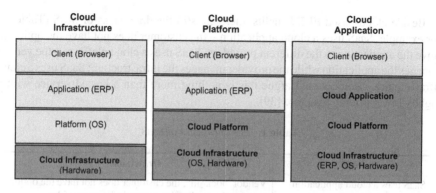

Fig. 1. The framework of cloud-based ERP (Source: Johnson [25])

Cloud-Based ERP Benefits and Drawbacks. Some benefits of the cloud-based ERP (Fig. 2): (1) Lower cost: Instead of being purchased entirely, cloud-based ERP deployments are paid for through a model of subscription, which typically includes not only software but also storage and support costs [53]. Hence, the initial capital cost required for deployment is significantly lower cost than for corny systems and operating costs. (2) Less staff: it needs less IT staffs and business analysis specialists as most of the ERP services including technical support, which is handled by the organizational outside experts [51]. (3) Increased innovation: it conducts through the open source software usage, all functions of ERP have benefited from the innovation acceleration that can be brought from cloud [47]. (4) Mobility and usability: the cloud allows users to access the ERP service through mobile devices [20], it increases the cloud-based ERP use inside and outside the corporation [15]. (5) Rapid deployment: A major limitation for both in-house and cloud-based ERP systems is that the system integrators and the vendors who regularly use existing templates must be configured and customized to accommodate the process and specific practice of the corporation [39]. (6) Flexibility and scalability: cloud-based ERP can be easily customized and expanded [15], it develops new solutions for the corporation to acquire functions and additional software without going through the delivery process of usual forbidden software [53].

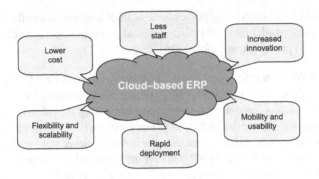

Fig. 2. Some benefits of cloud-based ERP

Beside cloud-based ERP benefits, there are still the drawbacks of SaaS (Table 1). For example, SaaS uses a cloud application – the customer does not have the option to move the application to the different providers; SaaS uses a cloud platform – the vendor manages the application while the provider manages the infrastructure; SaaS uses a cloud infrastructure – some would argue this is nothing more than a hosted service with a slightly lower pricing structure [39].

Table 1. The drawbacks of SaaS

SaaS	Drawback
SaaS uses a cloud application	Vendor "lock–in", the customer does not have the option to move the application to the different provider
SaaS uses a cloud platform	Coordination drawbacks, the vendor manages the application while the provider manages the infrastructure
SaaS uses a cloud infrastructure	Some would argue this is nothing more than a hosted service with a slightly lower pricing structure

Source: Nguyen et al. [39]

IS security issues are a huge challenge for the cloud-based ERP. Nevertheless, both software and hardware are applied by many solutions of security for the Internet platform, and the cloud has higher security standards than the Internet [31]. In addition, deliberate the relation between security risks and benefits, the enterprise can fully nominate the secure cloud applications [39]. Thus, the service providers of cloud-based solutions must provide the latest technology with a commitment to security.

2.2 Related Work

IS Success
IS success does not have a unified definition [12, 43, 48]. A typical definition of the IS success as "*information systems success ultimately corresponds to what DeLone and McLean label individual impact or organizational impact*" [21, p. 213], another

definition with organizational perspective as *"information system-impact of an information system as a measure at a point in time of the stream of net benefits from the information systems, to date and anticipated, as perceived by all key user groups"* [17, p. 381]. Whereby, Keen [26] provided that the scientific basis of imperfection in the studies of IS and enunciated that the dependent variable should continue research and clarify on the IS theories. In measuring the success of IS, there are many ways for the measurement. Which considered "information" as an IS output or message in a "communication", that can be measured at three levels (technical level, semantic level, and effectiveness) [10]. Suitably, the technical level is the system accuracy and efficiency with information's procedures, the semantic level is the information success in conveying meaning, and the effectiveness level is the information effect on the recipient [50]. Then, Mason [34] investigated "effectiveness" as "influence" and defined the effectiveness level is "hierarchy of events which take place at the receiving end of an information system which may be used to identify the various approaches that might be used to measure output at the influence level" [34, p. 277].

The original IS success model (original D&M) was proposed by DeLone and McLean [10] at three levels with six factors, as (1) system initialization: "system quality" measures of the information processing system itself, and "information quality" measures of an IS output; (2) system use: "use" is understood as recipient consumption of an IS output and "user satisfaction" is understood as the recipient consumption of an IS output; (3) system effectiveness: "individual impact" is the information effectiveness on the recipient behavior, and "organizational impact" is the information effectiveness on the firm performance [10]. DeLone and McLean [11] updated the original IS success model and proposed a new model of IS success (updated D&M). The main differences with the original IS success model, as adding "service quality" factor to reflect the service and support importance in the IS success; adding "intention to use" factor to measure user attitudes as a substitute for "use". The updated IS success models consists of six factors: three quality factors (information quality, system quality, and service quality), intention to use, use, user satisfaction, and net benefits [11, 12]. The concept of "intention to use" relates to the theory of technology adoption in TAM of Davis [9], UTAUT of Venkatesh et al. [55]). In addition, "intention to use" can replace "use" in some contexts, "intention to use" is the attitude – "use" is the behavior, attitude and behavior can be linked [11]. Because of difficult to measure "use", so many studies propose the attitude scale for the behavior scale. There are some works used the term IS to be tantamount to the IS success, others have used the IS effectiveness to cover the concepts of individual impact and organizational impact or net benefits [10–12].

Meanwhile, there are still the gaps in the IS works, especially, the dependent variable of the IS models. Hence, scholars have rummaged to find new factors and new relationships for the contribution to the theory of IS success. Whereas there are many studies about cloud computing (e.g., Badi et al. [5]), also about ERP (e.g., Ngai et al. [37]). Little is known on the adoption model and IS success model in the context of the cloud computing, especially, cloud-based ERP (except, e.g., Albar and Hoque [2]). In short, most of the related studies have not provided the integration model between IS success, trust and perceived risk of cloud-based ERP.

Trust and Perceived Risk

The trust is thought to be an action, an attitude or a tendency, a relationship or an option [3], also the degree a consumer confides in a trustee and feels secure to take any transaction with that specific service provider [30]. Trust is *"a defining feature of the major social and economic interactions in which uncertainty is present"* [42, p. 106]. The trust is the subjective belief that a party will fulfill its obligations according to the expectations of stakeholders as the goodwill [42]. Declarations such as "trust me" or "cloud security" which do not help much to increase the trust level of consumer confidence, unless the information is presented with the products or services [27].

The perceived risk theory was proposed by Bauer [6] for consumer behavior of IT sector which has been aware of the risks. Perceived risk is *"a combination of uncertainty plus seriousness of outcome involved"* [6, p. 13]. Perceived risk includes two perceived risks negatively related to the product or service and online transaction [41]. In which, product or service risk is the overall account to uncertainty or discomposure observed by a user in a conspicuous product or service when used to e-commerce [41], and online transaction risk is a possible the transaction risk, users can face when disclosed to e-commerce [41]. Thus, e-commerce adoption model (e-CAM) of Park et al. [41] is a typical model of perceived risk. More and more personal information and companies are placed in the cloud. The concern is how to ensure a safe environment [52].

There are many related works perceived risk about the online purchase (e.g., Jarvenpaa et al. [24]), online process (e.g., Gefen et al. [19]), cloud-based ERP (e.g., Lim et al. [32]). The perceived risk in this study is known as the opportunistic behavior related to the disclosure of organizational information submitted by the cloud-based ERP adoption. The risks include that information is misused and available to unknown individuals, companies, or government agencies [14]. Interestingly, several scholars have worked on both of trust and perceived risk. For example, Gefen et al. [19], Park et al. [41], so empirical evidence on the cloud-based ERP with trust and perceived risk are the suitable theory.

3   Research Model

3.1   Theoretical Framework

Based on the background of cloud-based ERP, the literature review of the D&M models, other IS theories such as the technology adoption (TAM, UTAUT), the theories of trust and perceived risk, and the related works, a model for cloud-based ERP success is built in Fig. 3. In which, the elements of trust and perceived risk are based on Bauer [6], Pavlou [42], e-CAM of Park et al. [41], the elements of system quality and information quality are based on the original D&M model of DeLone and McLean [10], the elements of IT service quality and net benefit are based on the updated D&M models of DeLone and McLean [11, 12], and intention to use is based on the D&M models of DeLone and McLean [10–13], TAM of Davis [9], UTAUT of Venkatesh et al. [55]. All elements and the relationships among them are exculpated as below:

System Quality (SYQ): is a measure of the expected characteristics of an information system [10]. The concept of ease of use is an aspect of system quality in evaluating the models of IS [44], so system quality is similar to easy to use in TAM of Davis [9]. System quality is considered for the success of D&M models in DeLone and McLean [10–12], Gable et al. [17], Seddon [48], which is the most prominent concepts in the theoretical models for the IS success.

Information Quality (INQ): is a measure of the accuracy, timeliness, completeness, relevance, and consistency of an information system [10]. The concept of information quality may vary by systems, there may be major differences in practice, so this variance should be amounted to in the empirical IS study [49]. Along with system quality, information quality is one of the most factors in the conceptual frameworks of the IS success, both of them have DeLone and McLean [10] as the foundational theory.

IT Service Quality (ISQ): is known that system user is received the support of the IS and IT support staff [11, 12]. According to DeLone and McLean [11], the related works indicated that the IT service quality contributes to individual impact, should be considered as a concept in the D&M model, it is concerned as the service quality. Therefore, with the IS success model that can be added IT service quality as a new concept in the IS success model [13].

Perceived Risk (PER): is the customer perception of negative consequences and uncertainty or outcomes related to specific behavior [6], also as a structure that reverberates the customer uncertainty emotions about the possible negative effect on the using new technology [41]. Perceived risk can be caused by lack of ability, reputation, and concern to protect the user privacy, so high risk can ultimately result that users having a negative attitude [7].

Trust (TRU): is conceived as a belief in the ability, benevolence, integrity, and predictability of the e-provider [19]. Trust is also the belief of the individual that cannot be sure of the outcome, or the other act appropriately responsibly [43]. In addition, trust is formed by two components as a perceived component and a behavior component expressed as the willingness or desire to follow a specific action [8]. System quality and information quality are the antecedents of the original and updated D&M models as in DeLone and McLean [10–12] and service quality is an antecedent of the updated D&M models as in DeLone and McLean [11, 12]. Notwithstanding, DeLone and McLean [13] indicated that the antecedents of the D&M models can have the positive impact on trust. Moreover, Cabanillas et al. [8] evidenced these relationships. Hence, the under cloud-based ERP, we propose hypotheses H1, H2, and H3:

- *Hypothesis H1: System quality has a positive impact on the trust.*
- *Hypothesis H2: Information quality has a positive impact on the trust.*
- *Hypothesis H3: IT service quality has a positive impact on the trust.*

Some scholars argued that the relationships between perceived risk and trust are parallel as in Featherman and Pavlou [16], serial as in Cabanillas et al. [8], and trust is a function of perceived risk as in Pavlou [42]. Furthermore, Kim and Benbasat [28] mentioned that the lower level of perceived risk is related to the higher level of trust in the IS. Thus, under the cloud-based ERP, we propose a hypothesis H4:

– *Hypothesis H4: Perceived risk has a negative impact on the trust.*

Intention to Use (ITU): is the attitude and belief of the user about the ability to use the IS, using the multi-attribute tool to measure the intention of the user [11]. Several works envisage the intention to use as the success of IS variable based on the related theories, such as the TAM of Davis [9], UTAUT of Venkatesh et al. [55] for the theoretical illumination. Hence, most of the important elements for the intention to use are the individual characteristics of using the IS. The influence of the trust and the perceived risk as in Pavlou [42] on the intention to use or behavioral intention as in the theories of trust and perceived risk. In addition, Featherman and Pavlou [16], Gefen et al. [19] have also confirmed this relationship. Thus, under the cloud-based ERP, we propose hypotheses H5 and H6:

– *Hypothesis H5: Perceived risk has a negative impact on the intention to use.*
– *Hypothesis H6: Trust has a positive impact on the intention to use.*

Net Benefit (NEP): is the degree to which IS are contributing to the success of individuals, groups, organizations, industries, and nations [11]. DeLone and McLean [11] collapses two factors "individual impact" and "organizational impact" in the original D&M model into a single variable "net benefit" in the updated D&M model as an outcome of IS success, it does not act the problem go forth. Specifically, the original D&M model of DeLone and McLean [10] and the updated D&M model of DeLone and McLean [11] specified the positive impact of intention to use or use on net benefit. Furthermore, the related works have confirmed this path as in Petter et al. [43], Seddon [48]. Hence, under the cloud-based ERP, we propose a hypothesis H7:

– *Hypothesis H7: Intention to use has a positive impact on the net benefit.*

3.2 Research Method

Research Process

There are two phases in this work: (1) a preliminary research with the method of qualitative, and (2) a formal analysis of the method of quantitative. Firstly, from the well-known theory of IS success, the literature review and the related works such as the concepts of trust and perceived risk and other related studies, a draft scale is established. Then, discussions and focus groups with the experts who are the professional person on the topic of cloud-based ERP, the accuracy contents of the scale is consolidated. Next, the final measurement uses for the formal research. A 5-point Likert: 1 – strongly disagree; 2 – disagree; 3 – undecided; 4 – agree; 5 – strongly agree, which measures the item assessment levels. In the measurement scale, there are four items of the system quality element; four items of the information quality element; four items of the IT service quality element; five items of the perceived risk element; four items of the trust element; three items of the intention to use element; and four items of the net benefit element. A convenient sampling method of data is investigated, and the questionnaires are sent to participants who have used the system of cloud-based ERP. Finally, the collected data are analyzed with the structural equation modeling technique by AMOS and SPSS applications. In this work, there are 182 valid samples out of 200 samples of 28 items.

Data Description

Age: the age groups of 16–25 and 26–35 are the plurality with 29 and 54%, respectively, followed by the age group of 36–45 up to 14%, and age group of over 45 is the lowest roundly 3%. *Gender:* it has a small difference of 43% female and 57% male. *Education:* 87% of the university degree, intermediate/college and postgraduate amount to five and six percents, respectively, and two percent of high school. *Job Position:* staff is the highest percent of 54%, there are 24% of respondents are the team leader, and manager, director, and others account for 15, five and one percents, respectively. *Experience:* below three years has the most percent with 40%, followed by 4–6 years and 7–9 years accounts for 37 and 20%, and the experience of over ten years is the lowest roundly three percent.

Cloud based-ERP: most of the respondents use Bitrix cloud-ERP with 54%, Ecount ERP amounts to 25%, similarities exist between Infor cloud suite and Teamcrop cloud-ERP is roundly nine percent respondents, the other kinds of cloud based-ERP is only three percent, a lower rate. The data sample description is manifested details in Table 2.

Table 2. Data description

	Frequency	Percentage		Frequency	Percentage
– Age			*– Gender*		
Ages 16–25	52	29	Male	104	57
Ages 26–35	98	54	Female	78	43
Ages 36–45	26	14	*– Job Position*		
Over age 45	6	3	Staff	98	54
– Education			Team leader	43	24
High school	3	2	Manager	29	16
Inter./College	9	5	Director	10	5
University degree	158	87	Others	2	1
Postgraduate	12	6	*– Cloud based–ERP*		
– Experience			Bitrix cloud–ERP	98	54
Below 3 years	72	40	Ecount ERP	45	25
4–6 years	68	37	Infor cloud suite	17	9
7–9 years	37	20	Teamcrop cloud–ERP	17	9
Over 10 years	5	3	Others	5	3

4   Research Result

4.1   Exploratory and Confirmatory Factor Analyses

Firstly, eliminate an item of system quality factor (SYQ_1) and an item of information quality factor (INQ_3) in reliability analysis, because of the correlation-item value of these factors is less than 0.60. Secondly, eliminate an item of system quality factor (SYQ_4) in the first Exploratory Factor Analysis (EFA), because of the factor loading is less than 0.50. Then, in the second EFA with EFA's factor loading of all items is between

0.648 and 0.887. Finally, the Confirmatory Factor Analysis (CFA) is taken to refine and assess the measurement scales. CFA on the overall measurement model displays the following results: Chi-square (χ^2)/dF equal to 1.315; p-value equal to 0.003; GFI equal to 0.903; TLI equal to 0.960; CFI equal to 0.971; RMSEA equal to 0.042. The CFA's factor loading of all items ranges from 0.621 to 0.884. Besides, the Average Variance Extracted (AVE) values between 0.509 and 0.754 (Table 3), so the scale of measurement reaches the convergence value.

Table 3. Mean and confirmatory factor analysis results

	Mean	SD	SYQ	INQ	ITQ	PER	TRU	ITU	NEB
SYQ	3.792	0.821	0.754*						
INQ	3.865	0.703	0.256	0.618*					
ITQ	3.952	0.692	0.368	0.345	0.509*				
PER	3.431	0.965	0.062	0.003	0.032	0.557*			
TRU	3.885	0.663	0.250	0.159	0.381	0.003	0.536*		
ITU	3.910	0.607	0.478	0.441	0.254	0.094	0.425	0.518*	
NEB	3.861	0.609	0.185	0.336	0.295	0.002	0.254	0.276	0.546*

*SD: Standard Deviation; * Average Variance Extracted (AVE)*

In addition, the value of AVE for each element is larger than the square correlation coefficient (r^2), detail in Table 3, so the scale of measurement is the discriminant value. After the EFA and CFA, the data are the coincidence for the next analysis – structural equation modeling.

4.2 Structural Equation Modeling

The Structural Equation Modeling (SEM) is effectuated by the estimation of ML (Maximum Likelihood). The model indexes indicate adequate fit with the Chi-square (χ^2)/dF equal to 1.247; p-value equal to 0.018; CFI equal to 0.907; TLI equal to 0.967; CFI equal to 0.977; RMSEA equal to 0.037. The standardized path coefficients of the model are shown in Table 4. In which, there are the positive effect of system quality, information quality, and IT service quality on the trust with the γ coefficient equal to 0.304, 0.426, and 0.243 (p-value < 0.005), respectively, so that supports the hypotheses H1, H2, and H3. Besides, the path from perceived risk to the trust is not statistical significance (p-value > 0.05), so the hypothesis H4 is rejected. However, perceived risk has a negative effect on the trust with the γ coefficient equal to −0.179 (p-value < 0.05), which in turn the hypothesis H5 is supported. Trust has a positive effect on the intention to use with the γ coefficient equal to 0.672 (p-value < 0.001), so the hypothesis H6 is strongly supported. It has strongly supported the hypothesis H7 by showing the affecting of intention to use on the net benefit with the γ coefficient equal to 0.607 (p-value < 0.001). The results of the SEM also provide the indexes of the Standard Error (SE) as the standard deviation of the sampling distribution of the paths, detail as in Table 4.

Table 4. Structural equation modeling results and hypothesis testing

H	Path	Estimate	SE	p–value	Result
H1	TRU ← SYQ	0.304	0.056	***	Supported
H2	TRU ← INQ	0.426	0.073	***	Supported
H3	TRU ← ITQ	0.243	0.048	0.002	Supported
H4	TRU ← PER	−0.028	0.035	0.728	Rejected
H5	ITU ← PER	−0.179	0.046	0.035	Supported
H6	ITU ← TRU	0.672	0.094	***	Supported
H7	NEP ← ITU	0.607	0.105	***	Supported

*SE: Standard Error; *** p–value < 0.001*

4.3 Discussion

The research results provide that all measurement scales of the variables of antecedents (system quality, information quality, IT service quality, perceived risk) – intermediates (trust, intention to use) – outcome (net benefit), which ensure reliability. The EFA and the CFA purvey that the measurement scale reaches the convergence value. Mainly, the SEM test and valid all paths and hypotheses. Specifically, the paths from system quality and information quality to the trust are relatively large, with the γ coefficients of 0.304 and 0.426, respectively, and the IT service quality has a positive impact on the trust with the γ coefficient of 0.243 which confirmed the work of DeLone and McLean [13]. Differently, under the cloud-based ERP, although a path from the perceived risk to the trust is not significant, because the data does not support this relationship, a negative path from the perceived risk to the intention to use is significant with the γ coefficient of −0.179 as in Pavlou [42]. Distinctly, the path of the trust with the intention to use is the largest coefficient (γ equal to 0.672) in the structural model. Which has strongly confirmed the works of Gefen et al. [19], Pavlou [42]. Finally, under the cloud-based ERP, the research model is accuracy with the original D&M model of DeLone and McLean [10], the updated D&M models of DeLone and McLean [11, 12], because the data has strongly supported the positive relationship between the intention to use and the net benefit with the γ coefficients of 0.607. In summary, six out of seven hypotheses are supported.

Interestingly, the results also externalize that when the trust is included, the factors of systems (system quality, information quality, IT service quality), the perceived risk are able to explain the intention to use nearly 79% (R^2 equal to 0.786). In the findings, they are comparable to the IT adoption model such as TAM of Davis [9] and UTAUT of Venkatesh et al. [55], and which explained about 45 and 56% of the variance in intention to use, respectively.

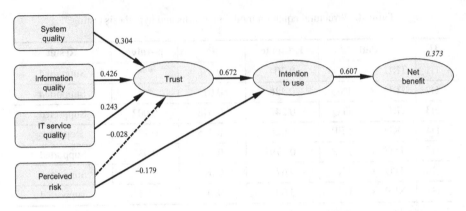

Fig. 3. The success model for cloud-based ERP

Besides, the antecedents and the intermediates are able to provide an overall deter-minant of the net benefit roughly 37% (R^2 equal to 0.372) in this empirical evidence on the context of cloud-based ERP. In which, did not concern on the related works of the IS success – the theory studies as in DeLone and McLean [10–12], or empirical studies as in Petter et al. [43], Seddon [48]. Interestingly, the predictors of IS success in the context of cloud-based ERP are empirically validated and theoretically significant. In Fig. 3, it illustrates the model for cloud-based ERP success, including the presentation of the paths of the model and also the hypotheses.

5   Conclusion and Future Work

This study approached the background of cloud-based ERP as the enterprise resource planning system in the cloud computing environment, including the infrastructures, and the benefits and drawbacks of cloud-based ERP. Distinctly, authors integrated the theo-retical exploration and confirmation of perceived risk and trust with the IS success into one model, this work proposes and investigates an empirical evidence of the IS success in the context of ERP in cloud computing. The model for cloud-based ERP success was empirically validated basic. Specifically, the determinants of system quality, informa-tion quality, and IT service quality have positively impact on trust. Perceived risk is positively related to the intention to use, and trust negatively related to the intention to use of cloud-based ERP. Moreover, intention to use of cloud-based ERP directly influ-ences the net benefit. Therefore, this work continues to contribute to the knowledge, exploring the theory of IS success and related theories as perceived risk and trust in the context of cloud-based ERP, and IS in general.

In future work, the authors may possible to add more the predictors of IS success for exculpating the net benefit, do literature the theory on the performance of IS with indi-vidual and organizational impacts. Furthermore, the demographics may be considered as moderating factors in the model of IS success.

Acknowledgments. The authors acknowledge the comments of three anonymous reviewers for this paper.

References

1. Ahmad, S., Bakar, A., Ahmad, N.: Social media adoption and its impact on firm performance: the case of the UAE. Int. J. Entrep. Behav. Res. (2018, in-press)
2. Albar, A., Hoque, M.: Factors affecting cloud ERP adoption in Saudi Arabia: An empirical study. Inf. Dev. (2018, in-press)
3. Alpern, K.: What do we want trust to be? Some distinctions of trust. Bus. Prof. Ethics J. **16**(1), 29–45 (1997)
4. Akter, S., Wamba, S., Ambra, J.: Enabling a transformative service system by modeling quality dynamics. Int. J. Prod. Econ. (2018, in-press)
5. Al-Badi, A., Tarhini, A., Al-Qirim, N.: Risks in adopting cloud computing: a proposed conceptual framework. In: Miraz, M.H., Excell, P., Ware, A., Soomro, S., Ali, M. (eds.) iCETiC 2018. LNICST, vol. 200, pp. 16–37. Springer, Cham (2018). https://doi.org/10.1007/978-3-319-95450-9_2
6. Bauer, R.: Consumer behavior as risk taking. In: AMA Proceedings (1960)
7. Benlian, A., Hess, T., Buxmann, P.: Drivers of SaaS-adoption-an empirical study of different application types. Bus. Inform. Syst. Eng. **5**, 357–369 (2009)
8. Cabanillas, F., Leiva, F., Fernandez, J.: Payment systems in new electronic environments: consumer behavior in payment systems via SMS. Int. J. Inform. Technol. Decis. Mak. **14**(2), 421–449 (2015)
9. Davis, F.: Perceived usefulness, perceived ease of use, and user acceptance of information technology. MIS Q. **13**(3), 319–340 (1989)
10. DeLone, W., McLean, E.: Information systems success: the quest for the dependent variable. Inform. Syst. Res. **3**(1), 60–95 (1992)
11. Delone, W., McLean, E.: The DeLone and McLean model of information systems success: a ten-year update. J. Manag. Inf. Syst. **19**(4), 9–30 (2003)
12. Delone, W., McLean, E.: Measuring e-commerce success: Applying the DeLone & McLean information systems success model. Int. J. Electron. Commer. **9**(1), 31–47 (2004)
13. Delone, W., Mclean, E.: Information systems success measurement. Found. Trends Inf. Syst. **2**(1), 1–116 (2016)
14. Dinev, T., Hart, P.: An extended privacy calculus model for e-commerce transactions. Inf. Syst. Res. **17**(1), 61–80 (2006)
15. Elmonem, M., Nasr, E., Geith, M.: Benefits and challenges of cloud ERP systems—a systematic literature review. Future Comput. Inf. J. **1**(2), 1–9 (2016)
16. Featherman, M., Pavlou, P.: Predicting e-services adoption: a perceived risk facets perspective. Int. J. Human-Comput. Stud. **59**(4), 451–474 (2003)
17. Gable, G., Sedera, D., Chan, T.: Re-conceptualizing information system success: the IS-impact measurement model. J. AIS **9**(7), 377–408 (2008)
18. Gangwar, H., Date, H., Ramaswamy, R.: Understanding determinants of cloud computing adoption using an integrated TAM-TOE model. J. Enterp. Inf. Manag. **28**(1), 107–130 (2015)
19. Gefen, D., Karahanna, E., Straub, D.: Trust and TAM in online shopping: an integrated model. MIS Q. **27**(1), 51–90 (2003)
20. Goel, M., Kiran, D., Garg, D.: Impact of cloud computing on ERP implementations in higher education. Int. J. Adv. Comput. Sci. Appl. **2**(6), 146–148 (2011)

21. Goodhue, D., Thompson, R.: Task-technology fit and individual performance. MIS Q. **19**(2), 213–236 (1995)
22. Gupta, S., Kumar, S., Singh, S., Foropon, C., Chandra, C.: Role of cloud ERP on the performance of an organization: contingent resource based view perspective. Int. J. Logistics Manag. **29**(2), 659–675 (2018)
23. Hashem, I., Yaqoob, I., Anuar, N., Mokhtar, S., Gani, A., Khan, S.: The rise of "big data" on cloud computing: review and open research issues. Inf. Syst. **47**, 98–115 (2015)
24. Jarvenpaa, S., Cantu, C., Lim, S.: Trust in virtual online environments. In: Handbook of the Psychology of the Internet at Work, pp. 103–130. Wiley, New York (2017)
25. Johnson, D.: ERP cloud news (2017). http://erpcloudnews.com
26. Keen, P.: MIS research: reference disciplines and a cumulative tradition. In: ICIS Proceedings, p. 9 (1980)
27. Khan, K., Malluhi, Q.: Establishing trust in cloud computing. IT Prof. **5**, 20–27 (2010)
28. Kim, D., Benbasat, I.: The effects of trust-assuring arguments on consumer trust in internet stores: application of Toulmin's model of argumentation. Inf. Syst. Res. **17**(3), 286–300 (2006)
29. Klaus, H., Rosemann, M., Gable, G.: What is ERP? Inf. Syst. Front. **2**(2), 141–162 (2000)
30. Komiak, S., Benbasat, I.: Understanding customer trust in agent-mediated electronic commerce, web-mediated electronic commerce, and traditional commerce. Inf. Technol. Manag. **5**(1), 181–207 (2004)
31. Lancon, F.: ERP in cloud computing and information security issues (2013). http://tuvanphanmem.vn
32. Lim, T.-M., Lee, A.S.-H., Yap, M.-K.: User Acceptance of SaaS ERP Considering Perceived Risk, System Performance and Cost. In: Kim, K., Joukov, N. (eds.) Information Science and Applications (ICISA) 2016. LNEE, vol. 376, pp. 53–63. Springer, Singapore (2016). https://doi.org/10.1007/978-981-10-0557-2_6
33. Marston, S., Li, Z., Bandyopadhyay, S., Zhang, J., Ghalsasi, A.: Cloud computing—the business perspective. Decis. Supp. Syst. **51**(1), 176–189 (2011)
34. Mason, R.: Measuring information output: a communication systems approach. Inf. Manag. **1**(4), 219–234 (1978)
35. Mell, P., Grance, T.: The NIST definition of cloud computing. In: NIST Special Publication, pp. 800–145. NIST (2011)
36. Mijac, M., Picek, R., Stapic, Z.: Cloud ERP system customization challenges. In: Information and Intelligent Systems, Varazdin, pp. 133–140 (2013)
37. Ngai, E., Law, C., Wat, F.: Examining the critical success factors in the adoption of enterprise resource planning. Comput. Ind. **59**(6), 548–564 (2008)
38. Nguyen, T.D., Nguyen, T.M., Cao, T.H.: Information systems success: a literature review. In: Dang, T.K., Wagner, R., Küng, J., Thoai, N., Takizawa, M., Neuhold, E. (eds.) FDSE 2015. LNCS, vol. 9446, pp. 242–256. Springer, Cham (2015). https://doi.org/10.1007/978-3-319-26135-5_18
39. Nguyen, T.D., Nguyen, T.T.T., Misra, S.: Cloud-based ERP solution for modern education in Vietnam. In: Dang, T.K., Wagner, R., Neuhold, E., Takizawa, M., Küng, J., Thoai, N. (eds.) FDSE 2014. LNCS, vol. 8860, pp. 234–247. Springer, Cham (2014). https://doi.org/10.1007/978-3-319-12778-1_18
40. Panorama: 2017 report on ERP systems and enterprise software. Panorama Consulting (2018). https://www.panorama-consulting.com
41. Park, J., Lee, D., Ahn, J.: Risk-focused e-commerce adoption model: a cross-country study. J. Glob. Inf. Technol. Manag. **7**, 6–30 (2004)

42. Pavlou, P.A.: Consumer acceptance of electronic commerce: integrating trust and risk with the technology acceptance model. Int. J. Electron. Commun. **7**(3), 101–134 (2003)

43. Petter, S., DeLone, W., McLean, E.: Information systems success: the quest for the independent variables. J. Manag. Inf. Syst. **29**(4), 7–62 (2013)

44. Rai, A., Lang, S., Welker, R.: Assessing the validity of IS success models: an empirical test and theoretical analysis. Inf. Syst. Res. **13**(1), 50–69 (2002)

45. Rich, D., Dibbern, J.: A team-oriented investigation of ERP post-implementation integration projects: how cross-functional collaboration influences ERP benefits. In: Piazolo, F., Felderer, M. (eds.) Innovation and Future of Enterprise Information Systems. LNISO, vol. 4. Springer, Heidelberg (2013). https://doi.org/10.1007/978-3-642-37021-2_10

46. Rouhani, S., Mehri, M.: Does ERP have benefits on the business intelligence readiness? An empirical study. Int. J. Inf. Syst. Change Manag. **8**(2), 81–105 (2016)

47. Schubert, P., Adisa, F.: Cloud computing for standard ERP systems: reference framework and research agenda. Fachbereich Informatik (2011)

48. Seddon, P.: A respecification and extension of the DeLone and McLean model of IS success. Inf. Syst. Res. **8**(3), 240–253 (1997)

49. Sedera, D., Gable, G.: A factor and structural equation analysis of the enterprise systems success measurement model. In: ICIS Proceedings, p. 36 (2004)

50. Shannon, C., Weaver, W.: A mathematical theory of communication. Bell Syst. Technol. J. **27**, 379–423 (1948)

51. Staehr, L.: Understanding the role of managerial agency in achieving business benefits from ERP systems. Inf. Syst. J. **20**(3), 213–238 (2010)

52. Subashini, S., Kavitha, V.: A survey on security issues in service delivery models of cloud computing. J. Netw. Comput. Appl. **34**(1), 1–11 (2011)

53. Utzig, C., Holland, D., Horvath, M., Manohar, M.: ERP in the cloud is it ready? Are you? Booz & Co., pp. 1–9 (2013)

54. Velte, A., Velte, T., Elsenpeter, R.: Cloud Computing: A Practical Approach. McGraw-Hill, New York (2010)

55. Venkatesh, V., Morris, M., Davis, F.: User acceptance of information technology: toward a unified view. MIS Q. **27**, 425–478 (2003)

Statistical Models to Automatic Text Summarization

Pham Trong Nguyen[✉] and Co Ton Minh Dang

Saigon University, Ho Chi Minh City, Vietnam
ptnguyen117@gmail.com, ctmdang@sgu.edu.vn

Abstract. This paper proposes statistical models used for text summarization and suggests models contributing to researches in text summarization issue. The evaluating experiment results methods have partially demonstrated synthesization technique's efficiency in automatic text summarization. Having been built and tested in real data, our system proves its accuracy.

Keywords: Text summarization · Statistical model
Natural language processing · Vietnamese

1 Introduction

In company with the amazing information development, the amount of everyday produced texts is tremendously increasing. It will take lots of labor and time to process these texts if one keeps doing it in manual, traditional way, let alone processing and summarizing many texts at a time. Facing with the progressively growing amount of texts as we know, the lack of new processing models will lead to difficulties. Automatic text summarization model comes into existence to meet this hard demand. A text of hundreds of sentences, even thousands will be automatically summarized in order to discover its central contents. This will help save a great deal of time and labor for people undertake the task. Some researching projects and automatically Vietnamese text processing models have been built and many approaches have been proposed, but none of which is optimal. Based on these results, this paper continues to study statistical models used for text summarization and suggest models contributing to researches in text summarization issue.

We organize the rest of this paper as follows: in Sect. 2, issue of Vietnamese text summarization issue is described; in Sect. 3, approaches relied on statistics in text summarization; in Sect. 4, presentation of experiment results, and finally, conclusion of the paper and discussion on possibilities for future work.

2 The Issue of Vietnamese Text Summarization

2.1 Approaches in Text Summarization

Two approaches to summarize texts are text abstractive summarization and text extractive summarization. In the first one, abstractive summarization, occur the changes

© Springer Nature Switzerland AG 2018
T. K. Dang et al. (Eds.): FDSE 2018, LNCS 11251, pp. 486–498, 2018.
https://doi.org/10.1007/978-3-030-03192-3_37

of sentence structures taken from the text to summarize. This approach bears high semantic quality. The semantic analysis in the text combining with natural language processing techniques gives birth to the results. This model consists in "paraphrasing" results obtained so that the 'new' sentences become clearer, easier to understand and consistent one with another. However, this model is being studied and has not been optimized yet. Having regard to text extractive summarization, we deal with the model extracting the whole sentences which have important data, touch upon the text's contents and putting them together to possess a shorter text which can always communicate the same main contents as the original text. Obviously, the second model cannot have a high semantic quality. The result possessed, namely the summarized text, is not a coherent whole owing to its sentences being relieved from different places in the original text. In whatever way, the ordinal quality in the result text must be the same as that in the original text. The problem of text extractive summarization will be presented in the paper:

Given the text T of n sentences and a number k > 0 find such a set A_T comprising t = k% × n sentences extracted from T that A_T bears T's most important data.

2.2   Automatic Text Summarization

Depending on Inderjeet Mani [1] automatic text summarization targets at extracting contents from an information source and presenting them in a concise and emotional form for users or a programme needing it. Models to automatic text summarization involve statistical, location, semantic net models. Statistical model: this one makes use of statistical data about the importance levels (weights) of terms, phrases, sentences or paragraphs. With the help of this, the system can reduce the number of objects needing examining and extract necessary language data units exactly. Statistical data are supplied by linguistic researches or machine learning from available sets. The techniques often used in statistical model [2, 3] are:

- *Suggested Phrase Model:* suggested phrases are those which make a sentence's weight identified clearly if they are present in this sentence. *For example*: underlining phrases: "in general...", "in particular...", "in the end...", "the content involves...", " the paper presents..."; redundant phrases: " it is rare...", " the paper does not mention...", " it is impossible...". If in a certain sentence exists an underlining suggested phrase, this sentence often takes part in those having important meanings. Other than suggested phrases whose significations are underlined, "redundant" phrases reveal the unnecessity of a sentence in the text's content. In this model, discovering suggested and redundant phrases helps identify sentences to select or omit during text extraction processes.
- *Statistical Model for Term Frequency:* This model based on the idea that the greater a term's frequency of presence in the original text and relevant ones, the greater its importance. Synthesizing terms' weights in a sentence determines its weight and decides weather or not to select it through text extraction processes. The techniques of statistical model for term frequency are as shown below:
  - Combining topical probability and general one: this is one of algorithms evaluating key terms relying on the combination of topical probability and general

one, it reaches a quite high exactitude TF.IDF (Term Frequency - Inverse Document Frequency).

– Term frequency set: this model aims at building terms pattern set, this set is constructed manually or by evaluating the number of a term's appearances in the original text and the others. The more considerable the number of its appearances in the original text, the more important the information provided by it; on the contrary, the more considerable the number of its appearances in the other texts, the less important the information provided by it.

- *Location Model:* It involves models identifying weights in line with the statistics of locations of terms, phrases, sentences in the text. In reference with each different text style, sentences taking place at the beginning of the text usually have more uniting quality than the ones in the middle or at the end. The important locations in the text are title, headline, the first parts and the last in paragraphs, illustrations, notes. This model's efficiency depends much on text style. Concerning some specific styles such as newspaper articles, scientific texts, which have highly coherent structure, this model proves effective using determining sentences' locations to obtain good results. However, texts of unstable structure put lots of limits on this model.

- *Semantic Net Model:* This model identifies important language data units by focusing on semantic relations between structure – grammar – semantics. A language data unit's weight is more significant if it bears more components relevant to other components. Evaluating relations is decided by semantic net or syntactic relations. Models often utilized in the semantic net model can be mentioned as model using relations between sentences, paragraphs; term series model; reference links model.

3 Approach Relying on Statistics of Text Summarization

3.1 Identifying Statistical Features

Location: this is model making use of sentence's location in a paragraph or in the whole text and relying on text's structure to find out sentences which convey important contents and express the whole text's content.

Term phrase: this model can be seen to focus on specific term phrases. Sentences or paragraphs containing specific term phrases are classified into two types: the first one has important content and the second one which is complementary explains the preceding sentences or the following sentences.

Term frequency: this model is the most used in automatic text summarization and gives rise to results of high exactitude. This model has the evaluation of every term's weight in a sentence as foundation for identifying the sentence's weight in the text, then from the results acquired selects the most important sentences revealing the original text's content.

Synthesizing Technique

The idea of using combined coefficients: this idea comes from the combination of 3 features of location, suggested term, and term frequency which identifies a sentence's weight in the text to summarize. Applied to real researches, weight identifications according to each feature disclose their own strengths and weaknesses: Location feature: If the text style is identified, the exactitude becomes pretty significant. However, this condition is not satisfied easily. Besides, the incoherent structure of the text will diminish the exactitude of automatic text summarization using location feature. Suggested phrase feature: here, the hardest thing to do is to identify suggested phrases precisely in the text to summarize. This task plays the decisive part of summarized text's quality. If suggested phrases are provided along with summarization demand, the probability of accurate summarization will be pretty satisfactory. Nevertheless, when asked to summarize a text entirely automatically, identifying suggested phrases would stay difficult and require labor and time but for this favourable condition. Term frequency feature: until now, several models have been proposed to evaluate weights of terms, sentences. Yet they provide accurate results only when applied to languages whose structure analysis is not too complicated like English, French... [4, 5] When dealing with Vietnamese, precision of separating terms presents such a challenge that one has difficulty overcoming it [6]. In addition, a sentence's length exerts an influence on its weight. Some long sentences, this means that they contain lots of terms, will obtain great weight. Despite their modest number of terms, some short sentences which transfer important contents cannot be neglected when summarizing. No model has been able to handle this problem.

The synthetization technique presented in the paper makes us expect to complement weaknesses of the above models and combine their strengths in order to improve the efficiency of statistically automatic text summarization.

Features will be identified as follows:

- Sentence location feature (DT1):
  Symbol K means the order of sentence s_k out of n sentences of text T. Sentence s_k's weight is symbolized by v(k).
  Formula to calculate DT1 is: $v(k) = (p_k)/N$
  Out of it, p_k is sentence s_k's weight, calculated according to sentence s_k's location in the text T and:

$$N = \sum_{k \in n} p_k$$

- Term phrase feature (DT2):
  Symbol D means (suggested, important) term phrase list in the text T. Given term phrase $d \in D$ and sentence s_k. Symbol of weight sentence s_k is $a(s_k)$.
  Formula to calculate DT2 is:

$$a(s_k) = \sum_{d \in s_k} u(d)$$

  Out of it u(d) is term phrase d's weight.
- Term frequency feature (DT3):
  Symbol p(w) means frequency of term w's appearances in T.
  Given term (w) and sentence s_k, symbol $b(s_k)$ is sentence s_k's weight.

Formula to calculate DT3 is:

$$b(s_k) = \sum_{w \in s_k} p(w)$$

- Synthetization technique:
  Given text T and s_k is the kth sentence out of n sentences in the text. Suppose that it is necessary to select c sentences and introduce them into the summarized text from T. If we only comply with DT1 feature, we will choose c sentences whose v values are the greatest. If we only comply with DT2 feature, we will choose c sentences whose a values are the greatest. If we only comply with DT3 feature, we will choose c sentences whose b value are the greatest.

  Synthetization technique:
  We will choose sentence i = 1, 2..., c on the condition that the coefficient th(i) = v(i) + a(i) + b(i)] has the greatest value.
  The importance of each feature in this technique is neither greater nor lesser than the others. Sentence's weight equals the total of all the features' weights, sentences whose weights are the greatest will be selected to be summarization results. Using term data corpus available to learn and form congruous coefficients α, β, γ.
  Select sentence i = 1, 2..., c; on the condition that the coefficient th1(i) = α * v(i) + β * a(i) + γ * b(i) has the greatest value. Coefficients function as increasing or decreasing features' importances in sentence's weight results. The greater a feature's coefficient, the greater its importance during sentence's weight evaluating. Likewise, the lesser a feature's coefficient, the less influence it has on sentence's weight evaluating.

3.2 Technique Identifying Coefficients α, β, γ

Building reference language data set:
  Reference language data set is built with the aim at having a reference when identifying statistical features in automatic text summarization. Apart from this, reference language data set is considered a foundation for comparing automatic text summarization program's results with summarized text's results by experts in order to estimate proposed text summarization technique's efficiency. Reference language data set comprises 3 principal components: original text, processed text (separated terns, separated sentences) and summarized text by language experts. Every original text will be summarized by experts at 3 levels: 10%, 20% and 30% of the number of sentences in the text.
  Reference language data set structure:

- Quantity: data set involves 100 texts, each of which has from 30 sentences to 70.
- Text sources: texts withdrawn from electronic newspaper VNExpress.
- Styles: withdrawn texts are classified into 10 styles.

  Number of experts summarizing texts: every text is summarized by 10 experts who work independently of each other. Every expert summarizes 1 text at 3 levels: 10%,

20% and 30% of the number of sentences in the text. Results of text summarization comes from synthesizing 10 experts' works. Sentences which are most selected by experts will take part in results of this text's summarization. For example, 10% of text sentences comprises 90 sentences. So, every expert offers 9 sentences as results. The final results consist of 9 sentences most selected (one selection by an expert for a sentence equals a vote.) Similarly working at different levels of 20% and 30%, we will obtain results expected.

Language data set components:

- 100 original texts which are not processed yet
- 100 processed texts: checked structure, checked orthography, separated terms, separated sentences.

Experts' summarization results set: every original text jointed 3 summarization results at each of summarization levels.

Building processes of language data reference set:

Step 1. Forming original texts: These original texts are automatically withdrawn from VNExpress's articles, address: www.vnexpress.net. Automatic program withdraws 10 texts non-duplicated from 10 different categories. Every text contains from 30 sentences to 70, jointed no pictures, links, notes, graphics.

Step 2. Text processing: Texts will be selected again in manual way. Texts have things listed, questions-answers will be left out. Separate terms, sentences of the text using VLSP tool which has generally been investigated and applied with high accuracy.

Step 3. Experts' text summarizing: Every text will be summarized by 10 experts at 3 levels: 10%, 20% and 30% of text sentences. Experts work independently of each other, each of whom offers a sentence selecting technique for every summarization level. Therefore, we will have 10 techniques for every summarization level.

Step 4. Checking and correcting summarization results by experts: Summarization results by experts are checked at the levels of quantity, content to make sure that they satisfy summarization level requirements. The quantity of selected sentences by experts will be checked by automatic program to make sure that it fits summarization requirements.

Step 5. Results synthesizing: Selected sentences in summarization levels by the majority of experts will be withdrawn by automatic program to form final summarization results of every text.

For example: Given text T of 51 sentences, summarization levels are as follows:
The 1st summarization level: 10%: 5 sentences. The 2nd summarization level: 20%: 10 sentences. The 3rd summarization level: 30%: 15 sentences.

The 1st summarization level result: selecting 5 sentences most selected by experts in 10 selecting methods at the 1st summarization level. The 2nd summarization level result: selecting 10 sentences most selected by experts in 10 selecting methods at the 2nd summarization level. The 3rd summarization level result: selecting 15 sentences most selected by experts in 10 selecting methods at the 3rd summarization level.

Identifying features' weight:

Weights according to sentence location feature (DT1): In some points of view, sentences which are located at the beginning of the text often have more weights than ones found in the following passages of the text. In the frame of the paper, we use language data reference set to identify weight in conformity with sentence location feature in the text. Weights according to suggested term phrase feature (DT2): Suggested term phrase can be a term or some ones being next to each other in a sentence and connected together. Suggested terms are either ones intimately relating to text's content or ones bearing uniting quality and insisting on signification of every sentence, paragraph in the text. Sentences in the text which contain suggested term phrases transfer important ideas of the text and can be introduced into summarized text. Suggested term phrases can be anteriorly selected or provided with summarization requirements. Every suggested term phrase in the sentence will be endowed with weight based on this suggested term phrase's number of appearances in the original text and its number of appearances in the reference text. Sentence's weight based on suggested term phrase feature will equal the total of all the suggested term phrases' weights in the sentence. Identifying suggested term phrases: every text must have suggested term phrase list. It is a man-made list. Suggested term phrases do not depend on summarization requirement levels. In the paper's frame, suggested term phrases being already provided are used to identify sentences' weights during automatic summarization processes.

Steps in identifying and evaluating suggested phrases' weights:

Step 1. identifying suggested phrases: Suggested phrases will be identified depending on selected sentences by experts in language data reference set. Terms and phrases in these sentences will be separated, then they will be suggested phrases, endowed with weights and used as base for evaluating sentence's weight during automatic summarization processes.

Step 2. Evaluating suggested phrases' weights:

- Calculating the number of phrase's appearances d in selected sentences from experts' summarization results at 3 levels. Symbolize it as nd.
- Calculating the number of phrase nd's appearances in 100 reference texts. Symbolize it as Nd.
- Calculating phrase d's weight according to formula: $u(d) = nd/Nd$
- The sentence's weight equals the total of suggested terms' weights in the sentence.
- The weight according to term frequency feature (DT3)
- Calculating term's weight in the sentence: term's weight in the sentence is identified on value TF.IDF (Term Frequency - Inverse Document Frequency).

$$\text{Weight}(w_i) = tf * idf$$

With:

$$tf = Ns(t)/\sum w$$

$$idf = \log(\sum d/(d:t \in d))$$

And:

Ns(t): Number of term t's appearances in the corpus f
$\sum w$: Sum of terms in the corpus f
$\sum d$ = Sum of corpus
d:t $\in$ d: Number of corpus containing term t

Identifying Coefficients α, β, γ

Given text T and summarization levels t1, t2, ..., tn. Given selected sentences at summarization levels t1: t1-1, t1-2, ..., t1-c. Given sentence t1-1:

- Classifying sentence t1-1 according to DT1 in all sentences of text T (Ex: class 2/36)
- Classifying sentence t1-1 according to DT2 in all sentences of text T (Ex: class 23/36)
- Classifying sentence t1-1 according to DT3 in all sentences of text T (Ex: class 12/36)

Therefore, the coefficients in t1-1 will respectively be:

α = 1 (the highest class); β = −1 (the lowest class); γ = 0 (the second highest class).

Respectively considering all sentences at summarization level t1, we will have all sentences' coefficients at summarization level t1.

Coefficients will be calculated as follows:

- Coefficients α, β, γ at selected levels equals arithmetical mean of coefficients according to t1-1, t1-2, ..., t1-c.
- Coefficients α, β, γ in the text equals arithmetical mean of coefficients according to summarization levels t1, t2, ..., tn.
- Coefficients α, β, γ in all the language data set equals arithmetical mean of coefficients in all the texts.
- Applied to the built language data set, the coefficients will be: α = 0.56, β = 0.27, and γ = 0.17.

4 Experiments

4.1 Materials to Experiment

Using automatic text summarization program combined with synthesizing technique, we summarize 100 texts of language data reference corpus. Afterwards, we compare between summarization program's results and 10 experts' summarization results.

Experiment 1: Automatic text summarization results applied to texts from language data corpus compared to experts' summarization results.

In regard to each summarization level of language data corpus, experts will propose different and non-duplicated selecting techniques. The program compares between summarization program's selected sentences and 10 experts' ones. It checks both of them to discover identical sentences selected by both two sides. The automatic text summarization's efficiency will be deduced from these results. This check is applied to summarization results with coefficients or results without coefficients.

Experiment 2: Automatic text summarization results evaluated by experts.

Every summarization level of each text in language data corpus will be evaluated at 3 levels: good, acceptable, and unacceptable. Experts give reviews of summarization results with coefficients and results without coefficients. With these reviews, we can evaluate the automatic text summarization's efficiency.

4.2 Experiment Results

Automatically result evaluating: Given text T, k sentences need to be selected from n sentences in the text T.

The 1st expert selects sentences a$(1, 1)$, a$(1, 2)$, ..., a$(1, k)$,

The 2nd expert selects sentences b$(2, 1)$, b$(2, 2)$, ..., b$(2, k)$,

...

The mth selects x$(m, 1)$, x$(m, 2)$, ..., x(m,k).

Symbols: T$* = \{$ a$(1, 1)$, a$(1, 2)$, ..., a$(1, k)$, b$(2, 1)$, b$(2, 2)$, ..., b$(2, k)$, ..., x$(m, 1)$, x$(m, 2)$, ..., x$(m, k)\}$ is the list of sentences selected by experts.

D $= \{$d(1), d(2), ..., d$(k)\}$ is the list of k sentences selected by the program.

Symbol x is the number belonging to D, but not to T* (Tables 1 and 2).

Table 1. Statistics of evaluated results with coefficients.

ID	Percentage of success	Quantity of texts		
		Summarization level of 10% of sentences total	Summarization level of 10% of sentences total	Summarization level of 10% of sentences total
1	Under 50%	11	00	00
2	From 50% to 75%	55	07	00
3	Over 75%	34	93	100
4	100%	17	54	64

Statistics reveal the high percentage of automatic summarization results' success, consider results of summarization model with coefficients:

- At summarization level of 10% of sentences total, only 11 out of 100 summarized texts reach 50% of success and 17 texts 100% of success.
- At summarization level of 20% of sentences total, none of texts reaches under 50% and 54 texts 100% of success.

Table 2. Statistics of evaluated results without coefficients

ID	Percentage of success	Quantity of texts		
		Summarization level of 10% of sentences total	Summarization level of 20% of sentences total	Summarization level of 30% of sentences total
1	Under 50%	14	00	00
2	From 50% to 75%	52	07	00
3	Over 75%	34	93	100
4	100%	16	50	63

- At summarization level of 30% of sentences total, none of texts reaches under 50% and 64 texts 100% of success.

This shows that the more the sentences in summarized results, the higher the percentage of success.

Besides, the percentage of success of summarization model with coefficients is better than that of summarization model without coefficients in:

- The number of sentences of success percentage reaching under 50% in model with coefficients lesser than that in model without coefficients, namely 11 compared to 14.
- The number of sentences of success percentage reaching 100% in model with coefficients greater than that in model without coefficients on all the summarization levels.

Apart from this, we need to take account of results evaluated by experts to be aware more clearly of automatic text summarization's efficiency of the program (Tables 3 and 4).

Table 3. Statistics of experts' results evaluating - with coefficients

ID	Evaluation levels	Quantity of texts		
		Summarization level of 10% of sentences total	Summarization level of 20% of sentences total	Summarization level of 30% of sentences total
1	Good	16	29	28
2	Acceptable	69	65	65
3	Unacceptable	15	06	07

Experts' Results Evaluating

Referring to experts' evaluation, automatic summarization results are pretty good, consider results in model with coefficients:

Table 4. Statistics of experts' results evaluating - without coefficients.

ID	Evaluation levels	Quantity of texts		
		Summarization level of 10% of sentences total	Summarization level of 20% of sentences total	Summarization level of 30% of sentences total
1	Good	11	22	16
2	Acceptable	80	75	82
3	Unacceptable	09	03	02

- At summarization level of 10% of sentences total, only 15 out of 100 summarized texts are unacceptable and 16 of them good.
- At summarization level of 20% of sentences total, only 6 out of 100 summarized texts are unacceptable and 29 of them good.
- At summarization level of 30% of sentences total, only 7 out of 100 summarized texts are unacceptable and 28 of them good (Tables 5 and 6).

Table 5. Comparison between automatic evaluation and experts' evaluation at summarization level 10% - with coefficients.

ID	File name	n	Automatic evaluation			Experts' evaluation		
			Level 10%			Level 10%		
			k	k − x	Success percentage (k − x)/k	Good	Acceptable	Unacceptable
1	family1	36	4	1	25.00%		X	
2	world0	33	3	1	33.30%		X	
3	world3	30	3	1	33.30%		X	
4	world4	34	3	0	0.00%		X	
5	law1	34	3	1	33.30%		X	
6	law4	35	4	1	25.00%		X	
7	education4	31	3	1	33.30%		X	
8	science3	32	3	1	33.30%		X	
9	entertaiment4	30	3	1	33.30%		X	
10	sport1	32	3	0	0.00%			X
11	digital3	32	3	1	33.30%		X	

This is consistent with automatic summarization results, the more the sentences in summarized results, the higher the percentage of success. However, in regard to summarization results in two models with coefficients and without coefficients, experts remark the difference in summarization levels:

- At summarization level of 10%: summarization results without coefficients are considered better than those with coefficients: the number of unacceptable texts in

Table 6. Comparison between automatic evaluation and experts' evaluation at summarization level 10% without coefficients.

ID	File name	n	Automatic evaluation			Experts' evaluation		
			Level 10%			Level 10%		
			k	k − x	Success percentage (k − x)/k	Good	Acceptable	Unacceptable
1	family1	36	4	1	25.00%	X		
2	world0	33	3	1	33.30%		X	
3	world3	30	3	1	33.30%		X	
4	world4	34	3	0	0.00%		X	
5	law1	34	3	1	33.30%		X	
6	law4	35	4	1	25.00%		X	
7	education4	31	3	1	33.30%		X	
8	education6	51	5	2	40.00%	X		
9	science3	32	3	1	33.30%		X	
10	business9	52	5	2	40.00%		X	
11	entertaiment4	30	3	1	33.30%	X		
12	sport1	32	3	0	0.00%			X
13	digital3	32	3	1	33.30%		X	
14	digital7	51	5	2	40.00%		X	

summarization model without coefficients is lower than that in summarization model with coefficients (namely 9 compared to 15). However, the number of good texts in summarization model with coefficients is higher than that in summarization model without coefficients (16 compared to 11).

- At other summarization levels, summarization model with coefficients is estimated better than summarization model without coefficients due to the fact that summarization model with coefficients has fewer unacceptable texts and more good texts than summarization model without coefficients.
- Examine the synthesization of automatic evaluation and experts' evaluation in order to make a final remark about automatic text summarization's efficiency of the program.

Synthetization of Automatic Evaluation and Experts' Evaluation:

In reference with summarization results without coefficients: at summarization level 10% of the total of sentences which have summarization results obtain success percentage under 50%.

According to experts' evaluation, only 1 out of 14 results is unacceptable and the rest acceptable and above.

The synthesization of automatic evaluation and experts' evaluation shows: automatic text summarization's efficiency of the program is pretty high. Summarization results with coefficients has better success than summarization results without coefficients.

5   Conclusion

The paper has investigated, studied statistical models in automatic text summarization. Based on models available, it proposes synthesization technique combining statistical models' features to build the program of automatic text summarization. The evaluating experiment results methods have partially demonstrated synthesization technique's efficiency in automatic text summarization. The paper has paved the way for building language data reference corpus; in spite of the modesty of data quantity, the base for building, referring, comparing obtained results has taken shape.

In the time coming, the language data reference corpus will need improving, expanding to upgrade the accuracy of identification of features' weights. Studying other machine learning will aim at identifying new coefficients, improving the efficiency of sentence's weight evaluating and producing more exact summarization results. We are also interested in developing applications for emotional analysis based on studies by Thien et al. [7–9].

Acknowledgements. This paper was supported by the research project CS2017-61 funded by Saigon University.

References

1. Mani, I.: Summarization Evaluation: An Overview. John Benjamins Publishing, Amsterdam (2001)
2. Nguyen, T.: Lac Hong research project: Xây dựng hệ thống rút trích các nội dung chính của văn bản khoa học dựa trên cấu trúc. Develop a system for extracting key contents of science texts. HCMC, Vietnam (2012)
3. Balabantara, R.C., et al.: Text summarization using term weights. Int. J. Comput. Appl. **38**(1), 10–14 (2012). (0975 – 8887)
4. Lin, C.Y.: Rouge: A Package for Automatic Evaluation of Summaries. Information Sciences Institute, University of Southern California (2004)
5. Hirohata, M., et al.: Sentence extraction-based presentation summarization techniques and evaluation metrics. Department of Computer Science, Tokyo Institute of Technology (2005)
6. Dang, C.T.M.: Modeling syntactic structures of vietnamese complex sentences. In: Silhavy, R., Silhavy, P., Prokopova, Z. (eds.) CoMeSySo 2018. AISC, vol. 859, pp. 81–91. Springer, Cham (2019). https://doi.org/10.1007/978-3-030-00211-4_9
7. Tran, T.K., Phan, T.T.: A hybrid approach for building a Vietnamese sentiment dictionary. J. Intell. Fuzzy Syst. **35**(1), 967–978 (2018)
8. Tran, T.K., Phan, T.T.: Mining opinion targets and opinion words from online reviews: Int. J. Inf. Technol. **9**(3), 239–249 (2017)
9. Tran, T.K., Phan, T.T.: Towards a sentiment analysis model based on semantic relation analysis. Int. J. Synth. Emot. (IJSE) **9**(2), 54–75 (2018)

Author Index